Lecture Notes in Mathematics

A collection of informal reports and seminars
Edited by A. Dold, Heidelberg and B. Eckmann, Zürich

Series: Department of Mathematics, University of Maryland, College Park
Adviser: J. K. Goldhaber

185

Several Complex Variables II
Maryland 1970

Proceedings of the International Mathematical
Conference, held at College Park, April 6–17, 1970

Edited by John Horváth, University of Maryland,
College Park, MD/USA

Springer-Verlag
Berlin · Heidelberg · New York 1971

ISBN 3-540-05372-7 Springer-Verlag Berlin · Heidelberg · New York
ISBN 0-387-05372-7 Springer Verlag New York · Heidelberg · Berlin

Offsetdruck: Julius Beltz, Weinheim/Bergstr.

CONTENTS

Griffiths, Phillip A. *Some transcendental methods in the study of algebraic cycles* 1

Gunning, R. C. *Analytic structures on the space of flat vector bundles over a compact Riemann surface* 47

Hervé, Michel. *Analytic continuation on Banach spaces* 63

Igusa, Jun-ichi. *On certain representations of semi-simple algebraic groups* 76

Kubota, Tomio. *Some number-theoretical results on real analytic automorphic forms* 87

Lelong, Pierre. *Recent results on analytic mappings and plurisubharmonic functions in topological linear spaces* 97

Royden, H. L. *Remarks on the Kobayashi metric* 125

Satake, Ichiro. *$\bar{\partial}$-cohomology spaces associated with a certain group extension* 138

Shimura, Goro. *Class fields over real quadratic fields in the theory of modular functions* 169

Siu, Yum-Tong. *An Osgood type extension theorem for coherent analytic sheaves* 189

Wolf, Joseph A. *Complex manifolds and unitary representations* 242

SOME TRANSCENDENTAL METHODS IN THE STUDY OF ALGEBRAIC CYCLES

Phillip A. Griffiths

0. *Introduction*

The algebraic cycles which lie in a smooth, projective algebraic variety, together with the various equivalence relations (such as numerical or algebraic equivalence) which may be defined on such cycles, constitute a purely algebro-geometric concept. One may, however, attempt to study these by transcendental methods, and historically much of the structure of algebraic cycles (such as the notion of linear equivalence of divisors as related to the Jacobian and Picard varieties) was discovered in this way. Conversely, the questions in complex function theory which arose initially from algebraic geometry are frequently quite interesting in themselves. For example, the first and second Cousin problems for divisors in $\mathbb{C}^n$ were very much motivated by consideration of the Jacobian varieties of curves.

In this paper I shall give an expository account of a few transcendental methods for studying algebraic cycles of intermediate dimension. The main technique to be discussed is the use of *intermediate Jacobians*, and concerning these I shall focus on two things:

(i) how these complex tori may be used to give an interesting equivalence relation on algebraic cycles, which is between algebraic and rational equivalence, and which may well lead to a good generalization of the classical Picard variety[0].

---

[0] This is the so-called *incidence equivalence relation* which was introduced somewhat obscurely in [9]. An application of this equivalence relation is given in [4], where it is shown that it allows us to reconstruct a non-singular cubic-threefold from the singular locus of the canonical theta divisor on its intermediate Jacobian.

(ii) how the intermediate Jacobians may serve to detect the quotient of homological modulo algebraic equivalence.

The developments concerning (i) and (ii) are as yet incomplete, and one purpose of this paper is to formulate precise problems concerning intermediate Jacobians whose solution would very nicely yield the structure of the Picard ring (cf. section 1) of an algebraic variety. For all but one of these problems we can give a plausibility argument, and in all cases we can formulate and prove some analogue of the problem for varieties defined over function fields.

A second technique to be briefly discussed is the notion of *positivity* for general algebraic cycles, and we shall propose a definition which is suggested by looking at the cohomology classes on a smooth, projective variety which are represented using de Rham's theorem by positive differential forms.

Most of the material given below has already appeared in [9], [10], and [11]; however, here I have tried to give clearer and more precise formulations than before. The discussion of intermediate Jacobians has some points of contact with Lieberman's paper [22], and a few of the results on normal functions have appeared only in preprint form. In discussing positivity for algebraic cycles, I have used Hartshorne's definition of an ample vector bundle [14], and the presentation has been influenced by recent results of Bloch and Geiseker [3] concerning the numerical positivity of such bundles. Finally, the study of algebraic cycles via transcendental methods which we have tried to present is, in spirit, very much akin to the treatment given by Kodaira in [17], [18], [19] for the structure of the divisors on a smooth, projective variety.

Table of Contents

1. Algebraic cycles and equivalence relations
2. Intermediate Jacobian varieties
3. Concerning Abel's theorem
4. Concerning the inversion theorem
5. Definition of normal functions
6. Some results about normal functions
7. Positive algebraic cycles

1. *Algebraic cycles and equivalence relations*

In this section we will give some concepts of a purely algebro-geometric nature. Then, in the remainder of the paper we will discuss some transcendental methods for studying these concepts.

We consider a complete, smooth, and projective algebraic variety V over the complex numbers, and we want to discuss the algebraic cycles, and the various equivalence relations among these, which lie on V (cf. Samuel [24]). For this we recall that an *algebraic cycle* on V is given by

$$Z = n_1 Z_1 + \dots + n_\mu Z_\mu \tag{1.1}$$

where the Z_j are irreducible algebraic subvarieties of V and the n_j are integers. We say that Z has codimension q if all the Z_j have codimension q, and will call Z *effective* if all n_j are non-negative. We may write

$$Z = Z_+ - Z_- \tag{1.2}$$

where Z_+ and Z_- are effective algebraic cycles, and this may be done in many different ways.

Two effective algebraic cycles Z and Z' are *strongly algebraically equivalent* if there is a connected algebraic variety T and an effective cycle $W \subset T \times V$ such that all intersections $W_t = (\{t\} \times V) \cdot W$ are defined and of the same dimension, and such that $Z = W_{t_1}$ and $Z' = W_{t_2}$ for $t_1, t_2 \in T$. A general algebraic cycle Z is *algebraically equivalent to zero*, written

$$Z \approx 0,$$

if we may write Z in the form (1.1) where Z_+ and Z_- are strongly algebraically equivalent. An equivalent definition is that we may write $Z = Z_+ - Z_-$ where Z_+ and Z_- are in the same (topological) component of the Chow variety of V.

We will say that Z is *rationally equivalent to zero*, written

$$Z \cong 0,$$

if Z is algebraically equivalent to zero as above with the parameter variety T being a rational variety. If we take all the algebraic cycles on V modulo rational equivalence, there results the graded *Chow ring*[1] (cf. [13])

$$C(V) = \bigoplus_{q=0}^{n} C_q(V)$$

where the grading is by codimension of cycles and where the product

$$C_p(V) \times C_q(V) \to C_{p+q}(V)$$

is induced by taking the intersection of cycles. Implicit in this definition is the assertion that any two algebraic cycles Z and W on V are rationally equivalent to cycles Z' and W' which intersect properly and the resulting product (1.2) should then be well-defined.

---

[1] A graded ring will be a graded, commutative ring with unit $R = \bigoplus_{q=0}^{n} R_q$ such that the multiplication $R_p \times R_q \to R_{p+q}$ is compatible with the grading.

All equivalence relations we shall consider will be weaker than rational equivalence, and will therefore generate naturally an ideal in $C(V)$. We also recall that a map

$$f\colon V \to W$$

between smooth, projective varieties V and W induces additive homomorphisms

$$(1.3)\qquad \begin{cases} f_*\colon C(V) \to C(W) \\ f^*\colon C(W) \to C(V), \end{cases}$$

which are related to the product (1.2) by the formula

$$(1.4)\qquad f_*(X)\cdot Y = f_*[X\cdot f^*(Y)] \qquad (X\in C(V),\ \ Y\in C(W)).$$

Geometrically, the map f_* means "pushing cycles forward by f" and f^* means "lifting cycles back under f".

The cycles which are algebraically equivalent to zero generate a graded ideal

$$A(V) = \bigoplus_{q=0}^{n} A_q(V),$$

which is preserved under the maps f_* and f^* in (1.3). The quotient

$$NS(V) = C(V)/A(V)$$

will be called the *Neron-Severi ring* of V. For $q = 1$, $NS_1(V)$ is just the group of divisors modulo algebraic equivalence, and is the usual Neron-Severi group of V.

There is a homomorphism

$$(1.5)\qquad h\colon C(V) \to H^{\text{even}}(V, \mathbb{Z})$$

which sends a cycle $Z\in C_q(V)$ into its *fundamental class* $h(Z)\in H^{2q}(V, \mathbb{Z})$, the latter being the Poincaré dual of the homology class carried by Z. The kernel of h is the graded ideal

$$H(V) = \bigoplus_{q=0}^{n} H_q(V)$$

of cycles which are homologous to zero,[2] written

$$Z \sim 0.$$

We note that $A(V) \subset H(V)$; i.e., algebraic equivalence implies homological equivalence.

A map $f\colon V \to W$ induces on cohomology the usual maps $f_*\colon H^*(V,\mathbb{Z}) \to H^*(W,\mathbb{Z})$ and $f^*\colon H^*(W,\mathbb{Z}) \to H^*(V,\mathbb{Z})$, and the homomorphism h in (1.5) is functorial with respect to these induced maps.

Before defining the last equivalence relation, I should like to comment that there are in general two methods of studying algebraic subvarieties of arbitrary codimension: (i) by the divisors which pass through the subvariety, and (ii) by the subvarieties which meet the given subvariety. The first method (*method of syzygies*) has a global version in vector bundles and $K(V)$; the second method (*method of incidence*) leads to an equivalence relation on algebraic cycles which will now be defined.

Let Z be an algebraic cycle of codimension q on V and consider an algebraic family $\{W_s\}_{s \in S}$ of effective algebraic cycles of dimension $q-1$ whose parameter space S is smooth and complete. To be precise, we assume given an effective cycle

$$W \subset S \times V$$

such that all intersections $W_s = (\{s\} \times V) \cdot W$ are defined and of dimension $q-1$. By changing Z in its rational equivalence class, we may assume that the intersection $W \cdot (S \times Z)$ is defined and that

---

[2] Using l-adic cohomology, $H(V)$ may be defined purely algebraically.

$$D_Z = \mathrm{pr}_S[W\cdot(S\times Z)]$$

is a divisor on S. This *incidence divisor* D_Z is the set of all points $s\in S$ such that W_s meets Z, and where the points are of course counted with multiplicities ([9]).

Example. The first interesting case is when $\dim V = 3$ and Z is a curve. Then the W_s are also curves, so that the incidence relation may be pictured something like this:

(Fig. 1)

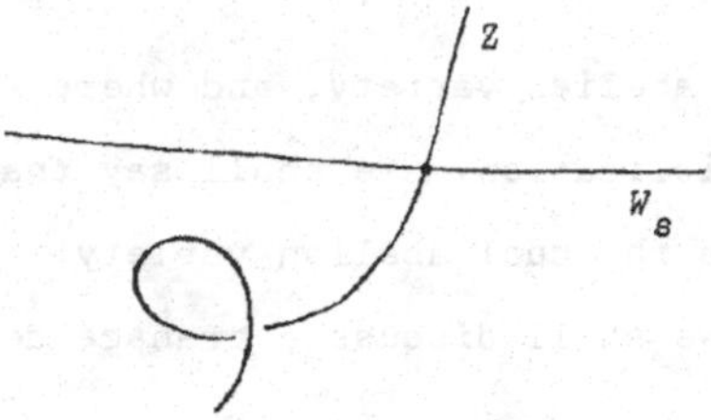

The linear equivalence class of the incidence divisor D_Z is well-defined by the rational equivalence class of Z, and we shall say that Z is *incidence equivalent to zero*, written $Z \overset{\sim}{=} 0$, if all such incidence divisors D_Z are linearly equivalent to zero on the parameter space S. The incidence equivalence relation again generates a graded ideal

$$I(V) = \bigoplus_{q=0}^{n} I_q(V),$$

and this ideal is preserved by the maps in (1.3).[3]

(1.6) *Definition. We define the Picard ring* Pic (V) *by*

$$\mathrm{Pic}\ (V) = C(V)/I(V),$$

and we also set $\mathrm{Pic}^0(V) = A(V)/I(V)$.

---

[3] The incidence equivalence relation has been defined here by playing off Z against cycles whose codimension is $\dim(Z) - 1$. We could try to play off Z against all cycles whose dimension is $\leqq \dim(Z) - 1$, but there is good evidence that this gives nothing beyond the case considered here.

As an initial justification for this definition, we observe that $\mathrm{Pic}_1(V)$ is the usual Picard variety of V because of the following easy

(1.7) *Lemma. If Z is a divisor on V, then Z is incidence equivalent to zero if, and only if, Z is linearly equivalent to zero.*

Let us define a *graded abelian variety* to be a direct sum

$$A = \bigoplus_{q=0}^{n} A_q$$

where each A_q is an abelian variety, and where A is made into a ring with a trivial multiplication. We shall say that A is *self-dual* if, up to isogeny, A_q is the dual abelian variety $\hat{A}_{n-q+1}$ to A_{n-q+1}. In section 2 below we shall discuss a transcendental proof of the

(1.8) *Proposition. $\mathrm{Pic}^0(V)$ is, in a natural way, a graded abelian variety.*[4]

We observe also that $\mathrm{Pic}^0(V)$ is functorially associated to V in that the maps in (1.3) induce maps between $A(V)$, $I(V)$ and $A(W)$, $I(W)$ which make the obvious diagrams commutative.

The outstanding question concerning $NS(V)$ and $\mathrm{Pic}\,(V)$ is the

Problem A. (i) *Is the Neron-Severi ring $NS(V)$ finitely generated?* (ii) *Is the graded abelian variety $\mathrm{Pic}^0(V)$ self-dual?*

The first part of this problem seems to be well known (I learned it from Mumford), and (1.8) together with the discussion in sections 2 and 3 below should lend credence to the second part, which at any

[4] Thus each graded component $\mathrm{Pic}^0_q(V)$ of $\mathrm{Pic}^0(V)$ is an abelian variety, and the induced product

$$A(V)/I(V) \otimes A(V)/I(V) \to A(V)/I(V)$$

is identically zero. I know of no algebro-geometric proof for either of these statements.

event is known in the following special cases:

a) $q = 1$ [5]

b) $\dim V = 2m - 1$ and $q = m$.

Observe that Problem A would very nicely give the structure of Pic (V) because of the exact sequence

$$(1.9) \qquad 0 \to \mathrm{Pic}^0(V) \to \mathrm{Pic}\,(V) \to NS(V) \to 0,$$

which results from the inclusions

$$(1.10) \qquad I(V) \subset A(V) \subset H(V) \subset C(V).$$

2. *Intermediate Jacobian varieties*

We want to discuss a transcendental method for studying the Chow ring $C(V)$ and its various ideals (1.10) introduced above. This is the use of the Jacobian variety $T(V)$ of the smooth, projective variety V, which will be a graded, self-dual complex torus functorially associated to V and whose definition we will now give.[6]

First we recall the *Hodge decomposition* [8]

$$(2.1) \qquad H^m(V,\mathbb{C}) = \bigoplus_{r+s=m} H^{r,s}(V) \qquad (H^{r,s}(V) = \overline{H^{s,r}(V)}),$$

---

[5] In this case $\mathrm{Pic}_1^0(V)$ is the identity component of the usual Picard variety of V and $\mathrm{Pic}_n^0(V)$ is isogeneous to the *Albanese variety* of V.

[6] Thus $T(V) = \bigoplus_{q=0}^{n} T_q(V)$ is a direct sum of complex tori $T_q(V)$ where $T_q(V)$ is isogeneous to the dual torus $\hat{T}_{n-q+1}(V)$ of $T_{n-q+1}(V)$ and where a map $f: V \to W$ generates induced maps $f_*: T(V) \to T(W)$ and $f^*: T(W) \to T(V)$ which satisfy the usual functoriality properties.

and the associated *Hodge filtration*

(2.2)
$$F^{m,p}(V,\mathbb{C}) = \bigoplus_{\substack{r+s=m\\ s\leq p}} H^{r,s}(V).$$

There is a useful method for describing the Hodge filtration in terms of the de Rham description of $H^m(V,\mathbb{C})$ together with the complex structure on V, which is the following: Let $A^{m,p}(V)$ be the vector space of global C^∞ differential forms on V which have total degree m and have type $(m,0) + \ldots + (m-p,p)$ (these are the same indices as in the definition of $F^{m,p}(V,\mathbb{C})$). If $Z^{m,p}(V)$ are the d-closed forms in $A^{m,p}(V)$, then by de Rham's theorem there is a natural map

$$Z^{m,p}(V)/dA^{m-1,p-1}(V) \to H^m(V,\mathbb{C}).$$

Because V carries a Kähler metric, this map turns out to be injective with image $F^{m,p}(V,\mathbb{C})$, and so there results an isomorphism

(2.3)
$$F^{m,p}(V,\mathbb{C}) \cong Z^{m,p}(V)/dA^{m-1,p-1}(V).$$

We shall use the following notations:

(2.4)
$$H^{2q-1}_{+}(V) = \bigoplus_{\substack{r+s=2q-1\\ s\leq q-1}} H^{r,s}(V) = F^{2q-1,q-1}(V,\mathbb{C})$$

$$H^{2q-1}_{-}(V) = \bigoplus_{\substack{r+s=2q-1\\ r\leq 2q-1}} H^{r,s}(V) \cong H^{2q-1}(V,\mathbb{C})/H^{2q-1}_{+}(V)$$

$$H^{\mathrm{odd}}_{+}(V) = \bigoplus_{q=1}^{n} H^{2q-1}_{+}(V)$$

$$H^{\mathrm{odd}}_{-}(V) = \bigoplus_{q=1}^{n} H^{2q-1}_{-}(V).$$

Observe the decomposition

(2.5)
$$H^{2q-1}(V,\mathbb{C}) = H^{2q-1}_{+}(V) + H^{2q-1}_{-}(V)$$

of $H^{2q-1}(V,\mathbb{C})$ into a direct sum of conjugate subspaces. The non-degenerate pairing

$$H_{-}^{2q-1}(V) \otimes H_{+}^{2n-2q+1}(V) \to H^{2n}(V,\mathbb{C}) \cong \mathbb{C}$$

induces a duality isomorphism

$$H_{+}^{2n-2q+1}(V) \cong \check{H}_{-}^{2q-1}(V). \tag{2.6}$$

Definition. The qth intermediate Jacobian $T_q(V)$ is the complex torus $H_{+}^{2q-1}(V)\backslash H^{2q-1}(V,\mathbb{C})/H^{2q-1}(V,\mathbb{Z}) \cong H_{-}^{2q-1}(V)/H^{2q-1}(V,\mathbb{Z})$. The Jacobian variety $T(V)$ is the direct sum $\bigoplus_{q=0}^{n} T_q(V)$ of the intermediate Jacobians.

From (2.5) together with the definition of the intermediate Jacobian there results an $\mathbb{R}$-*linear* isomorphism

$$T_q(V) \cong H^{2q-1}(V,\mathbb{R})/H^{2q-1}(V,\mathbb{Z}). \tag{2.7}$$

The complex Lie algebra of $T_q(V)$ is $H_{-}^{2q-1}(V)$ and, using (2.6) and (2.4), the holomorphic differentials on $T_q(V)$ are given by the isomorphisms

$$H^{1,0}(T_q(V)) \cong H_{+}^{2n-2q+1}(V) \cong F^{2n-2q+1,n-q}(V,\mathbb{C}). \tag{2.8}$$

The intermediate Jacobian defined above is closely related to, but not the same as, the intermediate Jacobian variety $J_q(V)$ introduced by Weil [25] and studied by Lieberman [22]. The main points of comparison are: (i) there is a natural $\mathbb{R}$-linear isomorphism between the two complex tori; (ii) $T_q(V)$ varies holomorphically with V whereas $J_q(V)$ does not, and (iii) $J_q(V)$ is an abelian variety whereas $T_q(V)$ has an r-convex polarization (cf. [8]). A propos the point (iii) just made, we recall from [8] the

(2.9) *Lemma. Let S be a complex sub-torus of* $T_q(V)$ *whose complex Lie algebra is contained in the subspace* $H^{q-1,q}(V)$ *of the Lie algebra of* $T_q(V)$. *Then the r-convex polarization is 0-convex on S, so that in particular S is naturally an abelian variety.*

We have defined the direct sum $\bigoplus_{q=0}^{n} T_q(V)$ to be the (total) Jacobian variety of V by analogy with the classical definition of the Jacobian of a curve as being the complex torus associated to the period matrix of the differentials of odd degree on the curve. On the other hand, the Picard variety Pic (V), as defined in section 1, is a purely algebro-geometric concept arising out of the notion of linear equivalence of divisors. The relation between these falls under the general heading of "Abel's theorem" and will be discussed in section 3. The fact that the Jacobian variety $T(V)$ is a graded, self-dual complex torus functorially attached to V (cf. footnote [6]) follows immediately from the definitions, the duality (2.6), and the functorial behavior which the induced maps f_* and f^* on cohomology have with respect to the Hodge filtration (2.2).

The algebro-geometric importance of the Jacobian variety rests in the fact that there is an *Abel-Jacobi homomorphism*

(2.10) $$\Phi: H(V) \to T(V)$$

whose definition is a generalization of the classical procedure for sending divisors of degree zero on a compact Riemann surface S into the Jacobian variety of S. To define Φ, it will suffice to give meaning to a symbol

(2.11) $$\langle\phi, Z\rangle \qquad (\phi \in H^{1,0}(T_q(V)), \quad Z \in H_q(V))$$

which has the properties of (i) being a complex number, (ii) being defined modulo periods, and (iii) being linear in each factor.

Using the isomorphism (2.8), we may consider ϕ as being a closed, C^{∞} differential form of type $(2n-2q+1, 0) + \dots + (n-q+1, n-q)$ and which is defined modulo exact forms $d\eta$ where η is of type $(2n-2q, 0) + \dots + (n-q+1, n-q-1)$. Furthermore, $Z \in H_q(V)$ is represented by an algebraic cycle of codimension q which is the boundary of a $(2n-2q+1)$-chain Γ on V.[7] We then let

(2.12) $$\langle\phi,\Gamma\rangle = \int_{\Gamma} \phi .$$

This symbol is well defined modulo periods since $\int_{\Gamma} d\eta = \int_{Z} \eta = 0$ if $\phi = d\eta$ as above. Furthermore, it is clearly bilinear, and so we may then use it to define the Abel-Jacobi mapping (2.10).

The mapping Φ has the following basic properties:

(2.13) Φ is holomorphic on the ideal $A(V)$ of cycles algebraically equivalent to zero on V.

(2.14) The image $\Phi_q[A_q(V)]$ of codimension q cycles which are algebraically equivalent to zero is a complex sub-torus $I_q^0(V)$ whose Lie algebra is a subspace of $H_-^{q-1,q}(V)$. It follows then from (2.9) that $I_q^0(V)$ is an abelian variety.[8]

---

[7] There is a foundational question here as to just what is meant by the equation "$\partial\Gamma = Z$". Moreover, we shall want to integrate over such "chains" Γ, and later on we shall want to let everything in sight depend on parameters. The foundational questions pose a significant problem which is resolved, using the theory of integral currents, in King's paper [16].

[8] The meaning of the notation $I_q^0(V)$ for the image in $T_q(V)$ of cycles algebraically equivalent to zero will be discussed in section 4 below.

(2.15) The Abel-Jacobi mapping Φ satisfies the hoped-for functorial properties. Thus, a holomorphic mapping $f\colon V \to W$ between smooth, projective varieties leads to commutative diagrams (cf. (1.3))[9]

$$\begin{array}{ccc} H(V) & \xrightarrow{\Phi_V} & T(V) \\ {\scriptstyle f_*}\downarrow & & \downarrow{\scriptstyle f_*} \\ H(W) & \xrightarrow[\Phi_W]{} & T(W) \end{array}$$

$$\begin{array}{ccc} H(W) & \xrightarrow{\Phi_W} & T(W) \\ {\scriptstyle f^*}\downarrow & & \downarrow{\scriptstyle f^*} \\ H(V) & \xrightarrow[\Phi_V]{} & T(V) \end{array}$$

(2.16) A noteworthy special case of (2.15) occurs when we consider a fixed cycle $Z \in C_q(V)$. Intersection with Z induces a map $Z\colon H_p(V) \to H_{p+q}(V)$, while cup-product with the fundamental class $h(Z) \in H^{2q}(V, \mathbb{Z})$ leads to a homomorphism $h(Z)\colon T_p(V) \to T_{p+q}(V)$, and we have a commutative diagram

$$\begin{array}{ccc} H_p(V) & \xrightarrow{\Phi_p} & T_p(V) \\ {\scriptstyle Z}\downarrow & & \downarrow{\scriptstyle h(Z)} \\ H_{p+q}(V) & \xrightarrow[\Phi_{p+q}]{} & T_{p+q}(V). \end{array}$$

Other basic properties of Φ occur when the complex structure of V is allowed to vary with parameters; these will be discussed in section 5 below.

[9] These properties, which are heuristically quite reasonable, were first proved by Lieberman [22] and then independently by myself, using residues, in [9]. They also follow from the results in [16].

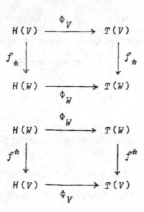

3. *Concerning Abel's theorem*

We continue discussing the Abel-Jacobi mapping (2.10), and in this section we are interested in the kernel of Φ; i.e., what is the equivalence relation on cycles given by Φ. In the classical case of divisors, the equivalence of Φ is linear equivalence,[10] from which it follows that

$$\Phi: \mathrm{Pic}_1^0(V) \to T_1(V)$$

is an isomorphism. In particular, this proves by transcendental methods that $\mathrm{Pic}_1^0(V)$, as defined algebro-geometrically in section 1 above, is an abelian variety. For general codimension, one-half of the above version of Abel's theorem can be proved:

(3.1) *Proposition* [9]. *If* $Z \in A(V)$ *is algebraically equivalent to zero and* $\Phi(Z) = 0$ *in* $T(V)$, *then* Z *is incidence equivalent to zero.*[11]

We want to restate (3.1) in a more suggestive manner. To do this, we let

$$K(V) = \bigoplus_{q=1}^{n} K_q(V)$$

be the kernel of the Abel-Jacobi mapping Φ on $A(V)$. It follows from (2.15) and (2.16) that $K(V)$ is a graded ideal in $A(V)$ which is functorially associated to V. The quotient

$$A(V)/K(V) \cong I^0(V)$$

[10] In the framework we are using, this result is proved by Kodaira in [18].

[11] In the case of curves, this proposition corresponds to what is ordinarily the "more difficult" half of Abel's theorem, which is the construction of a linear equivalence using differentials of the third kind.

is an abelian variety by virtue of (2.14). From (3.1) we have the inclusion

$$K(V) \subset I(V),$$

and this leads to the commutative diagram

(3.2)
$$\begin{array}{ccc} A(V) & \xrightarrow{\Phi} & I^0(V) \subset T(V) \\ & \searrow \quad \swarrow \Psi & \\ & \mathrm{Pic}^0(V) , & \end{array}$$

which is a direct sum of the commutative diagrams

$(3.2)_q$
$$\begin{array}{ccc} A_q(V) & \xrightarrow{\Phi_q} & I^0_q(V) \subset T_q(V) \\ & \searrow \quad \swarrow \Psi_q & \\ & \mathrm{Pic}^0_q(V) & \end{array}$$

for each $q = 1,\ldots,n$.

(3.3) *Corollary.* $\mathrm{Pic}^0_q(V)$ *is an abelian variety for all* $q = 1,\ldots,n$.

(3.4) *Corollary. The induced pairing* $A(V)/I(V) \otimes A(V)/I(V) \to A(V)/I(V)$ *is zero, so that* $\mathrm{Pic}^0(V) = \bigoplus_{q=1}^{n} \mathrm{Pic}^0_q(V)$ *is a graded abelian variety according to the definition in section* 1.

Using recent results of Deligne [5], the diagram $(3.2)_q$ can be further understood in one important special case:

(3.5) *Proposition. Suppose that* $\dim V = 2m - 1$ *is odd and consider the diagram* $(3.2)_m$. *Then* (i) *the kernel of* Ψ_m *is finite, and* (ii) *the auto-duality* $T_m(V) \cong \check{T}_m(V)$ *induces an auto-duality* $I_m(V) \cong \check{I}^0_m(V)$ *(both dualities are up to isogenies).*[12]

---

[12] See next page.

(3.6) *Corollary. In the diagram* $(3.2)_q$*, the mapping* Ψ_q *is an isogeny for* $q = 1, n$*, or in case* $q = m$ *and* $\dim V = 2m - 1$*. In both of these situations, the duality formula*

$$\mathrm{Pic}_q^0(V) \cong \check{\mathrm{Pic}}_{n-q+1}^0(V)$$

holds true (up to isogeny).

Problem B. (i) *Is the mapping* Ψ *in* (3.2) *an isogeny?* (ii) *Is the graded abelian variety* $I^0(V)$ *self-dual?*

An affirmative answer to this problem would settle part (ii) of problem A in section 1. Furthermore, it would identify the the equivalence relation on $A(V)$ induced by the Abel-Jacobi mapping as being the incidence equivalence relation (up to a finite group). As remarked in [9], both parts of problem B would have an affirmative answer if we knew the general Hodge conjecture [12].

Thus far we have only considered the kernel of Φ on the cycles algebraically equivalent to zero, whereas Φ is defined on the cycles which are homologous to zero. In this regard, let me propose

Problem C. *Is the induced mapping*

$$\Phi: H(V)/A(V) \to T(V)/I^0(V)$$

injective (up to a finite group)?

There are two bits of evidence for this problem. The first is that it would follow from a "relative version" of the Hodge conjecture, and the second is that one can prove an analogue of this problem over function fields. The precise statement of this latter result will be given in section 6 below.

---

[12] The first significant special case of this Proposition is when $m = 1$, which is the study of curves on a threefold. Applications of this are given in [4].

4. *Concerning the inversion theorem*

We continue to let V be a smooth, projective variety and denote by $\Phi: H(V) \to T(V)$ the Abel-Jacobi mapping defined in section 2. We further denote by $I(V)$ and $I^0(V)$ the images of

$$\begin{cases} \Phi: H(V) \to T(V) \\ \Phi: A(V) \to T(V), \end{cases}$$

so that $I(V)$ is the *group of invertible points* on the Jacobian variety $T(V)$ and $I^0(V)$ is the subgroup coming from cycles which are algebraically equivalent to zero. We recall from section 2 that $I^0(V)$ is an abelian subvariety of $T(V)$ whose Lie algebra is contained in the subspace $\bigoplus_{q=0}^{n} H^{q-1,q}(V)$ of $H^{\text{odd}}_{-}(V)$.[13] It is *not* possible to give any such (linear) restriction on the points in $I(V)$ ([10]).

The *inversion problem* is to describe the subgroups $I(V)$ and $I^0(V)$ in an a priori manner. For $I^0(V)$ there is a candidate, which is again suggested by the Hodge conjecture (cf. Lieberman [22]). To say what this is, we let $A_q(V)$ be the largest complex sub-torus of $T_q(V)$ whose *real* Lie algebra is a sub-Hodge structure of $H^{2q-1}(V,\mathbb{R})$ which is defined over $\mathbb{Q}$ and whose complexification is contained in $H^{q,q-1}(V) \oplus H^{q-1,q}(V)$. We let $A(V) = \bigoplus_{q=0}^{n} A_q(V)$ and observe from (2.14) the inclusion $I^0(V) \subset A(V)$ (cf. Grothendieck [12]).

Problem D. *Do we have the equality* $I^0(V) = A(V)$?

If this question is answered affirmatively, then it follows that problem B in section 3 is also answered in the affirmative. In particular, $\text{Pic}^0(V)$ as defined in section 1 would be a graded,

---

[13] In particular, except for $q = 1, n$ together witn a few other special cases, we cannot expect $\Phi_q: H_q(V) \to T_q(V)$ to be surjective.

self-dual abelian variety and the notion of "incidence equivalence" would be justified.

The question of finding the whole group $I(V)$ of invertible points in the Jacobian is, to me, the most mysterious problem regarding algebraic cycles. At least for other questions, such as the Hodge conjecture or problems A, B, and D above, we have a plausible answer which has yet to fail (although there are precious few non-trivial examples).

Problem E. *Describe the subgroup $I(V)$ of invertible points on $T(V)$.*

In explicit terms, given a basis $\phi_1,\ldots,\phi_g$ of $H^{1,0}(T_q(V)) \cong Z^{2n-2q+1,n-q}(V)/dA^{2n-2q,n-q-1}(V)$ we want to know which points $(z_1,\ldots,z_g)\in\mathbb{C}^g$ are solutions of the inversion equations

$$(4.1)\qquad \begin{cases} z_1 = \int_\Gamma \phi_1 \\ \vdots \qquad \vdots \\ z_g = \int_\Gamma \phi_g , \end{cases}$$

where Γ is a $2n-2q+1$ chain whose boundary is an algebraic cycle.

To add to the mystery surrounding this question, we shall see in section 6 below (cf. Corollary 6.14) that a knowledge of $I(V)$ would have strong implications regarding the Hodge conjecture, but I don't see any reason that the converse statement should be true.

Our final problem regarding $I(V)$ and $I^0(V)$ was originally suggested to me by Mumford:

Problem F. *Is the quotient group $I(V)/I^0(V)$ finitely generated?*

To motivate this problem, we observe that $I(V)/I^0(V)$ is a *countable* subgroup of $T(V)/I^0(V)$, and so we have a Mordell-Weil type of question. In this form, an analogue over function fields of problem F can be proved (cf. Proposition 6.8). If both problems D and F are answered in the affirmative, then we would have a positive answer to problem A. In fact, being very optimistic, if problems B-D and F could be answered affirmatively, then we could draw two conclusions: (i) that the purely algebro-geometric problem A has a positive answer, and (ii) that the use of intermediate Jacobians provides a very strong transcendental method for studying algebraic cycles.

5. *Definition of normal functions*

We want to discuss how the intermediate Jacobians vary with parameters. More precisely, we will consider a situation

(5.1) $$f: X \to S$$

where X and S are smooth, projective varieties and where the fibres $V_s = f^{-1}(s)$ are smooth, projective varieties for almost all $s \in S$. Letting $T(V_s)$ be the Jacobian variety of such V_s, we want to fit the $T(V_s)$ together and then discuss how the algebraic cycles on X relate to cross-sections of the resulting fibre space of commutative, complex Lie groups. In order to carry out this program we shall make the assumptions

(5.2) $$\begin{cases} \dim S = 1, \text{ and} \\ f \text{ has only non-degenerate critical points.} \end{cases}$$

The first assumption is not too serious, but the second one is much too restrictive and it should be possible to eliminate it entirely, especially in light of recent results of P. Deligne and W. Schmid (cf. problem G below).

We shall analyze the intermediate Jacobians along the fibres of $f: X \to S$ in three steps.

(a) *Local theory around a non-critical value.* We let Δ denote the unit disc $\{s \in \mathbb{C} : |s| \leqq 1\}$ and $\Delta^* = \Delta - \{0\}$ the corresponding punctured disc. Assume given a situation

$$f: W \to \Delta$$

where W is a complex manifold (with boundary) and f is a proper, smooth, and projective holomorphic mapping. Thus the fibres V_s are all smooth, projective algebraic varieties. Topologically, W is diffeomorphic to the product $V_0 \times \Delta$ so that we may identify all of the cohomology groups $H^*(V_s, \mathbb{C})$ with $H^*(V_0, \mathbb{C})$[14]. When this is done, the Hodge filtration

$$F^{m,p}(V_s, \mathbb{C}) = H^{m,0}(V_s) + \ldots + H^{m-p,p}(V_s)$$

gives a subspace of $H^m(V_0, \mathbb{C})$; this subspace has the two basic properties:

(i) $F^{m,p}(V_s, \mathbb{C})$ varies holomorphically with $s \in \Delta$;

(ii) the *infinitesimal bilinear relation* [8]

$$\frac{d}{ds}\{F^{m,p}(V_s, \mathbb{C})\} \subset F^{m,p+1}(V_s, \mathbb{C}) \tag{5.3}$$

---

[14] To be precise, there is a retraction $r: W \to V_0$ and an inclusion $i: V_s \to W$. The composite map on cohomology

$$H^*(V_0, \mathbb{C}) \xrightarrow{r^*} H^*(W, \mathbb{C}) \xrightarrow{i^*} H^*(V_s, \mathbb{C})$$

is an isomorphism and gives the identification in which we are interested.

is satisfied.[15]

From (i) it follows that we may canonically construct an analytic fibre space of complex tori

$$\text{(5.4)} \qquad \tilde{\omega}: T(W/\Delta) \to \Delta$$

with fibres $\tilde{\omega}^{-1}(s) = T(V_s)$. The holomorphic vector bundle of complex Lie algebras associated to (5.4) will be denoted by

$$\pi: \mathbb{L} \to \Delta.$$

The fibres $\mathbb{L}_s$ of π are given by

$$\mathbb{L}_s \cong H_{-}^{\text{odd}}(V_s).$$

Letting $O(T(W/\Delta))$ denote the group of holomorphic cross-sections of the fibre space (5.4), the exponential mapping gives an exact sheaf sequence

$$0 \to \Lambda \to O(\mathbb{L}) \xrightarrow{\exp} O(T(W/\Delta)) \to 0.$$

The sheaf Λ is a locally constant sheaf of $\mathbb{Z}$-modules, which is easily identified by the isomorphism (cf. (2.7))

$$\Lambda \cong R_{f_*}^{\text{odd}}(\mathbb{Z}),$$

where $R_{f_*}^{\text{odd}}(\mathbb{Z})$ is the *Leray direct image sheaf* for the constant sheaf $\mathbb{Z}$ on W relative to the map $f: W \to S$.

---

[15] These two statements mean that we can choose vectors $e_1(s),\ldots,e_l(s)$ in $H^m(V_0, \mathbb{C})$ with the properties that (i) the $e_j(s)$ vary holomorphically with $s \in \Delta$ and $e_1(s) \wedge \ldots \wedge e_l(s) \neq 0$; (ii) the $e_j(s)$ give a basis for $F^{m,p}(V_s, \mathbb{C})$; and (iii) the vectors $de_j(s)/ds$ lie in $F^{m,p+1}(V_s, \mathbb{C})$.

We want now to discuss the implications of (5.3) on the fibre space (5.4). For this, we construct over Δ a holomorphic vector bundle

$$\mathbb{J} = \bigoplus_{q=1}^{n} \mathbb{J}_q$$

where the fibre $(\mathbb{J}_q)_s$ of $\mathbb{J}_q$ is given by

$$(\mathbb{J}_q)_s = H^{2q-1}(V_s, \mathbb{C})/F^{2q-1,q}(V_s, \mathbb{C}).$$

Then (5.3) implies that there is a homomorphism

$$\text{(5.5)} \qquad D\colon \mathcal{O}(T(W/\Delta)) \to \Omega^1(\mathbb{J}),$$

and the subsheaf of sections ν of $T(W/\Delta)$ which satisfy the equation

$$\text{(5.6)} \qquad D\nu = 0$$

will be denoted by $Hom\ (\Delta, T(W/\Delta))$. Thus we have

$$0 \longrightarrow Hom\ (\Delta, T(W/\Delta)) \longrightarrow \mathcal{O}(T(W/\Delta)) \longrightarrow \Omega^1(\mathbb{J}).$$

We will now explain the geometric meaning of the condition (5.6). For this, we let Z be an analytic cycle on W such that all intersections $Z_s = V_s \cdot Z$ are defined and induce an algebraic cycle Z_s which is homologous to zero on V_s. Using the Abel-Jacobi mappings

$$\Phi_{V_s} : H(V_s) \to T(V_s),$$

we may define a cross-section ν_Z of the fibre space (5.4) by the rule

$$\nu_Z(s) = \Phi_{V_s}(Z_s).$$

(5.7) *Proposition* [8]. *The cross section* ν_Z *is holomorphic and satisfies* $D\nu_Z = 0$.

(b) *Local theory around a non-degenerate critical value.* Now we assume given a situation

$$f\colon W \to \Delta$$

where W is a complex manifold and f is a proper, projective holomorphic mapping which is smooth except that $s = 0$ is a non-degenerate critical value for f. Thus the fibres V_s $(s \neq 0)$ are smooth, projective varieties while V_0 has an isolated, ordinary double point around which the mapping f has the local form

$$(z_1)^2 + \ldots + (z_{n+1})^2 = s$$

for suitable holomorphic coordinates $z_1,\ldots,z_{n+1}$ on W.

We let $W^* = f^{-1}(\Delta^*) = W - V_0$ so that $f\colon W^* \to \Delta^*$ is a differentiable fibre bundle. Consequently the fundamental group $\pi_1(\Delta^*, s_0)$ acts on the cohomology $H^*(V_{s_0}, \mathbb{Z})$, and we let $I^*(V_{s_0}, \mathbb{Z})$ be the subspace on which $\pi_1(\Delta^*, s_0)$ acts trivially (these are the so-called *local invariant cycles*). There is an obvious restriction map

$$H^*(W, \mathbb{Z}) \xrightarrow{r^*} I^*(V_{s_0}, \mathbb{Z}), \tag{5.8}$$

which turns out to be an isomorphism in our case where V_0 has ordinary double points.[16] From this it follows that the stalk of the Leray direct image sheaf for $\mathbb{Z}$ on W is given by

$$R^*_{f_*}(\mathbb{Z})_0 = I^*(V_{s_0}, \mathbb{Z}).$$

---

[16] This follows from an analysis of the topology of the degeneration $V_s \to V_0$ given by Lefschetz [21]. It has recently been proved by P. Deligne that r^* in (5.8) is *surjective over* $\mathbb{Q}$ where V_0 has arbitrary singularities; this is the *local invariant problem* which is discussed in §15 of [7].

The analysis of how the Hodge filtration $\{F^{m,p}(V_s, \mathbb{C})\}$ of $H^m(V_s, \mathbb{C})$ behaves as $s \to 0$ is not too difficult in the situation at hand, and this analysis leads to the construction of a *generalized Jacobian* $T(V_0)$ with the following properties [10]:

(i) $T(V_0)$ is a commutative complex Lie group which fits into an exact sequence

$$1 \to \mathbb{C}^* \to T(V_0) \to T(\tilde{V}_0) \to 0,$$

where $\tilde{V}_0$ is the standard desingularization of V_0.[17]

(ii) There exists a complex-analytic fibre space

$$\tilde{\omega}: T(W/\Delta) \to \Delta$$

of abelian complex Lie groups such that the fibres $\tilde{\omega}^{-1}(s) = T(V_s)$ for *all* $s \in \Delta$. Furthermore, letting $\pi: \mathbb{L} \to \Delta$ be the holomorphic vector bundle of complex Lie algebras, we have again the exponential sheaf sequence

$$0 \longrightarrow \Lambda \longrightarrow \mathcal{O}(\mathbb{L}) \xrightarrow{\exp} \mathcal{O}(T(W/\Delta)) \longrightarrow 0$$

where, because of the isomorphism (5.8), the isomorphism

$$\Lambda \cong R^{\text{odd}}_{f_*}(\mathbb{Z})$$

of Λ with the Leray direct image sheaf holds just as before.

(iii) If Z is an analytic cycle on W such that all intersections

[17] The "standard desingularization" of V_0 is obtained by blowing up the double point on V_0 to obtain a smooth, projective variety $\tilde{V}_0$ containing a non-singular quadric which may be contracted to yield the singular point on V_0.

$Z_s = Z \cdot V_s$ are defined and such that Z_s is homologous to zero on V_s for $s \neq 0$, then the Abel-Jacobi maps

$$\Phi_{V_s} : H(V_s) \to T(V_s) \qquad (s \neq 0)$$

induce a holomorphic cross-section ν_Z of $T(W/\Delta)$ over all of the disc Δ.

Before going on, I should like to discuss briefly the possibility of extending (i)-(iii) to the case of general $f: W \to \Delta$ where V_0 is allowed to have arbitrary singularities. The main tools used in (i)-(iii) were the isomorphism (5.8), which related the topology of V_{s_0} and W by means of the action of $\pi_1(\Delta^*, s_0)$ on $H^*(V_{s_0}, \mathbb{C})$, and the knowledge of how the Hodge filtration on $H^*(V_s, \mathbb{C})$ behaves as $s \to 0$. Now the former has been done in general by Deligne and Katz, with one conclusion being the local invariant cycle theorem over $\mathbb{Q}$ as discussed in footnote [16]. The latter has recently been done by W. Schmid, who has in particular verified the conjecture of Deligne as given in §9 of [7]. Thus, it seems that perhaps the time is ripe to work on the

Problem G. *Analyze the behavior of the Jacobian varieties along the fibres of* $f: W \to \Delta$ *where* V_0 *has arbitrary singularities. In particular, can we define a generalized Jacobian* $T(V_0)$ *so that the analogues of* (i)-(iii) *above will remain valid?*

A side condition on this problem is given by remark 6.9 below.

(c) *Global theory and definition of normal functions.* We assume now that we are given $f: X \to S$ where S is a compact Riemann surface, X is a smooth, projective algebraic variety, and f is a holomorphic mapping which has only non-degenerate critical points. Thus, localizing around a point $s_0 \in S$, we find either the situation in (a) or (b) above according as to whether s_0 is a regular or critical value of f. We denote the critical values of f by $\{s_1,\ldots,s_N\}$ and set $S^* = S - \{s_1,\ldots,s_N\}$, $X^* = X|S^*$. Then $f\colon X^* \to S^*$ is topologically a fibre bundle and so the fundamental group $\pi_1(S^*,s_0)$ acts on the cohomology $H^*(V_{s_0},\mathbb{Z})$.

Example. To see how to construct such a situation, we take an arbitrary smooth, projective variety X' embedded in a projective space $\mathbb{P}_m$. In $\mathbb{P}_m$ we consider a general pencil $|\mathbb{P}_{m-1}(s)|_{s\in\mathbb{P}_1}$ of linear hyperplanes, and we let $V_s = X'\cdot\mathbb{P}_{m-1}(s)$ be the residual intersection of the hyperplane $\mathbb{P}_{m-1}(s)$ with X'. If we blow up X' along the base locus $V_0\cdot V_\infty$ of the pencil $|V_s|_{s\in\mathbb{P}_1}$, then we obtain a smooth, projective algebraic variety X together with an obvious mapping $f\colon X \to \mathbb{P}_1$.[18] The point $s_0 \in \mathbb{P}_1$ is a critical value for f if, and only if, the hyperplane $\mathbb{P}_{m-1}(s_0)$ is tangent to X'. In this case, the singular points of V_{s_0} occur along the locus of tangency, and to say that s_0 is a non-degenerate critical value for f means that V_{s_0} should have one isolated, ordinary double point at the place of tangency of $\mathbb{P}_{m-1}(s_0)$.[19] If all critical values for f are of this sort, then we shall say that $|V_s|_{s\in\mathbb{P}_1}$ is a *Lefschetz pencil* on X', and in this case the resulting fibration $f\colon X \to \mathbb{P}_1$ is of the type we want to consider.

Returning to the general case of $f\colon X \to S$ where f has non-degenerate critical points, we may combine the results of (a) and (b) to construct a complex-analytic fibre space of abelian complex Lie

---

[18] Set theoretically, X is the disjoint union $\bigcup_{s\in\mathbb{P}_1} V_s$ of the hyperplane sections of X', and the map $f\colon X \to \mathbb{P}_1$ sends $x \in V_s$ onto $s \in \mathbb{P}_1$.

[19] There is a nice discussion of these matters in the paper of Andreotti-Frankel [1] and in the exposés of Katz [15].

groups

$$\tilde{\omega}: T(X/S) \to S$$

with $\tilde{\omega}^{-1}(s) = T(V_s)$ the Jacobian, or generalized Jacobian if s is a critical value for f, of V_s. The Lie algebras along the fibres of $\tilde{\omega}$ give a holomorphic vector bundle

$$\pi: \mathbb{L} \to S,$$

and we have the exponential sheaf sequence

$$(5.9) \qquad 0 \to \Lambda \to \mathcal{O}(\mathbb{L}) \to \mathcal{O}(T(X/S)) \to 0$$

where the sheaf Λ is described by the isomorphism

$$(5.10) \qquad \Lambda \cong R^{\text{odd}}_{f_*}(\mathbb{Z})$$

with the Leray direct image sheaf. We shall continue to denote by $Hom(S,T(X/S))$ the subsheaf of $\mathcal{O}(T(X/S))$ of sections ν which satisfy the condition $D\nu = 0$ as explained in (a).

Definition. The group Hom $(S,T(X/S))$ *of global sections of* $Hom(S,T(X/S))$ *will be called the group of normal functions associated to* $f: X \to S$.

We now want to relate the algebraic cycles on X to these normal functions. For this we let $C(F)$ be the ideal in the Chow ring $C(X)$ of all rational equivalence classes of algebraic cycles on X which lie in a fibre of $f: X \to S$. Similarly, we let $H^{\text{even}}(F,\mathbb{Z})$ be the ideal in $H^{\text{even}}(X,\mathbb{Z})$ which is given by the Poincaré dual of the image

$$H_{\text{even}}(V_s,\mathbb{Z}) \to H_{\text{even}}(X,\mathbb{Z}) \qquad (s \in S^*).$$

The quotients by these two ideals will be denoted by

$$\begin{cases} C(X/S) = C(X)/C(F), & \text{and} \\ H^{\text{even}}(X/S,\mathbb{Z}) = H^{\text{even}}(X,\mathbb{Z})/H^{\text{even}}(F,\mathbb{Z}). \end{cases}$$

The fundamental class mapping (1.5) and the restriction mapping $H^{\text{even}}(X,\mathbb{Z}) \xrightarrow{r^*} H^{\text{even}}(V_s,\mathbb{Z})$ induce maps

(5.11)
$$\begin{cases} H^{\text{even}}(X/S,\mathbb{Z}) \xrightarrow{r^*} H^{\text{even}}(V_s,\mathbb{Z}), \quad \text{and} \\ \qquad\qquad\qquad \rho^* \\ C(X/S) \longrightarrow H^{\text{even}}(X/S,\mathbb{Z}) \longrightarrow H^{\text{even}}(V_s,\mathbb{Z}). \end{cases}$$

Definition. The ideal Prim (X/S) *in* $C(X,S)$ *which is given by the kernel of* ρ^* *in* (5.11) *will be called the ring of primitive algebraic cycles for* $f\colon X \to S$.

Similarly, we will denote by $\text{Prim}^{\text{even}}(X/S,\mathbb{Z})$ the kernel of r^* in (5.11) To understand the importance of primitive cycles, we refer to the example of a Lefschetz pencil discussed above. On X', there is a famous theorem of Lefschetz [21] which states that, over $\mathbb{Q}$, every cohomology class is the sum of a primitive class together with a class supported on a hyperplane section. In other words, the primitive cycles are the "building blocks" for all of the homology of X'.

Now we can come to the main point. Referring to (5.7) there will be a homomorphism

(5.12) $$\nu\colon \text{Prim}\,(X/S) \to \text{Hom}\,(S,T(X/S))$$

which assigns to each primitive algebraic cycle Z on X the cross-section ν_Z given as follows: By changing Z in its rational equivalence class, we may assume that all intersections $Z_s = Z\cdot V_s$ are defined and induce algebraic cycles which are homologous to zero on V_s $(s \in S^*)$. Then we have the formula

(5.13) $$\nu_Z(s) = \Phi_{V_s}(Z_s)$$

where $\Phi_{V_s} : H(V_s) \to T(V_s)$ is the Abel-Jacobi mapping as given in section 2.

In the next section we will give some results and open problems concerning this homomorphism (5.12). In the case where $\dim X = 2$ and the base S is the projective line $\mathbb{P}_1$, the normal functions were introduced by Poincaré (1910) and used by Lefschetz [21] to give a complete analysis of the curves lying on an algebraic surface, including the results that: (i) homological and algebraic equivalence are the same for curves on a surface; (ii) the Abel-Jacobi map

$$\Phi\colon H_1(X) \longrightarrow \mathrm{Pic}_1^0(X)$$

is surjective (existence theorem for the Picard variety of an algebraic surface); and (iii) a homology class $\Gamma \in H_2(X, \mathbb{Z})$ is carried by an algebraic curve if, and only if, Γ is of type (1,1) (Lefschetz theorem). The proofs of these results were based on the formal properties of normal functions together with the Jacobi inversion theorem for the Jacobians $T(V_s)$. In the next section, we shall point out that the formal properties mostly go through, but, as discussed in section 4, the inversion theorem is completely missing and this is the hangup in trying to understand algebraic cycles on X by means of normal functions.

6. *Some results about normal functions*

We retain the notations of section 5, so that we are studying a situation

$$f\colon X \to S$$

where f has only non-degenerate critical points. In this paragraph we shall make the additional assumption that the base S is a projective line $\mathbb{P}_1$. With this assumption, the Leray spectral sequence for the constant sheaf $\mathbb{Z}$ on X and the mapping f degenerates at the E_2 term (cf. [1]), from which we may draw the following conclusions:

(i) There is a filtration on $H^{\text{even}}(X,\mathbb{Z})$ whose associated graded module is the direct sum

$$H^2(S, R^{\text{even}}_{f_*}\mathbb{Z}) + H^1(S, R^{\text{odd}}_{f_*}\mathbb{Z}) + H^0(S, R^{\text{even}}_{f_*}\mathbb{Z}).$$

(ii) The ideal $H^{\text{even}}(F,\mathbb{Z})$ of $H^{\text{even}}(X,\mathbb{Z})$ is $H^2(S, R^{\text{even}}_{f_*}\mathbb{Z})$, and the subgroup of $H^{\text{even}}(X,\mathbb{Z})$ of cycles which restrict to zero on the fibres of f is $H^2(S, R^{\text{even}}_{f_*}\mathbb{Z}) + H^1(S, R^{\text{odd}}_{f_*}\mathbb{Z})$.

iiii) From (i) and (ii) there results the isomorphism

$$\text{Prim}^{\text{even}}(X/S,\mathbb{Z}) \cong H^1(S, R^{\text{odd}}_{f_*}\mathbb{Z}). \tag{6.1}$$

We wish to study the group Hom $(S, T(X/S))$ of normal functions. For this we consider the cohomology sequence of the exponential sheaf sequence (5.9) together with the isomorphisms (5.10) and (6.7) to arrive at our basic diagram [20]

$$\begin{array}{ccccccc}
0 \to H^0(S,\mathcal{O}(\mathbb{L}))/H^0(S,\Lambda) & \to & H^0(S,\mathcal{O}(T(X/S))) & \xrightarrow{\delta} & \text{Prim}^{\text{even}}(X,\mathbb{Z}) \\
\uparrow & & \uparrow & & \| \\
0 \longrightarrow \text{Fix}\,(T(X/S)) & \longrightarrow & \text{Hom}\,(S,T(X/S)) & \xrightarrow{\delta} & \text{Prim}^{\text{even}}(X,\mathbb{Z}). \\
\uparrow & & \uparrow & & \\
0 & & 0 & &
\end{array} \tag{6.2}$$

[20] The subgroup Fix $(T(X/S))$ is defined to be the group of sections μ of $H^0(S,\mathcal{O}(\mathbb{L}))/H^0(S,\Lambda)$ whose projections into $\mathcal{O}(T(X/S))$ satisfy the equation $D\mu = 0$ as explained above (5.5).

We want to relate the diagram (6.2) to the normal functions which arise from algebraic cycles. For this we consider the diagram

$$\begin{array}{ccc} \mathrm{Hom}\,(S, T(X/S)) & \xrightarrow{\ \delta\ } & \mathrm{Prim}^{\mathrm{even}}(X, \mathbb{Z}) \\ & \nwarrow^{\nu} \quad {}^{h}\nearrow & \\ & \mathrm{Prim}\,(X/S) & \end{array} \tag{6.3}$$

which arises from (6.2), (5.12), and (5.11). Speaking geometrically, ν assigns to a primitive algebraic cycle Z on X the normal function given by (5.13), and h assigns to Z its homology class in $H^{\mathrm{even}}(X, \mathbb{Z})/H^{\mathrm{even}}(F, \mathbb{Z})$.

(6.4) *Proposition* [10]. *The diagram* (6.3) *is commutative, so that the homology class* $h(Z) \in \mathrm{Prim}^{\mathrm{even}}(X, \mathbb{Z})$ *may be computed from the corresponding normal function* ν_Z.

To give the geometric interpretation of $\mathrm{Fix}\,(T(X/S))$, we observe from (2.15) that the inclusions $V_s \subset X$ induce a map

$$T(X) \xrightarrow{\ r^*\ } \mathrm{Fix}\,(T(X/S)). \tag{6.5}$$

Moreover, in case X arises from X' by the method of Lefschetz pencils as discussed in the example of section 5, (2.15) leads to a commutative diagram

$$\begin{array}{ccc} T(X) & \xrightarrow{\ r^*\ } & \mathrm{Fix}\,(T(X/S)) \\ & \nwarrow \quad {}^{\rho^*}\nearrow & \\ & T(X'). & \end{array} \tag{6.6}$$

(6.7) *Proposition* [8]. *The restriction mapping* r^* *in* (6.5) *is onto. Moreover, in case* X *arises from* X' *by Lefschetz pencils, the mapping* ρ^* *in* (6.6) *is an isomorphism.*

There are two geometric conclusions which may be drawn from (6.4) and (6.7). The first is:

(6.8) (*Mordell-Weil for intermediate Jacobians*). The group $Hom\ (S, T(X/S))$ of normal functions is an extension of a finitely generated group by the "trace" or "fixed part" of the Jacobians along the fibres of the mapping $f\colon X \to S$.[21]

(6.9) *Remark.* This result has been proved in [8] when S is a curve of any genus but where always f is assumed to have only non-degenerate critical values. Referring to problem G in section 5 where we asked for an analysis of how the Jacobians $T(V_s)$ behave as s tends to a critical value s_0 for a general mapping $f\colon X \to S$, it should be the case that this analysis will lead to a proof of (6.8) for arbitrary mappings f.

The result (6.8) may be thought of as an analogue of problem F in section 4 over function fields.

The second consequence of (6.4) and (6.7) is

(6.10) (*Induction principle for Lefschetz pencils*). Suppose that X arises from X' by the method of Lefschetz pencils. Then the study of the (rational equivalence classes of) primitive algebraic cycles on X' as regards the homology class $h(Z')$ of such a cycle Z', or in case Z' is homologous to zero, the point $\Phi_{X'}(Z')$ in the Jacobian variety of X', may be done by studying the Jacobians of the hyperplane sections of X' and using the method of normal functions.

---

[21] By definition, the *fixed part* of the family $\{T(V_s)\}_{s \in S}$ of Jacobians along the fibres of $f\colon X \to S$ is the complex torus given by the image of the restriction mapping $T(X) \to T(V_s)$. The finitely generated group is identified in a special case by the exact sequence (6.13) below.

As mentioned at the end of section 5, this induction principle (6.10) may be thought of as giving the generalization of most of the formal properties possessed by the Jacobians of the curves in a Lefschetz pencil on an algebraic surface. However, the existence theorems are missing because we don't understand the group $I(V_s)$ of invertible points in $T(V_s)$, a state of affairs which is made even more frustrating by the following:

Let $f: X \to S$ be as above and define the group

$$\mathrm{Prim}^{\mathrm{Hodge}}(X, \mathbb{Z})$$

to be the subgroup of $\mathrm{Prim}^{\mathrm{even}}(X, \mathbb{Z})$ which comes from the subgroup $\bigoplus_{q=0}^{n} H^{q,q}(X)$ of $H^{\mathrm{even}}(X, \mathbb{C})$. Obviously the diagram (6.3) may be refined to a new diagram

$$\text{(6.11)} \qquad \begin{array}{ccc} \mathrm{Hom}\,(S, T(X/S)) & \xrightarrow{\ \delta\ } & \mathrm{Prim}^{\mathrm{even}}(X, \mathbb{Z}) \\ \Big\uparrow \nu & & \Big\uparrow i = \text{inclusion map} \\ \mathrm{Prim}\,(X/S) & \xrightarrow{\ h\ } & \mathrm{Prim}^{\mathrm{Hodge}}(X, \mathbb{Z}). \end{array}$$

(6.12) *Proposition. Suppose that X arises from X' by the method of Lefschetz pencils and where X' is a smooth hypersurface in a projective space $\mathbb{P}_{2m+1}$. Then the images of i and δ in (6.11) coincide, so that we have the exact sequence*

$$\text{(6.13)} \qquad 0 \to \mathrm{Fix}\,(T(X/S)) \to \mathrm{Hom}\,(S, T(X/S)) \to \mathrm{Prim}^{\mathrm{Hodge}}(X, \mathbb{Z}) \to 0.$$

This proposition gives in a very special case the structure of the finitely generated group in the Mordell-Weil theorem (6.8). For the problem of constructing algebraic cycles, there is the following

(6.14) *Corollary. Let X' be a smooth hypersurface in $\mathbb{P}_{2m+1}$ and $\Gamma \in H^{m,m}(X') \cap \mathrm{Prim}^{2m}(X',\mathbb{Z})$ a primitive, integral homology class of type (m,m). Then Γ comes from an algebraic cycle if, and only if, the corresponding normal function $\nu_\Gamma \in \mathrm{Hom}\,(S,T(X/S))$ satisfies the inversion property that $\nu_\Gamma(s) \in I(V_s)$ for all $s \in S^*$.*

Thus, at least for smooth hypersurfaces in projective space, the construction of algebraic cycles in a given homology class is thrown back to the inversion problem as discussed in section 4.

Because of the lovely result by Gherardelli [6] on the intermediate Jacobian of the cubic threefold, we have the following

(6.15) *Corollary. Let V be a smooth hypersurface of degree three in $\mathbb{P}_5$ (V is a cubic fourfold). Then a class $\Gamma \in H^{2m}(V,\mathbb{Z})$ is algebraic if, and only if, Γ is of type (m,m) (for any m).*

7. *Positive algebraic cycles*

(a) *Preliminary comments on vector bundles.* Let V be a smooth, projective algebraic variety and $\mathbb{E} \to V$ an algebraic vector bundle of rank r. We denote by $P(\mathbb{E})$ the projective bundle of the dual $\check{\mathbb{E}}$ of $\mathbb{E}$, and shall use the notation $\mathbb{L} \to P(\mathbb{E})$ to denote the tautological line bundle over $P(\mathbb{E})$. The isomorphism of cohomology

$$H^k(V,\mathcal{O}(\mathbb{E}^{(\mu)})) \cong H^k(P(\mathbb{E}),\mathcal{O}(\mathbb{L}^{(\mu)}))$$ [22]

may serve to eliminate confusion between bundles and their duals. Following Hartshorne [14], we say that $\mathbb{E} \to V$ is *ample* if the

---

[22] $\mathbb{E}^{(\mu)}$ is the μth symmetric power of $\mathbb{E}$, and thus the fibre $\mathbb{E}_x^{(\mu)}$ is the vector space of homogeneous forms of degree μ on the projective space $P(\mathbb{E})_x$.

tautological line bundle $\mathbb{L} \to P(\mathbb{E})$ is ample, in the usual sense of the word for line bundles. An equivalent formulation is that the *vanishing theorem*

$$H^k(V, O(\mathbb{E}^{(\mu)}) \otimes S) = 0 \qquad (k > 0, \quad \mu \geqq \mu_0(S))$$

should hold for every coherent sheaf S on V.

We shall also use the notion of very ample, which deals with the vector space $H^0(\mathbb{E})$ of holomorphic cross-sections of $\mathbb{E} \to V$. First, we recall that $\mathbb{E}$ is said to be *generated by its sections* if the restriction mappings

$$H^0(\mathbb{E}) \to \mathbb{E}_x \qquad (x \in V)$$

are surjective for all points x. In this case there is an exact bundle sequence

$$0 \to \mathbb{F} \to V \times H^0(\mathbb{E}) \to \mathbb{E} \to 0$$

where the fibre $\mathbb{F}_x = \{\sigma \in H^0(\mathbb{E}) : \sigma(x) = 0\}$. For $\sigma \in \mathbb{F}_x$, the differential $d\sigma(x) \in \mathbb{E}_x \otimes \check{\mathbb{T}}_x(V)$ is well defined, and $\mathbb{E}$ is *very ample* if $\mathbb{E}$ is generated by its sections and if we also have the surjection

$$\mathbb{F} \xrightarrow{d} \mathbb{E} \otimes \check{\mathbb{T}}(V) \longrightarrow 0.$$

Geometrically, $\mathbb{E}$ is very ample if it is induced by a holomorphic immersion

$$f : V \to \text{Grassmannian}$$

where the image $f(V)$ is "sufficiently twisted". We recall that:

$$\begin{cases} \mathbb{E} \text{ very ample} \implies \mathbb{E} \text{ ample,}^{23} \quad \text{and} \\ \mathbb{E} \text{ ample} \implies \mathbb{E}^{(\mu)} \text{ very ample for } \mu \geqq \mu_0. \end{cases}$$

For an algebraic vector bundle $\mathbb{E} \to V$ of rank r the *total Chern class*

$$c(\mathbb{E}) \in C(\mathbb{E})$$

may be defined [13]. Writing

$$c(\mathbb{E}) = c_0(\mathbb{E}) + c_1(\mathbb{E}) + \dots + c_n(\mathbb{E}),$$

we have $c_0(\mathbb{E}) = 1 \cdot V$ and $c_k(\mathbb{E}) = 0$ for $k > r$. Let $I = (i_0, \dots, i_r)$ be an r-tuple of non-negative integers and $|I| = i_1 + 2i_2 + \dots + ri_r$. We define the *Chern monomials*

$$c_I(\mathbb{E}) = c_1(\mathbb{E})^{i_1} \cdots c_r(\mathbb{E})^{i_r} \in C_{|I|}(\mathbb{E}).$$

From [11] we recall that there is a set $\mathbb{P}^+(r)$ of polynomials $P(c_1, \dots, c_r)$ with rational coefficients, called *positive polynomials*, which has the properties:

(i) The polynomials in $\mathbb{P}^+(r)$ form a graded, convex cone over $\mathbb{Q}^+$;

(ii) all monomials c_I are in $\mathbb{P}^+(r)$, but for $r > 1$ these do not generate $\mathbb{P}^+(r)$; and

(iii) if $\mathbb{E} \to V$ is a very ample vector bundle, then any positive polynomial $P(c_1(\mathbb{E}), \dots, c_r(\mathbb{E}))$ is *numerically positive* in the sense that the intersection number

---

[23] It is not the case that $\mathbb{E}$ is ample if it is induced from a holomorphic immersion $f : V \to$ Grassmannian, unless of course the rank of $\mathbb{E}$ is one.

(7.1) $$\deg\,[P(c_1(\mathbb{E}),\ldots,c_r(\mathbb{E}))\cdot Z] > 0$$

for every effective cycle Z of complementary dimension to $P(c_1(\mathbb{E}),\ldots,c_r(\mathbb{E}))$.

Remarks. For $r = 1$, the only positive polynomials of degree k are the obvious ones

$$\lambda c_1^k \qquad (\lambda \in \mathbb{Q}^+).$$

In this case, it is a theorem of Moishezon-Nakai [23] that the inequality (7.1) is both necessary and sufficient for $\mathbb{E} \to V$ to be ample.

For $r > 1$, there will be positive polynomials such as $c_1^2 - c_2$ which are not positive linear combinations of Chern monomials.

The positivity property (7.1) may well be true if we only assume that $\mathbb{E} \to V$ is ample. For the Chern classes themselves, this has recently been given a very nice proof by Bloch and Geiseker [3].

In order to state our main problem about ample vector bundles, we first give the

Definition. $\mathbb{E}$ *is numerically positive if, for each irreducible subvariety* W *of* V *and each quotient bundle* $\mathbb{Q}$ *of* $\mathbb{E}/W$*, the numerical positivity property* (7.1) *is valid.*

In [11] it was proved, by differential-geometric methods, that a sufficiently ample bundle is numerically positive in the above sense. Contrary to what mistakenly appeared in [11], the converse result does not seem to have been proved except in special cases; e.g., the result when $\dim V = 1$ is due to Hartshorne. I should like to offer my personal apology for any confusion which may have arisen from this mixup.

Problem H. *Find necessary and sufficient numerical conditions in order that a bundle* $\mathbb{E} \to V$ *should be ample. More specifically, is it true that* $\mathbb{E}$ *is ample if, and only if, it is numerically positive in the above sense?*

(b) *Positivity of divisors.* Let D be an effective divisor on a smooth, projective variety V. Any of the following four equivalent conditions may be taken as the definition of what it means for D to be *positive*:[24]

(i) There is a very ample line bundle $\mathbb{L} \to V$ such that D is a positive multiple of $c_1(\mathbb{L})$ in the rational Chow ring $C(V) \otimes_{\mathbb{Z}} \mathbb{Q}$.

(ii) For any effective algebraic q-cycle on V, the intersection number

$$D^{(q)} \cdot Z > 0.$$

(This is the Moishezon-Nakai criterion mentioned above.)

(iii) Denoting by $I(D)$ the ideal sheaf of D, we have the vanishing theorem

$$H^k(V, I(D)^{\mu} \otimes S) = 0 \qquad (k > 0, \quad \mu \geqq \mu(S)) \tag{7.2}$$

for any coherent sheaf S on V.

(iv) The complement $V - D$ is an *affine algebraic variety*, and is therefore *strongly-pseudo-convex* in the sense of complex function theory. (It is not the case that $V - D$ affine $\Longrightarrow$ D ample.)

The condition (iv) implies the vanishing theorem

$$H^k(V - D, S) = 0 \qquad (k > 0) \tag{7.3}$$

[24] We shall give the definition of positivity for non-effective divisors in a little while. It may be that the adjective "ample" should be used rather than "positive" in the present context, but we prefer to stick to the older terminology.

in either the algebraic or analytic category. Conversely, the vanishing theorem (7.3) implies either that $V-D$ is an affine algebraic variety or is strongly-pseudo-convex according to the category in which we are given the result.

I should now like to mention two uses of positive (effective) divisors for problems in algebraic geometry. Both of these results are proved using the vanishing theorem (7.2), so that one might say that whereas the conditions (i) or (ii) are the more appealing geometrically, it perhaps is (iii) which is the most useful technically.

(a) If D is a positive divisor, then there is an integer μ_0 such that the set of all effective divisors which are algebraically equivalent in the strong sense to μD $(\mu \geqq \mu_0)$ generates the identity component of the Picard variety of V.[25]

(b) If $\{V_s\}_{s\in\Delta}$ is any variation of the complex structure of $V = V_0$ given by a situation

$$f\colon W \to \Delta$$

as discussed in section 4(a), if $\Gamma \in H^2(V_s, \mathbb{Z}) \cong H^2(V_0, \mathbb{Z})$ is the homology class of a divisor on V_s for all $s \in \Delta$, and if $h(D) = \Gamma$ in $H^2(V, \mathbb{Z})$, then for $\mu \geqq \mu_0$ there will exist divisors D_s' on V_s which vary holomorphically with $s \in \Delta$ and which specialize to μD at $s = 0$.[26]

---

[25] More precisely, we let $\Lambda(\mu)$ be the (normalized) component of the the Chow variety of V which contains the effective divisor μD. Then for $\mu \geqq \mu_0$ the Abel-Jacobi mapping

$$\Phi : \mathrm{Alb}(\Lambda(\mu)) \to T_1(V)$$

is an isomorphism. This is proved in Kodaira [19].

[26] This property might be stated as saying that "sufficiently positive divisors are stable under variation of structure".

We now discuss briefly the notion of positivity for non-effective divisors. For this we observe that either of conditions (i) or (ii) still makes sense and may be used to define what it means for a general divisor D to be positive. With this definition, there are two additional properties of positive divisors which I should like to mention:

(c) If D is a (not necessarily effective) divisor which satisfies (i) or (ii) above, then for $\mu \geqq \mu_0$ the divisor μD will be linearly equivalent to an effective divisor.

(d) If D is a positive divisor and D' is any divisor, then the divisor

$$\mu D + D'$$

will be positive for $\mu \geqq \mu(D')$.

(c) *Positivity of general cycles.* In view of the usefulness of positive divisors for problems in algebraic geometry, it is of importance to have a definition for positive algebraic cycles of any codimension. I should like to propose such a definition which is based on the incidence relation discussed in section 1.

To motivate this, we first give a transcendental condition which should certainly guarantee that an algebraic cycle be positive by any reasonable definition. Thus, suppose that Z is an algebraic cycle of codimension q on V and let $h(Z) \in H^{2q}(V, \mathbb{C})$ be the fundamental class of Z. The transcendental condition we have in mind is that, using the de Rham isomorphism, $h(Z)$ should be given by a *positive* (q,q) form Θ. This means locally

$$\Theta = (\sqrt{-1})^{q(q-1)/2} \{ \sum_{\alpha} \theta_{\alpha} \wedge \overline{\theta}_{\alpha} \} \tag{7.4}$$

where the θ_{α} are $(q,0)$ forms such that, for any set $\tau_1, \ldots, \tau_q$ of linearly independent $(1,0)$ tangent vectors, the contraction

$$< \theta_\alpha, \tau_1 \wedge \ldots \wedge \tau_q > \neq 0$$

for at least one index α. If this condition is satisfied, we shall say that Z *is positive in the differential-geometric sense.*[27]

(7.5) *Example.* Z is positive in the differential-geometric sense if, in the rational Chow ring $C(V) \otimes_{\mathbb{Z}} \mathbb{Q}$, Z is a positive polynomial in the Chern classes of a very ample vector bundle (cf. [11]).

We want to draw an algebro-geometric corollary of the notion of positive in the differential-geometric sense. For this we suppose that $\{W_s\}_{s \in S}$ is an algebraic family of effective algebraic $(q-1)$ cycles on V and assume that this family is *effectively parametrized* in the sense that

$$W_s = W_{s'} \Longrightarrow s = s'.$$

Then from [4] we have the easy

(7.6) *Proposition. If Z is positive in the differential-geometric sense, then the incidence divisor D_Z is positive on S.*

With (7.6) as motivation, we propose the

(7.7) *Definition. The algebraic cycle $Z \in C_q(V)$ is positive if, for any algebraic family $\{W_s\}_{s \in S}$ of effective algebraic $(q-1)$-cycles as above, the incidence divisor D_Z is positive on S.*

Using the Moishezon-Nakai numerical criterion for positivity of divisors recalled above, we want to give an equivalent numerical

---

[27] A divisor is positive in the algebro-geometric sense if, and only if, it is positive in the differential-geometric sense (Kodaira). If Z is smooth and is positive in the differential-geometric sense, then the *normal bundle* $\mathbb{N} \to Z$ is ample according to the definition in section 7(a).

formulation of (6.5). For this we consider the effective algebraic correspondences

$$W \subset S \times V$$

where $\dim [W \cdot \{s\} \times V] = q - 1$ for all $s \in S$. We shall use the notation V^k for the product $\underbrace{V \times \ldots \times V}_{k}$, Z^k for the diagonal cycle on V^k induced by Z on V, and $W_s = W \cdot \{s\} \times V$. Given a subvariety T of S, we define the effective algebraic cycle $\Delta_k(T)$ on the product V^k by the rule

$$\Delta_k(T) = \{(x_1, \ldots, x_k) \in V^k : \text{all } x_j \in W_s \text{ for some } s \in T\}.$$

Then $\dim [\Delta_k(T)] = k(q-1) + \dim T$ and $\operatorname{codim} [Z^k] = kq$. Using Moishezon-Nakai it follows that

(7.8) *Proposition. The cycle* $Z \in C_q(V)$ *is positive if, and only if, the numerical condition*

$$\deg [Z^k \cdot \Delta_k(T)] > 0$$

is satisfied for all k-dimensional subvarieties of S.

(7.9) *Corollary. The positive cycles form a graded, convex cone* $P(V) = \bigoplus_{q=0}^{n} P_q(V)$ *in the Chow ring* $C(V)$.

The analogues of the properties (i)-(iv) in section 7(b) for positive divisors are these:

(i) In $C(V) \otimes_{\mathbb{Z}} \mathbb{Q}$, Z is a positive linear combination of positive polynomials in the Chern classes of ample vector bundles.

(ii) The numerical condition (7.8) is satisfied.

(iii) In case Z is effective, we denote by $I(Z)$ the ideal sheaf of Z. Then we have the vanishing theorem

$$H^k(V, I(Z)^\mu \otimes S) = 0 \qquad (k \geqq q, \quad \mu \geqq \mu(S))$$

for any coherent sheaf S on V.

(iv) Again assuming Z is effective, the complement $V - Z$ is everywhere strongly $(q-1)$-convex and is strongly $(n-q-1)$-concave outside a compact set in $V - Z$.[28] In particular, for any coherent analytic sheaf S on $V - Z$ we have

$$(7.10) \qquad \begin{cases} H^k(V - Z, S) = 0 & (k > q) \\ \dim H^k(V - Z, S) < \infty & (k < q, \quad S \text{ locally free}). \end{cases}$$

The vanishing-finite-dimensionality theorem (7.10) is due to Andreotti-Grauert [2] and has been stated analytically as I do not know if the corresponding algebraic theorem has been proved.[29]

Concerning these four properties, the main facts which I know are these:

Property (ii) is true if Z is positive in the differential-geometric sense (cf. Example (7.5)).

If Z is smooth and is the top Chern class of a very ample vector bundle, then all of the properties (i)-(iv) are true [11].

Problem I. *Are properties* (i)-(iv) *true for positive linear combinations of Chern classes of very ample vector bundles?*

---

[28] In case Z is smooth, the $(q-1)$-convexity is roughly supposed to mean that the normal bundle of Z should be positive plus the assumption that Z should meet every subvariety of complementary dimension. The $(n-q-1)$-concavity is always present because Z is locally the zeroes of q holomorphic functions.

[29] To be specific, suppose that Z is smooth and the normal bundle $\mathbb{N} \to Z$ is ample. Then is it true that

$$\dim H^k(V - Z, S) < \infty \qquad (k > q)$$

for any coherent algebraic sheaf S on $V - Z$ and where the cohomology is in the Zariski topology? The corresponding analytic result is true by using (7.10).

The reason for stating this problem is two-fold: First, it may not be too difficult to prove it using the methods of [11] together with the analysis of the singularities of the basic Schubert cycles associated to a very ample vector bundle which has recently been given by Kleiman and Landolfi [20]. Secondly, in the rational Chow ring $C(V) \otimes_{\mathbb{Z}} \mathbb{Q}$ every cycle Z is a linear combination of these basic Schubert cycles, so that problem H might suffice for many applications of positive subvarieties to problems in algebraic geometry. In this connection we shall close by mentioning the

Problem J. *Which analogues of the properties* (a)-(d) *in section* 7(b) *for positive divisors are true for positive algebraic cycles?*

Bibliography

1. A. Andreotti and T. Frankel, *The second Lefschetz theorem on hyperplane sections*, Global analysis (papers in honor of K. Kodaira), 1-20.

2. A. Andreotti and H. Grauert, *Théorèmes de finitude pour la cohomologie des espaces complexes*, Bull. Soc. Math. France 90 (1962), 193-259.

3. S. Bloch and D. Geiseker, *The positivity of the Chern classes of an ample vector bundle*, to appear.

4. C. H. Clemens and P. Griffiths, *Intermediate Jacobian varieties and the cubic threefold*, to appear.

5. P. Deligne, *Théorie de Hodge*, Inst. Hautes Etudes Sci. Publ. Math., to appear.

6. F. Gherardelli, *Un'osservazione sulla varietà cubica di* P^4, Rend. Sem. Mat. Fis. Milano 37 (1967), 3-6.

7. P. Griffiths, *Periods of integrals on algebraic manifolds: summary of main results and discussion of open problems*, Bull. Amer. Math. Soc. 76 (1970), 228-296.

8. ———, *Periods of integrals on algebraic manifolds* I, II, III, Amer. J. Math. 90 (1968), 568-626 and 805-865, and to appear.

9. ———, *Some results on algebraic cycles on algebraic manifolds*, Bombay colloquium on algebraic geometry (1968), 93-191.

10. ———, *On the periods of certain rational integrals* I, II, III, Ann. of Math. (2) 90 (1969), 460-495 and 498-541, and to appear.

11. P. Griffiths, *Hermitian differential geometry, Chern classes, and positive vector bundles*, Global analysis (papers in honor of K. Kodaira), 185-251.

12. A. Grothendieck, *Hodge's general conjecture is false for trivial reasons*, Topology 8 (1969), 299-303.

13. ———, *La théorie deś classes de Chern*, Bull. Soc. Math. France 86 (1958), 137-154.

14. R. Hartshorne, *Ample vector bundles*, Inst. Hautes Etudes Sci. Publ. Math. 29 (1966), 319-350.

15. N. Katz, *Etude cohomologique des pinceaux de Lefschetz*, Séminaire de Géométrie Algébrique No. 7, Exposé XVIII.

16. J. King, *The currents defined by analytic varieties*, Acta Math., to appear.

17. K. Kodaira, *On Kähler varieties of restricted type*, Ann. of Math. (2) 60 (1954), 28-48.

18. ———, *Green's forms and meromorphic functions on compact analytic varieties*, Canad. J. Math. 3 (1951), 108-128.

19. ———, *Characteristic linear systems of complete continuous systems*, Amer. J. Math. 78 (1956), 716-744.

20. S. Kleiman and J. Landolfi, *Singularities of special Schubert varieties*, to appear.

21. S. Lefschetz, *L'analysis situs et la géométrie algébrique*, Gauthier-Villars, Paris, 1924.

22. D. Lieberman, *Higher Picard varieties*, Amer. J. Math. 90 (1968), 1165-1199.

23. Y. Nakai, *A criterion of an ample sheaf on a projective scheme*, Amer. J. Math. 85 (1963), 14-26.

24. P. Samuel, *Relations d'equivalence en géométrie algébrique*, Proc. International Cong. Math. 1958, 470-487.

25. A. Weil, *On Picard varieties*, Amer. J. Math. 74 (1952), 865-893.

Princeton University, Princeton, New Jersey 08540

Remark added in proof. A recent monograph by R. Hartshorne entitled *Ample subvarieties of algebraic varieties* (Springer-Verlag Lecture Notes #156) treats many questions closely related to our section 6.

ANALYTIC STRUCTURES ON THE SPACE OF FLAT VECTOR BUNDLES OVER A COMPACT RIEMANN SURFACE

R. C. Gunning

1. *Introduction*

Consider a compact Riemann surface M of genus $g > 1$. The set of flat complex line bundles over M is by definition the cohomology group $H^1(M, \mathbb{C}^*)$, where $\mathbb{C}^*$ is the multiplicative group of non-zero complex numbers; and this cohomology group can be identified in a familiar manner with the group $\mathrm{Hom}(\pi_1(M), \mathbb{C}^*) \cong (\mathbb{C}^*)^{2g}$, hence has the structure of a complex Lie group of complex dimension $2g$. The exact sequence of sheaves of Abelian groups

$$0 \longrightarrow \mathbb{C}^* \xrightarrow{i} \mathcal{O}^* \xrightarrow{dl} \mathcal{O}^{1,0} \longrightarrow 0,$$

where $\mathcal{O}^*$ is the multiplicative sheaf of germs of holomorphic nowhere-vanishing functions, $\mathcal{O}^{1,0}$ is the additive sheaf of germs of holomorphic differential forms of type (1,0), and dl is the mapping which takes a germ $f \in \mathcal{O}^*$ to the germ $\frac{1}{2\pi i}\, d \log f \in \mathcal{O}^{1,0}$, gives rise to an exact cohomology sequence containing the segment

$$0 \longrightarrow \Gamma(M, \mathcal{O}^{1,0}) \xrightarrow{\delta} H^1(M, \mathbb{C}^*) \xrightarrow{i} H^1(M, \mathcal{O}^*).$$

Two flat line bundles are called analytically equivalent when they determine the same complex analytic line bundle, or what is the same thing, when they have the same image in $H^1(M, \mathcal{O}^*)$; and the latter exact sequence shows that the analytic equivalence classes of flat line bundles are just the cosets of the Lie subgroup $\delta\Gamma(M, \mathcal{O}^{1,0}) \subset H^1(M, \mathbb{C}^*)$.

All the equivalence classes are thus complex analytic submanifolds of $H^1(M, \mathbb{C}^*)$, analytically equivalent to $\mathbb{C}^g$; and $H^1(M, \mathbb{C}^*)$ is a complex analytic fibre bundle over the quotient space $H^1(M, \mathbb{C}^*)/\delta\Gamma(M, \mathcal{O}^{1,0})$ under this equivalence relation. The quotient space is of course the Picard variety of the Riemann surface M.

In an analogous manner, the set of flat complex vector bundles of rank n over M is by definition the cohomology set $H^1(M, GL(n, \mathbb{C}))$; this set does not possess a natural group structure when $n > 1$, since the general linear group $GL(n, \mathbb{C})$ is not then an Abelian group. Two flat vector bundles are called analytically equivalent when they determine the same complex analytic vector bundle; and there is then the problem of describing this equivalence relation on the set $H^1(M, GL(n, \mathbb{C}))$. The present paper is a discussion of this problem, correcting and extending the treatment of the same topic in §9 of the notes [2].

As a preliminary observation, note that it is possible to introduce, in addition to the set of flat vector bundles $H^1(M, GL(n, \mathbb{C}))$, the set of special flat vector bundles $H^1(M, SL(n, \mathbb{C}))$. Any flat vector bundle can be represented in a unique manner as the tensor product of a flat line bundle and a special flat vector bundle; and two flat vector bundles are analytically equivalent if and only if their factors in this representation are correspondingly analytically equivalent. Since the analytic equivalence of flat line bundles can be taken as a known relation, it follows that it is possible to restrict the discussion henceforth to the case of special flat vector bundles. Of course it is further possible to introduce the set $H^1(M, PL(n, \mathbb{C}))$ of projective flat vector bundles, where the projective linear group $PL(n, \mathbb{C})$ is the quotient of the general linear group $GL(n, \mathbb{C})$ by its center, or equivalently, the quotient of the special linear group $SL(n, \mathbb{C})$ by its center. Imbedding $\mathbb{Z}/n\mathbb{Z} = \mathbb{Z}_n$ in $\mathbb{C}^*$ as the subgroup consisting of the n-th roots of unity, the elements of $H^1(M, \mathbb{Z}_n)$ can be viewed as flat complex line bundles over M; and under tensor product, the group $H^1(M, \mathbb{Z}_n)$ acts as a finite group of transformations of the set $H^1(M, SL(n, \mathbb{C}))$. Note that $H^1(M, PL(n, \mathbb{C}))$ splits naturally into n components, which can be put in one-to-one correspondences with one another; and that the quotient space $H^1(M, SL(n, \mathbb{C}))/H^1(M, \mathbb{Z}_n)$ can in turn be put in one-to-one correspondence with each of these components. This follows quite directly from an examination of the exact sequence

$$0 \longrightarrow \mathbb{Z}_n \longrightarrow SL(n, \mathbb{C}) \longrightarrow PL(n, \mathbb{C}) \longrightarrow 0,$$

so the details will be omitted. Since analytic equivalence commutes with these correspondences, it follows again that it is really sufficient to restrict the discussion henceforth to the case of special flat vector bundles.

Perhaps it should be mentioned, before beginning the detailed discussion, that the set of analytic equivalence classes of flat vector bundles of rank n can be described, as a subset of the set of complex analytic vector bundles of rank n, by Weil's theorem, [7]; and that the further subset consisting of stable vector bundles, or equivalently consisting of irreducible unitary flat bundles, can be given a complex analytic structure, as in the work of M. S. Narasimhan and C. S. Seshadri, [4,6]. There seems at present no further interpretation of this quotient space, though. On the other hand, at least some of the individual fibres in this equivalence do have a further interpretation; for example, the unique analytic equivalence class in $H^1(M, PL(2, \mathbb{C}))$ containing the maximally unstable bundles consists precisely of those projective bundles arising from projective coordinate coverings of the Riemann surface M, and thus has some significance in studying the uniformizations of the surface M, [3].

This indicates that there may well be more interest in the space $H^1(M, SL(n,\mathbb{C}))$ and its decomposition into equivalence classes than in the quotient space itself.

2. *Analytic structure of the space of flat bundles*

There is a natural one-to-one correspondence between the cohomology set $H^1(M, SL(n,\mathbb{C}))$ and the quotient space $\mathrm{Hom}(\pi_1(M), SL(n,\mathbb{C}))/SL(n,\mathbb{C})$, where $SL(n,\mathbb{C})$ acts as a group of transformations on $\mathrm{Hom}(\pi_1(M), SL(n,\mathbb{C}))$ by inner automorphisms, [1]; actually of course the elements of the center of $SL(n,\mathbb{C})$ act trivially as inner automorphisms, so this quotient space can be written equivalently as $\mathrm{Hom}(\pi_1(M), SL(n,\mathbb{C}))/PL(n,\mathbb{C})$. The fundamental group $\pi_1(M)$ can be described canonically as a group with $2g$ generators $\sigma_1,\ldots,\sigma_g,\tau_1,\ldots,\tau_g$ and one defining relation $[\sigma_g,\tau_g]\cdots[\sigma_1,\tau_1] = 1$, where $[\sigma,\tau] = \sigma\tau\sigma^{-1}\tau^{-1}$. Introducing the complex analytic mapping $F\colon SL(n,\mathbb{C})^{2g} \to SL(n,\mathbb{C})$ defined by $F(S_1,\ldots,S_g,T_1,\ldots,T_g) = [S_g,T_g]\cdots[S_1,T_1]$, and the complex analytic subvariety $R = \{(S_1,\ldots,S_g,T_1,\ldots,T_g) \in SL(n,\mathbb{C})^{2g} \mid F(S_1,\ldots,S_g,T_1,\ldots,T_g) = I\}$, the mapping which associates to an element $\rho \in \mathrm{Hom}(\pi_1(M), SL(n,\mathbb{C}))$ the point $(\rho(\sigma_1),\ldots,\rho(\sigma_g),\rho(\tau_1),\ldots,\rho(\tau_g)) \in R$ clearly identifies $\mathrm{Hom}(\pi_1(M), SL(n,\mathbb{C}))$ with the complex analytic variety R, and can thus be used to impose the structure of a complex analytic variety on the set $\mathrm{Hom}(\pi_1(M), SL(n,\mathbb{C}))$. It is quite easy to verify that this complex analytic variety structure is independent of the choice of a presentation of the group $\pi_1(M)$; but this is not needed here, so one presentation will be selected and the set $\mathrm{Hom}(\pi_1(M), SL(n,\mathbb{C}))$ will be identified with the complex analytic subvariety $R \subset SL(n,\mathbb{C})^{2g}$. This subvariety has singularities whenever $n > 1$, $g > 1$; of course, a similar construction can be carried out for any finitely presented group, and this leads to an interesting class of singularities to be investigated. These singularities do not seem to play a significant role in the present investigation, though; so introduce the subset $R_0 \subset R$ consisting of those points at which the mapping F is regular, noting that R_0 is a complex analytic manifold of complex dimensions $(n^2-1)(2g-1)$.

Theorem 1. *The manifold R_0 is the dense open subset of R consisting of those representations of the group $\pi_1(M)$ having only scalar commutants. The tangent space to the manifold R_0 at a representation ρ can be identified with the space $Z^1(\pi_1(M), \mathrm{Ad}_0\rho)$ of cocycles of the group $\pi_1(M)$ with coefficients in the $\pi_1(M)$-module of $n \times n$ matrices of trace zero under the group representation* $\mathrm{Ad}\,\rho$. (Here it is assumed that $n > 1$, $g > 1$.)

Proof. The differential of the mapping F in the natural coordinates was calculated explicitly in [2, page 182] for the case of the full general linear group; the calculation is the same for the special linear group, except that the natural coordinates in the Lie algebra are $n \times n$ matrices of trace zero. Consequently the kernel of the differential dF_ρ of the mapping F at the representation ρ can be identified with the vector space $Z^1(\pi_1(M), \mathrm{Ad}_0\rho)$; and the condition that F be regular at ρ is just that

$$\dim Z^1(\pi_1(M), \mathrm{Ad}_0\rho) = \dim SL(n, \mathbb{C})^{2g} - \dim SL(n, \mathbb{C}) = (n^2-1)(2g-1).$$

However, recalling [2, page 133], note that

$$\dim Z^1(\pi_1(M), \mathrm{Ad}_0\rho) = (n^2-1)(2g-1) + q,$$

where q is the largest integer such that the identity representation of degree q is contained in the adjoint representation to $\mathrm{Ad}_0\rho$; thus q is the dimension of the space of matrices A of trace zero such that $\rho(\gamma)A = A\rho(\gamma)$ for all $\gamma \in \pi_1(M)$. Consequently F is regular at ρ precisely when only scalar matrices (and thus no matrices of trace zero) commute with the representation ρ, and the proof is thereby concluded.

Remark. This proof is basically the same as the first part of the proof of Theorem 26 in [2, pages 184-185]; but the last step of the latter proof, the assertion that a representation is irreducible precisely when it has only scalar commutants, is of course sheer nonsense, and the statements of Theorem 26 and several of the later results must be modified accordingly.

Theorem 2. The subset $R_0 \subset R$ *is mapped onto itself under inner automorphisms by any element of* $PL(n, \mathbb{C})$*; and the quotient space* $V = R_0/PL(n, \mathbb{C})$ *has the structure of a complex analytic manifold of dimension* $2(n^2-1)(g-1)$*, such that the natural projection* $R_0 \to V$ *is a complex analytic principal* $PL(n, \mathbb{C})$ *bundle. The tangent space to the manifold* V *at a point corresponding to a representation* $\rho \in R_0$ *can be identified with the cohomology group* $H^1(\pi_1(M), \mathrm{Ad}_0\rho)$ *of the group* $\pi_1(M)$ *with coefficients in the* $\pi_1(M)$*-module of* $n \times n$ *matrices of trace zero under the group representation* $\mathrm{Ad}\,\rho$. (Here it is also assumed that $n > 1$, $g > 1$.)

Proof. The subset $R_0 \subset R$ consists of those $2g$-tuples of matrices having only scalar commutants, by Theorem 1; and this subset is obviously preserved by inner automorphisms. Indeed, the group $PL(n, \mathbb{C})$ acts freely (no transformation in $PL(n, \mathbb{C})$ other than the identity has any fixed points) as a complex analytic transformation group on R_0. The crucial step in the next part of the proof consists in showing that any point $\rho \in R_0$ has an open neighborhood U in R_0 such that the set $\{T \in PL(n, \mathbb{C}) | \mathrm{Ad}(T)U \cap U \neq \emptyset\}$ has compact closure in $PL(n, \mathbb{C})$; and for this purpose it suffices to show that whenever U is an open neighborhood of ρ such that $\overline{U}$ is compact and contained in R_0, then $\tilde{U} = \{T \in PL(n, \mathbb{C}) | \mathrm{Ad}(T)\overline{U} \cap \overline{U} \neq \emptyset\}$ is already compact in $PL(n, \mathbb{C})$. Note that $T \in \tilde{U}$ precisely when there are points $\sigma, \tau \in \overline{U}$

such that $\tau = \mathrm{Ad}(T)\sigma = T\sigma T^{-1}$; here σ, τ are $2g$-tuples of matrices in $SL(n,\mathbb{C})$, and the last equation must hold for corresponding components. Now the set $\{(\sigma,\tau)\in\overline{U}\times\overline{U}\mid \tau = T\sigma T^{-1}$ for some $T\in PL(n,\mathbb{C})\}$ is a closed and hence compact subset of $\overline{U}\times\overline{U}$, since the condition that (σ,τ) belongs to this subset is an algebraic condition on the matrix components; and for each point (σ,τ) in this subset there is a unique element $T\in PL(n,\mathbb{C})$, depending continuously on (σ,τ), such that $\tau = T\sigma T^{-1}$. Thus the set $\tilde{U}$ is the continuous image of a compact set, hence is itself compact, as desired. Having made this observation, it is a standard argument in the general theory of transformation groups that the quotient space $V = R_o/PL(n,\mathbb{C})$ is a complex manifold and the natural projection $R_o \to V$ is a fibration; a detailed proof is given in [2, pages 193-195], where the tangent space to V is also demonstrated to have the desired form. That then serves to complete the proof.

The topological and analytical properties of the manifold V have not yet been studied in detail; but the corresponding properties of the spaces of unitary or stable bundles have been treated by Newstead, [5].

3. *Analytic structures of spaces of analytically equivalent flat bundles*

Fixing a special flat vector bundle $\rho\in H^1(M,SL(n,\mathbb{C}))$, consider next the problem of describing the set of all special flat vector bundles analytically equivalent to ρ. The abstract description of this analytic equivalence class is most easily handled in terms of the complex analytic vector bundle represented by ρ; to simplify matters, this description will be carried out here only for bundles of rank 2.

Consider therefore a complex analytic vector bundle Φ, of rank 2 and determinant 1, on the Riemann surface M; and suppose that Φ admits a flat representative. By Weil's theorem, the latter condition is equivalent to the assertion that either Φ is analytically indecomposable or Φ is analytically equivalent to a direct sum of two line bundles, each having Chern class zero, [2, 7]. In terms of a coordinate covering $U = \{U_\alpha\}$ of the surface M, the bundle Φ can be represented by holomorphic matrices of the form

$$(1) \qquad \Phi_{\alpha\beta}(z) = \begin{pmatrix} \phi_{\alpha\beta}(z) & \tau_{\alpha\beta}(z) \\ 0 & \phi_{\alpha\beta}^{-1}(z) \end{pmatrix}$$

in each intersection $U_\alpha \cap U_\beta$; the holomorphic functions $\phi_{\alpha\beta}(z)$ represent a complex analytic line bundle ϕ contained as a subbundle of Φ, and it can be assumed that the Chern class of ϕ is the maximum of the Chern classes of all complex analytic line bundles contained in Φ. If the Chern class $c(\phi) \geqq 0$ and the bundle Φ is indecomposable, then the line bundle ϕ is uniquely determined, and $c(\phi) \leqq g-1$; and if the bundle Φ is decomposable, then by Weil's theorem $c(\phi) = 0$. If $c(\phi) < 0$ the bundle Φ is called *stable*; in this case, the line bundle ϕ is not necessarily uniquely determined, but at any rate $c(\phi) \geqq -g$. (Proofs of these assertions can be found in [2], among other places.)

The special flat bundles analytically equivalent to Φ can of course be described by cocycles of the form $\rho_{\alpha\beta} = F_\alpha(z)\Phi_{\alpha\beta}(z)F_\beta(z)^{-1}$, for some holomorphic matrices $F_\alpha(z)$ of determinant 1 in each neighborhood U_α, after passing to a refinement of the covering U if necessary. Conversely any set of holomorphic matrices $F_\alpha(z)$ of determinant 1 for which the expressions $\rho_{\alpha\beta}$ are constants describes a special flat vector bundle analytically equivalent to Φ; and two sets of such matrices $F_\alpha(z)$, $G_\alpha(z)$ describe the same flat bundle if and only if $G_\alpha(z) = C_\alpha F_\alpha(z)T_\alpha(z)$, where $C_\alpha \in SL(2,\mathbb{C})$ are constants and $T_\alpha(z)$ are holomorphic matrices of determinant 1 such that

$$T_\alpha(z)\Phi_{\alpha\beta}(z) = \Phi_{\alpha\beta}(z)T_\beta(z)$$

in each intersection $U_\alpha \cap U_\beta$. (As always, it may be necessary to pass to a refinement of the covering U; this comment will be understood, but not mentioned specifically henceforth.) A set of holomorphic matrices $T_\alpha(z)$ of determinant 1 satisfying this last condition will be called an *automorphism* of the bundle Φ; and the set of all such automorphisms will be denoted by Aut(Φ). Note that if $T^1 = \{T_\alpha^1(z)\}$ and $T^2 = \{T_\alpha^2(z)\}$ are two elements of Aut(Φ), then $T^1T^2 = \{T_\alpha^1(z)T_\alpha^2(z)\}$ is also contained in Aut(Φ); it is easy to see that with this operation Aut(Φ) is a complex Lie group in the usual sense.

It is an easy calculation to see that the expressions $\rho_{\alpha\beta}$ as above are constants precisely when the holomorphic matrix differential forms $\lambda_\alpha(z) = F_\alpha(z)^{-1}dF_\alpha(z)$ satisfy

$$(2) \qquad d\Phi_{\alpha\beta}(z) + \lambda_\alpha(z)\Phi_{\alpha\beta}(z) - \Phi_{\alpha\beta}(z)\lambda_\beta(z) = 0 \quad \text{in} \quad U_\alpha \cap U_\beta ;$$

these differential forms have trace zero, and for any matrix differential forms $\lambda_\alpha(z)$ of trace zero there are holomorphic functions $F_\alpha(z)$ of determinant 1 such that $\lambda_\alpha(z) = F_\alpha(z)^{-1}dF_\alpha(z)$. A set of holomorphic matrix differential forms $\lambda_\alpha(z)$ of trace zero satisfying (2) will be called a *connection* for the bundle Φ; and the set of all connections will be denoted by $\Lambda(\Phi)$. The set $\Lambda(\Phi)$ is not empty, since by assumption the bundle Φ has at least one flat representative; and the difference of any two connections is an element of $\Gamma(M,\mathcal{O}^{1,0}(\mathrm{Ad}_o\Phi))$. Thus the set $\Lambda(\Phi)$ can be identified with the complex analytic manifold $\mathbb{C}^m$, where $m = \dim \Gamma(M,\mathcal{O}^{1,0}(\mathrm{Ad}_o\Phi))$. Note that if $\{\lambda_\alpha(z)\} = \{F_\alpha(z)^{-1}dF_\alpha(z)\} \in \Lambda(\Phi)$, and if $G_\alpha(z) = C_\alpha F_\alpha(z)T_\alpha(z)$ for some constants C_α and automorphism $T = \{T_\alpha(z)\} \in \mathrm{Aut}(\Phi)$, then

$$\{G_\alpha(z)^{-1}dG_\alpha(z)\} = \{T_\alpha(z)^{-1}\lambda_\alpha(z)T_\alpha(z) + T_\alpha(z)^{-1}dT_\alpha(z)\} \in \Lambda(\Phi).$$

The Lie group Aut(Φ) thus acts as a group of complex analytic transformations on the complex analytic manifold $\Lambda(\Phi)$, by defining

$$\lambda_\alpha^T(z) = T_\alpha(z)^{-1}\lambda_\alpha(z)T_\alpha(z) + T_\alpha(z)^{-1}dT_\alpha(z)$$

for any $\{\lambda_\alpha(z)\}\in\Lambda(\Phi)$ and $T = \{T_\alpha(z)\}\in \mathrm{Aut}(\Phi)$; and there is a one-to-one correspondence between the set of special flat vector bundles analytically equivalent to the bundle Φ and the quotient space $\Lambda(\Phi)/\mathrm{Aut}(\Phi)$. This description leads quite readily to a number of interesting results.

Note first that considering $\Lambda(\Phi)$ and Aut(Φ) as complex manifolds, $\dim \Lambda(\Phi) - \dim \mathrm{Aut}(\Phi) =$

$$\dim \Gamma(M,\mathcal{O}^{1,0}(\mathrm{Ad}_o\Phi)) - \dim \Gamma(M,\mathcal{O}(\mathrm{Ad}_o\Phi)) = (n^2-1)(g-1),$$

by the Riemann-Roch formula; in particular, in the case $n = 2$ under consideration here, the difference in dimensions is $3(g-1)$. Consequently $\dim \Lambda(\Phi) \geqq 3(g-1)$, so that the manifold $\Lambda(\Phi)$ is certainly non-trivial whenever $g > 1$. The dimension of the manifold $\Lambda(\Phi)$ does vary with Φ, depending on the number of automorphisms of the bundle Φ; but if the group Aut(Φ) acts freely and properly as a group of transformations of the manifold $\Lambda(\Phi)$, so that the quotient space $\Lambda(\Phi)/\mathrm{Aut}(\Phi)$ is itself a complex manifold, then $\dim \Lambda(\Phi)/\mathrm{Aut}(\Phi) = 3(g-1)$ is independent of Φ.

This raises the question of the nature of the fixed points of transformations in Aut(Φ). Note that the scalar automorphisms form a finite subgroup of Aut(Φ) that acts trivially on the manifold $\Lambda(\Phi)$; hence it is more relevant to consider the quotient group P Aut(Φ) of Aut (Φ) by the scalar automorphisms, noting that P Aut(Φ) is a Lie group of the same dimension as Aut(Φ).

Theorem 3. *A connection* $\lambda = \{\lambda_\alpha(z)\}\in\Lambda(\Phi)$ *is a fixed point under a transformation* $T = \{T_\alpha(z)\}\in P$ Aut(Φ) *other than the identity if and only if the special flat vector bundle corresponding to* λ *has non-scalar commutants.*

Proof. The connection $\{\lambda_\alpha(z)\} \in \Lambda(\Phi)$ is a fixed point under the transformation $\{T_\alpha(z)\} \in \mathrm{Aut}(\Phi)$ precisely when $dT_\alpha(z) = T_\alpha(z)\lambda_\alpha(z) - \lambda_\alpha(z)T_\alpha(z)$. Selecting holomorphic matrices $F_\alpha(z)$ of determinant 1 such that $\lambda_\alpha(z) = F_\alpha(z)^{-1}dF_\alpha(z)$, the cocycle $\rho_{\alpha\beta} = F_\alpha(z)\Phi_{\alpha\beta}(z)F_\beta(z)^{-1}$ represents the special flat vector bundle corresponding to λ. Note that the holomorphic matrices $C_\alpha(z) = F_\alpha(z)T_\alpha(z)F_\alpha(z)^{-1}$ are non-scalar matrices, since by hypothesis the matrices $T_\alpha(z)$ are non-scalar; and that $C_\alpha(z)\cdot\rho_{\alpha\beta} = \rho_{\alpha\beta}\cdot C_\beta(z)$. A simple calculation shows that the functions $C_\alpha(z)$ are constants precisely when

$$0 = dC_\alpha(z) = F_\alpha(z)[dT_\alpha(z) + \lambda_\alpha(z)T_\alpha(z) - T_\alpha(z)\lambda_\alpha(z)]F_\alpha(z)^{-1} ;$$

and that suffices to conclude the proof.

Theorem 4. For any complex vector bundle Φ *of rank 2 and determinant 1 which admits flat representatives, the subset* $\Lambda_o(\Phi) \subseteq \Lambda(\Phi)$, *consisting of those connections which are not fixed points of any transformation of* $P\,\mathrm{Aut}(\Phi)$ *other than the identity, is a dense open subset of the manifold* $\Lambda(\Phi)$. *The quotient space* $\Lambda_o(\Phi)/P\,\mathrm{Aut}(\Phi)$ *has the structure of a complex analytic manifold of complex dimension* $3(g-1)$ *such that the natural projection* $\Lambda_o(\Phi) \to \Lambda_o(\Phi)/P\,\mathrm{Aut}(\Phi)$ *is a complex analytic principal* $P\,\mathrm{Aut}(\Phi)$ *bundle.*

Proof. Writing the bundle Φ in the form (1), it is a straightforward calculation to determine the group $\mathrm{Aut}(\Phi)$ and the manifold $\Lambda(\Phi)$ quite explicitly; that the subset $\Lambda_o(\Phi)$ is as desired, and that the group $P\,\mathrm{Aut}(\Phi)$ acts freely and properly (in the sense used in the proof of Theorem 2) on the manifold $\Lambda_o(\Phi)$, then follow by inspection. Since the details are given in [2, pages 212-222], there is no need to repeat them here.

Corollary. Any special flat vector bundle is analytically equivalent to a bundle having only scalar commutants.

Proof. This is an immediate consequence of Theorems 3 and 4.

Some remarks are perhaps in order here, before passing on to the next stage of the discussion. The earlier comment, that flat vector bundles with non-scalar commutants (or equivalently, representations of $\pi_1(M)$ with non-scalar commutants) do not seem to play an important role in this discussion, can be considered as being justified by the above Corollary; that has the great advantage that the discussion can to a great extent be limited to the complex manifolds $V = R_0/PL(2,\mathbb{C})$ and $\Lambda_0(\Phi)/P\ \mathrm{Aut}(\Phi)$. In the detailed proof of Theorem 4, it is noted that for a stable vector bundle Φ necessarily $P\ \mathrm{Aut}(\Phi) = 1$; consequently all stable special flat vector bundles have only scalar commutants. Furthermore the space $\Lambda_0(\Phi)/P\,\mathrm{Aut}(\Phi)$ as a complex analytic manifold is in this case analytically the same as the manifold $\mathbb{C}^{3(g-1)}$. For an indecomposable unstable vector bundle, on the other hand, it follows that the group $P\ \mathrm{Aut}(\Phi)$ is just the additive group $\mathbb{C}^m$, where $m = \dim \Gamma(M,\mathcal{O}(\phi^2))$; but an examination of the proof of Theorem 4 easily shows that, except in the special case that $c(\phi) = 0$, none of these transformations has any fixed points. Thus all special flat vector bundles representing unstable bundles with $c(\phi) > 0$ also have only scalar commutants. Since the group $P\ \mathrm{Aut}(\phi)$ is non-trivial, although the quotient $\Lambda(\Phi)/P\ \mathrm{Aut}(\Phi)$ is a complex manifold of complex dimension $3(g-1)$, it may not be the case that this manifold is just the manifold $\mathbb{C}^{3(g-1)}$; this is a matter yet to be settled. One special case of interest is that in which the line bundle ϕ has Chern class $c(\phi) = g-1$; actually it is necessary then that ϕ^2 be the canonical bundle, so that $P\ \mathrm{Aut}(\Phi) = \mathbb{C}^g$. In this case, it is easy to see that $\Lambda(\Phi)/P\ \mathrm{Aut}(\Phi) = \mathbb{C}^{3(g-1)}$; the projective bundles analytically equivalent to Φ are those arising from the projective coordinate coverings of the surface M, as discussed in [3]. Finally, when the bundle Φ is analytically decomposable, it is not necessarily the case that the $3(g-1)$-dimensional complex manifold $\Lambda_0(\Phi)/P\,\mathrm{Aut}(\Phi)$ is analytically, or even topologically, equivalent to $\mathbb{C}^{3(g-1)}$.

4. *Decomposition of the space of flat bundles into analytic equivalence classes*

Each analytic equivalence class $\Lambda_0(\Phi)/P\,\mathrm{Aut}(\Phi)$ of special flat vector bundles with only scalar commutants has been given the structure of a complex analytic manifold of dimension $3(g-1)$; and at the same time, each such class can be imbedded as a subset of the complex analytic manifold $V = R_0/PL(2,\mathbb{C})$ of dimension $6(g-1)$. This naturally leads to the question of the relationship between these two complex structures; and that question is answered in part by the following result.

Theorem 5. *For any complex vector bundle Φ of rank* 2 *and determinant* 1, *the inclusion mapping $\Lambda_0(\Phi)/P\,\mathrm{Aut}(\Phi) \to V$ is a non-singular complex analytic mapping between these two complex manifolds. The tangent space to the image at a point $\rho \in V$ can be identified with the* $3(g-1)$-*dimensional subspace of $H^1(\pi_1(M),\mathrm{Ad}_0\rho)$ consisting of the period classes of the Prym differentials $\Gamma(M,\mathcal{O}^{1,0}(\mathrm{Ad}_0\rho))$.*

Proof. Select a connection $\{\lambda_\alpha(z)\} \in \Lambda(\Phi)$, and let $\theta_i(z) = \{\theta_{i\alpha}(z)\}$ be a basis of the vector space $\Gamma(M,\mathcal{O}^{1,0}(\mathrm{Ad}_0\Phi))$ of 2×2 matrices of holomorphic differential forms $\theta_{i\alpha}(z)$ of trace zero in the various neighborhoods U_α such that $\theta_{i\alpha}(z) = \Phi_{\alpha\beta}(z)\theta_{i\beta}(z)\Phi_{\alpha\beta}(z)^{-1}$ in $U_\alpha \cap U_\beta$. Then local coordinates $\{t_i\}$ can be introduced in an open neighborhood Δ of the given connection in $\Lambda(\Phi)$ by

$$\{t_i\} \to \{\lambda_\alpha(z) + \sum_i t_i\theta_{i\alpha}(z)\}.$$

If the neighborhood Δ is sufficiently small, then by the Cauchy-Kowalewsky theorem there are holomorphic matrix functions $F_\alpha(z,t)$ of determinant 1 in $U_\alpha \times \Delta$ such that

$$(3) \qquad \frac{\partial}{\partial z} F_\alpha(z,t) = F_\alpha(z,t)\cdot[\lambda_\alpha(z) + \sum_i t_i\theta_{i\alpha}(z)];$$

and then $\rho_{\alpha\beta}(t) = F_\alpha(z,t)\Phi_{\alpha\beta}(z)F_\beta(z,t)^{-1}$ is a cocycle representing the flat bundle corresponding to the parameter value $\{t_i\}$ in Δ. It is clear that the mapping $\Lambda(\Phi) \to R$ is complex analytic, and hence that the induced mapping $\Lambda_o(\Phi)/P\ \mathrm{Aut}(\Phi) \to V$ is also complex analytic.

The tangent space to the manifold V at the representation ρ corresponding to the given connection $\{\lambda_\alpha(z)\}$ has been identified with the cohomology group $H^1(\pi_1(M),\mathrm{Ad}_o\rho)$; it is more convenient though, in dealing with the Riemann surface M, to consider this cohomology group in the form $H^1(M,\mathrm{Ad}_o\rho)$, where now $\mathrm{Ad}_o\rho$ denotes the flat vector bundle over M defined by the coordinate transition functions $\mathrm{Ad}\,\rho_{\alpha\beta}$ acting on the space of matrices of trace zero. (The equivalence of these two representations is discussed in [2].) Now the image in V of the curve through $\{\lambda_\alpha(z)\}\in\Lambda(\Phi)$ described by varying the single parameter t_i has as tangent vector at the representation ρ the vector

$$\{\rho_{\alpha\beta}(t)^{-1}\frac{\partial}{\partial t_i}\rho_{\alpha\beta}(t)|_{t=0}\} \in H^1(M,\mathrm{Ad}_o\rho);$$

and recalling the above form for the cocycle $\{\rho_{\alpha\beta}(t)\}$ an easy calculation shows that this vector is given by

$$(4) \qquad \rho_{\alpha\beta}(t)^{-1}\frac{\partial}{\partial t_i}\rho_{\alpha\beta}(t)|_{t=0} = \rho_{\alpha\beta}^{-1}G_{i\alpha}(z)\rho_{\alpha\beta} - G_{i\beta}(z)$$

where $G_{i\alpha}(z) = (\frac{\partial}{\partial t_i}F_\alpha(z,t))F_\alpha(z,t)^{-1}|_{t=0}$. However, upon differentiating this expression for the holomorphic matrix functions $G_{i\alpha}(z)$ and using formula (3), it follows readily that $dG_{i\alpha}(z) = F_\alpha(z)\theta_{i\alpha}(z)F_\alpha(z)^{-1}$;

but these differential forms satisfy $dG_{i\alpha}(z) = \rho_{\alpha\beta} dG_{i\beta}(z) \rho_{\alpha\beta}^{-1}$, hence are a basis for the space of Prym differentials $\Gamma(M, \mathcal{O}^{1,0}(\mathrm{Ad}_0\rho))$, and the cocycle (4) represents the period class of the corresponding Prym differential. Therefore the image in $H^1(M, \mathrm{Ad}_0\rho)$ of the mapping on the tangent spaces induced by the inclusion mapping $\Lambda_0(\Phi)/P\,\mathrm{Aut}(\Phi) \to V$ is precisely the space of period classes of the Prym differentials $\Gamma(M, \mathcal{O}^{1,0}(\mathrm{Ad}_0\rho))$; and since this latter space has dimension $3(g-1)$, as demonstrated in [2], it follows that the inclusion mapping is non-singular, and the proof is thereby concluded.

It follows from this theorem that the period classes of the Prym differentials describe an analytic involutive distribution in the tangent bundle of the manifold V, and that the integral submanifolds of this distribution are precisely the analytic equivalence classes of special flat vector bundles over the Riemann surface M. It is not at present clear whether this result is more useful in studying the decomposition of the manifold V into these analytic equivalence classes than in studying the period classes of the Prym differentials; for even in the case of scalar Prym differentials, the period classes are as yet rather mysterious.

Bibliography

1. R. C. Gunning, *Lectures on Riemann surfaces*, Princeton University Press, Princeton, N. J., 1966.

2. ———, *Lectures on vector bundles over Riemann surfaces*, Princeton University Press, Princeton, N. J., 1967.

3. ———, *Special coordinate coverings of Riemann surfaces*, Math. Ann. 170 (1967), 67-86.

4. M. S. Narasimhan and C. S. Seshadri, *Holomorphic vector bundles on a compact Riemann surface*, Math. Ann. 155 (1964), 69-89.

5. P. E. Newstead, *Topological properties of some spaces of stable bundles*, Topology 6 (1967), 241-262.

6. C. S. Seshadri, *Space of unitary vector bundles on a compact Riemann surface*, Ann. of Math. (2) 85 (1967), 303-336.

7. A. Weil, *Généralization des fonctions abéliennes*, J. Math. Pures. Appl. 17 (1938), 47-87.

Princeton University, Princeton, New Jersey 08540

ANALYTIC CONTINUATION ON BANACH SPACES

Michel Hervé

1. *Introduction and notation*

Most results in this talk are due to Gerard Coeuré [2], and his more detailed proofs will shortly appear in the Annales de l'Institut Fourier.

We consider a *manifold spread over a Banach space*, i.e., a triple (X,p,E), where X is a topological space, E a complex Banach space, p a local homeomorphism $X \to E$; we moreover assume that *X is an open subset of a connected manifold*, that *E is separable*, and those assumptions imply that X too is separable.

Given $x \in X$, if an open ball $\beta[p(x),r]$ in E, with center $p(x)$, radius r, is the homeomorphic image under p of an open neighborhood of x, this neighborhood is uniquely determined, it will be denoted by $B(x,r)$ and considered as the open ball with center x, radius r, in X. That situation holds for sufficiently small r, and there is a biggest radius r for which it holds, say $\delta(x)$; $\delta(x)$ is the analogue of Cartan-Thullen's [1] "Randdistanz", and a continuous function of x: $|\delta(x) - \delta(x')| \leqq \|p(x) - p(x')\|$ for $x' \in B[x,\delta(x)]$.

Let $A(X)$ be the algebra of complex valued analytic functions on X: $f \in A(X)$ if and only if $f \circ p_x^{-1}$ is analytic [4] on $\beta[p(x),\delta(x)]$ $\forall x \in X$, with the notation $p_x = p|B[x,\delta(x)]$. Then $f \circ p_x^{-1}$ has an absolutely convergent generalized homogeneous polynomial expansion on $\beta[p(x),\delta(x)]$:

$$(1) \qquad f \circ p_x^{-1}[p(x) + a] = \sum_{n \in \mathbf{N}} f_n(x,a), \quad a \in E, \ \|a\| < \delta(x),$$

with $f_0(x,a) = f(x)$, while for $n > 0$, $a \mapsto f_n(x,a)$ is a continuous n-homogeneous polynomial function [4] given by

$$(2)\quad f_n(x,a) = \frac{1}{2\pi}\int_0^{2\pi} e^{-in\theta}\, f\circ p_x^{-1}[p(x)+ae^{i\theta}]d\theta, \quad a\in E,\ \|a\| < \delta(x).$$

Given $x \in X$, each term $f_n(x,a)$ in the expansion (1) is bounded for bounded a; on account of this and (2), the three sets of radii $r \in]0,\delta(x)[$ such that

(i) (1) converges normally for $\|a\| \leqq r$,

(ii) (1) converges uniformly for $\|a\| \leqq r$,

(iii) f is bounded on $B(x,r)$,

have the same upper bound $\rho(x,f) \in]0,\delta(x)]$, which is a continuous function of x: $|\rho(x,f) - \rho(x',f)| \leqq \| p(x) - p(x')\|$ for $x' \in B[x,\rho(x,f)]$.

If E has a finite dimension, the equality $\rho(x,f) = \delta(x)$ $\forall x \in X$ and $f \in A(X)$ is well known; if E has an infinite dimension, that equality still holds for special functions f, for instance $f = u \circ p$, $u \in E'$, but $\inf_{f\in A(X)} \rho(x,f) = 0$ $\forall x \in X$, and therefore a simultaneous analytic continuation of all functions $\in A(X)$ cannot be achieved by the classical Weierstrass method.

2. *A topology on* $A(X)$ is defined as follows, by considering the open balls in X on which a function $\in A(X)$ is bounded.

Definition 1. *Given a lower semi-continuous function* τ *on* X, $0 < \tau \leqq \delta$, $A_\tau(X) = \{f \in A(X) \mid \rho(x,f) \geqq \tau(x) \quad \forall x \in X\}$; *the topology on* $A_\tau(X)$ *is defined by the seminorms* $f \mapsto \sup_{B(x,r)} |f|$, $x \in X$, $0 < r < \tau(x)$, *and the topology on* $A(X)$ *is the inductive limit of the topologies on the* $A_\tau(X)$.

Comments. Algebraically speaking: the $A_\tau(X)$ are an increasingly directed family of subalgebras of $A(X)$, which is their union.

$f \mapsto \sup_{B(x,r)} |f|$ is a norm when X is connected; since X is separable, $A_\tau(X)$ with those seminorms is a Fréchet space.

If E has a finite dimension: for any τ, $A_\tau(X) = A(X)$ and has the topology of compact convergence.

If E has an infinite dimension: $\tau' \leqq \tau$ implies $A_\tau(X) \hookrightarrow A_{\tau'}(X)$, and for each τ a $\tau' \leqq \tau$ can be chosen so that the topology on $A_\tau(X)$ is genuinely finer than that induced by $A_{\tau'}(X)$; as to the topology on $A(X)$, it is finer than that of compact convergence, and genuinely finer by Example 3 below.

Examples. 1) $A_\delta(X) \ni u \circ p$ for any $u \in E'$.

2) If $A_\tau(X) \ni f$, then $A_\tau(X)$ also contains $x \mapsto f_n(x,a)$ or f_n^a, and the linear map $f \mapsto f_n^a$ is continuous from $A_\tau(X)$ to $A_\tau(X)$, since relation (2) above implies

$$(3) \quad \sup_{B(x,r)} |f_n^a| \leqq \left(\frac{\|a\|}{\alpha}\right)^n \sup_{B(x,r+\alpha)} |f| \quad \text{for } \alpha > 0, \quad r+\alpha < \tau(x).$$

3) Given $x \in X$, $f \mapsto \sup_{\|a\| \leq 1} |f_1(x,a)|$ is a seminorm on $A(X)$, which by formula (3) is continuous for the topology we have chosen, but not for the topology of compact convergence. In fact, if $f = u \circ p$ with $u \in E'$, then $f_1(x,a) = u(a)$ [and, by the way, $f_n(x,a) = 0 \;\; \forall n > 1$], $\sup_{\|a\| \leqq 1} |f_1(x,a)| = \|u\|_{E'}$; if E has an infinite dimension, a sequence $(u_k) \subset E'$, $\|u_k\| = 1 \;\; \forall k$, can be chosen converging to 0 at each point $\in E$, and therefore uniformly on each compact set $\subset E$.

Theorem 1. *Let X_0 be an open connected subset of a* <u>*connected*</u> *manifold X, such that any $f \in A(X_0)$ has a (unique) continuation $\bar{f} \in A(X)$. Then:*

a) *the 1-1 map* $f \mapsto \overline{f}$ *is a homeomorphism between* $A(X_o)$ *and* $A(X)$ *with the topology in Definition* 1;

b) *any analytic map* g *from* X_o *into a Banach space* F *has a (unique) analytic continuation* $\overline{g}: X \to F$.

Proof. a) A linear map from $A(X)$ to $A(X_o)$ is continuous if it maps continuously each subspace $A_\tau(X)$ of $A(X)$ into a suitably chosen subspace $A_{\tau_o}(X_o)$ of $A(X_o)$: this is obvious for the restriction map $\overline{f} \to f$, with $\tau_o = \inf(\tau, \delta_o)$, where δ_o is the Randdistanz for X_o. The same idea of proof will now be used, with more difficulty, for the continuation map $f \mapsto \overline{f}$: let f belong to a given subspace $A_{\tau_o}(X_o)$.

Let $x \in X$ be such that $f \mapsto \overline{f}(x)$ is a continuous map: since this map is a homomorphism from the Fréchet algebra $A_{\tau_o}(X_o)$ to $\mathbb{C}$, there exists a finite union U of open balls $B(x_i, r_i)$, $x_i \in X_o$, $0 < r_i < \tau_o(x_i)$, such that $|\overline{f}(x)| \leqq \sup_U |f| \quad \forall f \in A_{\tau_o}(X_o)$. If $\alpha > 0$ and $r_i + \alpha < \tau_o(x_i)$ for each index i, by formula (3) we also have

$$|\overline{f}_n(x,a)| = |\overline{f_n^a}(x)| \leqq \left(\frac{\|a\|}{\alpha}\right)^n \sup_{U'} |f| \quad \forall f \in A_{\tau_o}(X_o),$$

with $U' = \bigcup_i B(x_i, r_i + \alpha)$; then, by formula (1):

$$(4) \quad |\overline{f}(x')| \leqq 2 \sup_{U'} |f| \quad \forall f \in A_{\tau_o}(X_o) \quad \text{for } x' \in B[x, \delta(x)] \cap B\left(x, \frac{\alpha}{2}\right).$$

This shows:

(i) that all linear forms $f \mapsto \overline{f}(x')$, $x' \in B[x,\delta(x)] \cap B\left(x, \frac{\alpha}{2}\right)$, are equicontinuous on $A_{\tau_o}(X_o)$;

(ii) that $\{r \in]0, \delta(x)[\mid \overline{f} \text{ bounded on } B(x,r) \ \forall f \in A_{\tau_o}(X_o)\}$ is a non empty set of radii, therefore has an upper bound $\tau(x) \in]0, \delta(x)]$.

Since X is connected, and the limit of a sequence of continuous linear forms on the Fréchet space $A_{\tau_o}(X_o)$ is continuous again: from (i) follows that $f \mapsto \overline{f}(x)$ is continuous on $A_{\tau_o}(X_o)$ $\forall x \in X$; then from (ii) follows that τ is a continuous function on X, defining a subalgebra $A_\tau(X)$ into which $A_{\tau_o}(X_o)$ is mapped by $f \mapsto \overline{f}$. Finally this linear mapping is continuous, since $\{f \in A_{\tau_o}(X_o) \mid |\overline{f}| \leqq 1$ on $B(x,r)\}$ is a barrel in the Fréchet space $A_{\tau_o}(X_o)$, $\forall x \in X$ and $r \in]0,\tau(x)[$.

b) Let $\rho(x,g)$ be defined in just the same way as $\rho(x,f)$ was defined in section 1 for a complex valued function f, and $\tau_o(x) = \rho(x,g)$: $u \circ g \in A_{\tau_o}(X_o)$ $\forall u \in F'$, the linear map $u \mapsto u \circ g$ is continuous from the Banach space F' to $A_{\tau_o}(X_o)$, hence also to $A(X_o)$. By part a) of the theorem, the linear map $u \mapsto \overline{u \circ g}$ = (Analytic continuation of $u \circ g$ to X) is continuous from F' to $A(X)$ for the topology of compact convergence, in particular an element $\overline{g}(x) \in F''$ is defined by $\langle u,\overline{g}(x)\rangle = \overline{u \circ g}(x)$ $\forall x \in X$; $\overline{g}(x) = g(x) \in F$ $\forall x \in X_o$, and formula (4) above entails $\| \overline{g}(x')\| \leqq 2 \sup_{U'} \|g\|$ for $x' \in B[x,\delta(x)] \cap B\left(x,\frac{\alpha}{2}\right)$.

Since $\overline{g}$ is locally bounded, $\overline{g} \circ p_x^{-1}$ will be analytic on $\beta[p(x),\delta(x)]$ if it is G-analytic [4], or if $v \circ \overline{g} \circ p_x^{-1}$ is analytic on $\beta[p(x),\delta(x)]$ $\forall v \in F'''$ with $\|v\| \leqq 1$. Now $v \circ \overline{g} \circ p_x^{-1}$ belongs to the closure, for the topology of pointwise convergence, of the set of functions $\overline{u \circ g} \circ p_x^{-1}$, $u \in F'$, $\|u\| \leqq 1$, which by (4) are locally bounded together, hence equicontinuous, on $\beta[p(x),\delta(x)]$.

Thus $\overline{g}$ is an analytic map from X to F'', and it only remains to show that $\overline{g}(X) \subset F$. If it were not so, there would exist $v \in F'''$, $v \equiv 0$ on F, $v \not\equiv 0$ on $\overline{g}(X)$, which imply $v \circ \overline{g} \in A(X)$, $v \circ \overline{g} \equiv 0$ on X_o, $v \circ \overline{g} \not\equiv 0$ on X, a contradiction to the connectedness of X.

Remarks. a) was proved with a repeated use of the fact that all subspaces $A_\tau(X)$, hence also their inductive limit $A(X)$, are barreled spaces. On the contrary, $A(X)$ with the topology of compact convergence is not barreled if E has an infinite dimension: in fact, for that topology, the seminorm in Example 3 is not continuous, but l.s.c. since each linear form $f \mapsto f_1(x,a)$ is continuous by formula (2). Therefore [3] it is dubious whether a result similar to a) holds with the topology of compact convergence on $A(X_0)$ and $A(X)$.

b) is meaningful without any topology on $A(X_0)$ and $A(X)$, and yet was proved with the help of the topology in Definition 1, because this one makes $f \leftrightarrow \bar{f}$ a homeomorphism between $A(X_0)$ and $A(X)$; if the topology of compact convergence had the same property, then b) could be proved in a simpler way [3].

In the situation of Theorem 1, we shall say that X is an *extension* of X_0 ; we now proceed to construct an extension of a given manifold X, with the usual tool in such problems (already used in the proof of Theorem 1 a), namely:

3. *Homomorphisms from the algebra $A(X)$ to $\mathbb{C}$.*

Let h be a continuous linear form on $A(X)$: for each subspace $A_\tau(X)$ exist a constant $c(\tau) > 0$ and a finite union $U(\tau)$ of open balls $B(x_i, r_i)$, $x_i \in X$, $0 < r_i < \tau(x_i)$, such that $|h(f)| \leqq$ $\leqq c(\tau) \sup_{U(\tau)} |f| \quad \forall f \in A_\tau(X)$.

If moreover h is a homomorphism, then $c(\tau) = 1$ for each τ since each $A_\tau(X)$ is an algebra. If h is a special homomorphism $\dot{x}$: $f \mapsto f(x)$, $x \in X$, then for each τ we may take $U(\tau) = B(x,r)$, $0 < r < \inf[\tau(x), \delta(x) - \gamma(\dot{x})]$, with a given $\gamma(\dot{x})$, $0 < \gamma(\dot{x}) < \delta(x)$; with that choice of $U(\tau)$, the Randdistanz δ remains $> \gamma(\dot{x})$ on $U(\tau)$ for each τ. We only consider homomorphisms or linear forms with the same property.

Definition 2. *For each subspace $A_\tau(X)$, let $U(\tau)$ be a finite union of open balls $B(x_i, r_i)$, $x_i \in X$, $0 < r_i < \tau(x_i)$: the family of sets $U(\tau)$ is a* <u>*regular family of carriers*</u> *for a linear form h*

on $A(X)$, *if* $|h(f)| \leq \sup_{U(\tau)} |f|$ $\forall f \in A_\tau(X)$, *and there exists a constant* $\gamma(h) > 0$ *which is* $\leq \delta$ *on* $U(\tau)$ *for each* τ.

Notation. $H(X)$ for the set of homomorphisms $h \not\equiv 0$ with a regular family of carriers; $L(X)$ for the set of linear forms h with the same property. The notation $f_n^a(x) = f_n(x,a)$ was already used in section 2, Example 2, and $f_0^a = f$.

Proposition 2. *Let* $h \in L(X)$. a) *For any* $a \in E$ *and* $n \in \mathbb{N}$, $L(X)$ *also contains the linear form* $h_n^a : f \mapsto h(f_n^a)$ *on* $A(X)$; $h_0^a = h$, $h_n^0 \equiv 0$ $\forall n \geq 1$. b) *More precisely, given a compact set* K *in* E *which is contained in the open ball* $\beta[0,\gamma(h)]$, *a same regular family of carriers* $V(\tau)$ *can be chosen for all* h_n^a, $a \in K$, $n \in \mathbb{N}$. c) *The series* $\sum_{n\in\mathbb{N}} h_n^a(f) = h^a(f)$ *converges absolutely for* $\|a\| < \gamma(h)$, $f \in A(X)$, *and defines* $h^a \in H(X)$, $h^0 = h$, *if* $h \in H(X)$.

Proof. a) will follow from b) since $h_n^{\lambda a} = \lambda^n h_n^a$ $\forall a \in E$, $\lambda \in \mathbb{C}$.

b) Let K be a compact balanced set in E, and $a \in K$ imply $\|a\| < \alpha < \gamma(h)$: $X_o = \{x \in X \mid \delta(x) > \alpha\}$ is an open subset of X, with a Randdistanz $\delta_o \geq \delta - \alpha$; X_o contains the closed set $Y = \{x \in X \mid \delta(x) \geq \gamma(h)\}$, and $Y \supset U(\tau)$ for each τ.

For $x \in X_o$: p_x^{-1} maps the compact subset $p(x) + K$ of $\beta[p(x),\delta(x)]$ onto a compact subset $K(x) \ni x$ of $B[x,\delta(x)]$; since a neighborhood of $K(x)$ contains $K(x')$ for x' sufficiently near x, $\tau_o(x) = \inf[\delta(x) - \alpha, \inf_{K(x)} \tau]$ defines a l.s.c. function $\tau_o \leq \tau$ for X_o, and a l.s.c. function $\tau' \leq \tau$ for X is τ_o on Y, τ on $X-Y$, depending only on τ.

Given τ, hence τ_o and τ' : $U(\tau')$ is a finite union of open balls $B(x_i,r_i)$, $x_i \in X_o$, $0 < r_i < \tau_o(x_i)$, so $f_0 \mapsto \sup_{U(\tau')} |f_0|$ is a continuous seminorm on $A_{\tau_o}(X_o)$; since the restriction of h to $A_{\tau'}(X) \cap A_{\tau_o}(X_o)$ is a linear form $f_0 \mapsto h(f_o)$ satisfying

$|h(f_0)| \leqq \sup_{U(\tau')} |f_0|$, by the Hahn-Banach theorem there exists a linear form $f_0 \mapsto h_0(f_0)$ on $A_{\tau_o}(X_o)$ satisfying $|h_0(f_0)| \leqq \sup_{U(\tau')} |f_0|$ $\forall\, f_0 \in A_{\tau_o}(X_o)$, and $h_0(f|X_o) = h(f)$ $\forall f \in A_{\tau'}(X)$ such that $f|X_o \in A_{\tau_o}(X_o)$.

Now let $f \in A_\tau(X)$: then $f_0 = f|X_o \in A_{\tau_o}(X_o)$ since $\tau_o \leqq \tau$ on X_o, $(f_0)^a_n = f^a_n|X_o \in A_{\tau_o}(X_o)$ and $f^a_n \in A_{\tau'}(X)$ (section 2, Example 2). Given $a \in K$, let $g_\theta(x) = f \circ p_x^{-1}[p(x) + ae^{i\theta}]$, $x \in X_o$: the generalized translation $x \mapsto y = p_x^{-1}[p(x) + ae^{i\theta}] \in K(x)$ maps an open ball $B(x,r)$, $0 < r < \tau_o(x) \leqq \tau(y)$, onto the open ball $B(y,r)$. Thus g_θ is bounded on $B(x,r)$ and, from the boundedness of f on $B(y,r+\beta)$, $\beta > 0$, $r+\beta < \tau_o(x)$, follows that $\sup_{B(x,r)} |g_{\theta'} - g_\theta| \to 0$ as $\theta' \to \theta$.

Since $\theta \mapsto g_\theta$ is a continuous map from $[0,2\pi]$ to $A_{\tau_o}(X_o)$, $G_n = \frac{1}{2\pi} \int_0^{2\pi} e^{-in\theta}\, g_\theta d\theta \in A_{\tau_o}(X_o)$; since $f_0 \to f_0(x)$, $x \in X_o$, and $f_0 \mapsto h_0(f_0)$ are continuous linear forms on $A_{\tau_o}(X_o)$, we have

$$(5) \qquad G_n(x) = \frac{1}{2\pi} \int_0^{2\pi} e^{-in\theta}\, g_\theta(x) d\theta = f_n(x,a) \quad \text{for } x \in X_o$$

by formula (2), and

$$(6) \qquad |h_0(G_n)| = \Big| \frac{1}{2\pi} \int_0^{2\pi} e^{-in\theta}\, h_0(g_\theta) d\theta \Big| \leqq \frac{1}{2\pi} \int_0^{2\pi} \sup_{U(\tau')} |g_\theta| d\theta \leqq \sup_{V(\tau)} |f|$$

where $V(\tau)$ is obtained as follows: for each index i in $U(\tau') = \bigcup_i B(x_i,r_i)$, $x_i \in X_o$, $0 < r_i < \tau_o(x_i)$, the compact set $K(x_i)$ is contained in a finite union of open balls with centers $\in K(x_i)$ and a same radius $\alpha_i > 0$ such that $r_i + \alpha_i < \tau_o(x_i)$; $V_i(\tau)$ is the union of the open balls with the same centers and radius $r_i + \alpha_i < \inf_{K(x_i)} \tau$, and $V(\tau) = \bigcup_i V(\tau_i)$.

Formula (5) means that $G_n = f_n^a | X_0$; since $f_n^a \in A_{\tau'}(X)$ and $f_n^a | X_0 \in A_{\tau_0}(X_0)$, from (6) we get

$$|h_n^a(f)| = |h(f_n^a)| = |h_0(G_n)| \leq \sup_{V(\tau)} |f| \quad \forall f \in A_\tau(X),$$

with $V(\tau)$ depending only on τ and K; finally the Randdistanz δ is at least $\delta(x_i) - \sup_{a \in K} \|a\|$ on $K(x_i)$, and $r_i + \alpha_i < \tau_0(x_i) \leq \delta(x_i) - \alpha$, hence $\delta \geq \alpha - \sup_{a \in K} \|a\|$ on each $V(\tau)$.

c) Given $a \in E$ with $\|a\| < \gamma(h)$, choose $r \in]1, \gamma(h)/\|a\|[$: by b) there exists a same regular family of carriers $V(\tau)$ for all linear forms $h_n^{ra} = r^n h_n^a$, hence $|h_n^a(f)| \leq \frac{1}{r^n} \sup_{V(\tau)} |f| \; \forall f \in A_\tau(X)$.

Since the series $\sum_{n \in \mathbb{N}} h_n^a(f)$ converges absolutely to $h^a(f)$, the family of complex numbers $\sum_{n,p \in \mathbb{N}} h_n^a(f) h_p^a(g)$ is summable to $h^a(f) h^a(g)$ $\forall f$ and $g \in A(X)$. If h is a homomorphism, $(fg)_m^a = \sum_{n+p=m} f_n^a \cdot g_p^a$ implies $h_m^a(fg) = \sum_{n+p=m} h_n^a(f) h_p^a(g)$, hence $h^a(fg) = \sum_{n,p \in \mathbb{N}} h_n^a(f) h_p^a(g)$, and h^a is a homomorphism from $A(X)$ to $\mathbb{C}$, with $|h^a(f)| \leq \frac{r}{r-1} \sup_{V(\tau)} |f| \; \forall f \in A_\tau(X)$.

Examples. 1) If h is a special homomorphism $\dot{x} : f \mapsto f(x)$, $x \in X$, the series in c) coincides with the expansion (1), which converges absolutely for $\|a\| < \delta(x)$, and then $\dot{x}^a = \dot{y}$, $y = p_x^{-1}[p(x) + a]$; i.e., the subset $\{\dot{x}^a \mid \|a\| < \delta(x)\}$ of $H(X)$ is the image of the open ball $B[x, \delta(x)]$ under the map $\phi : x \mapsto \dot{x}$.

2) Let X_0 be an open connected subset of X, on which the

Randdistanz δ has a lower bound $\alpha > 0$, Y an extension of X, and Y_o an extension of X_o which is an open subset of Y. If $f \in A(X)$ is continued into $\bar{f} \in A(Y)$, then $f|X_o$ is continued into $\bar{f}|Y_o$; if $f \in A_\tau(X)$, then $f|X_o \in A_{\tau_o}(X_o)$ with $\tau_o = \inf(\tau, \delta_o)$, where δ_o is the Randdistanz for X_o, and by the proof of Theorem 1 a) the homomorphism $f|X_o \mapsto \bar{f}(y)$, $y \in Y_o$, has the following property: there exists a finite union $U(\tau)$ of open balls $B(x_i, r_i)$, $x_i \in X_o$, $0 < r_i < \tau_o(x_i)$, such that $|\bar{f}(y)| \leqq \sup_{U(\tau)} |f|$. Since $\delta \geqq \alpha$ on $U(\tau) \subset X_o$, the $U(\tau)$ are a regular family of carriers for the homomorphism $f \mapsto \bar{f}(y)$ from $A(X)$ to $\mathbb{C}$.

Example 2 shows that $H(X)$ will contain at least some extensions of X if we obtain:

4. *$H(X)$ as a manifold spread over E.*

The necessary topology on $H(X)$ is suggested by Example 1 above, and chosen in a) below.

Proposition 3. a) *The sets* $V(h,\alpha) = \{h^a \mid \|a\| < \alpha\}$ *are a fundamental system of connected neighborhoods of* $h \in H(X)$, $0 < \alpha \leqq \gamma(h)$ (see Definition 2 and Proposition 2), *for a Hausdorff topology on $H(X)$, which is finer than the topology of pointwise convergence.*

b) *For that topology, the map* $\phi : x \mapsto \dot{x}$ *is a local homeomorphism from X to $H(X)$, hence its image* $\phi(X) = \dot{X}$ *is open in $H(X)$.*

c) *Let $E(X)$ be the union of all connected components of $H(X)$ meeting $\dot{X}$ (one only if X is connected): there exists a local homeomorphism $\hat{p}$ from $E(X)$ to E, such that* $p = \hat{p} \circ \phi$.

Proof. a) Given $a \in E$ with $\|a\| < \alpha \leqq \gamma(h)$, choose $r > 1$ (as in the proof of Proposition 2 c)) so that all linear forms $h_n^{ra} = r^n h_n^a$ have a same regular family of carriers $V(\tau)$: for a given τ, $V(\tau)$ is a finite union of open balls $B(x_i,r_i)$, $x_i\in X$, $0 < r_i < \tau(x_i)$; if $\beta > 0$ is chosen such that $r_i + \beta < \tau(x_i)$ $\forall i$, for $f\in A_\tau(X)$ on account of (3) we have

$$(7)\qquad |h_n^a(f_p^b)| \leq \frac{1}{r^n} \sup_{V(\tau)} |f_p^b| \leq \frac{1}{r^n}\left(\frac{\|b\|}{\beta}\right)^p \sup_{V'(\tau)} |f|$$

with $V'(\tau) = \bigcup_i B(x_i,r_i+\beta)$. Then, for $\|b\| < \beta$ the family of complex numbers $h_n^a(f_p^b)$, n and $p\in\mathbb{N}$, is summable to $\sum_{p\in\mathbb{N}} h^a(f_p^b) = (h^a)^b(f)$ if $\|b\| < \gamma(h^a)$, and also to $h^{a+b}(f)$ if $\|a+b\| < \gamma(h)$, since $f_m^{a+b} = \sum_{n+p=m} (f_p^b)_n^a$ implies $h_m^{a+b}(f) = \sum_{n+p=m} h_n^a(f_p^b)$.

Thus we have proved the relation $h^{a+tb}(f) = (h^a)^{tb}(f)$ for given a and $b\in E$, and $|t| \leq$ some number depending on τ, hence on f. Now each term $h_n^{a+tb}(f) = h(f_n^{a+tb})$ is a polynomial in t, and the series $\sum_{n\in\mathbb{N}} h_n^{a+tb}(f)$ converges normally for

$$t\in T = \{t\in\mathbb{C} \mid \|a+tb\| \leq \gamma(h)/r^2\}, \qquad r > 1 :$$

in fact $K = \{r(a+tb) \mid t\in T\}$ is a compact subset of E contained in the open ball $\beta[0,\gamma(h)]$, by Proposition 2 b) all linear forms $h_n^{r(a+tb)} = r^n h_n^{a+tb}$, $t\in T$, $n\in\mathbb{N}$, have a same regular family of carriers $W(\tau)$, hence $|h_n^{a+tb}(f)| \leq \frac{1}{r^n}\sup_{W(\tau)}|f|$ $\forall f\in A_\tau(X)$. Thus $h^{a+tb}(f)$ is a holomorphic function of t for $\|a+tb\| < \gamma(h)$, similarly $(h^a)^{tb}(f)$ is a holomorphic function of t for $\|tb\| < \gamma(h^a)$, and they coincide for t satisfying both inequalities, which no longer depend on f.

Finally, for $\|b\| < \beta = \inf[\alpha - \|a\|, \gamma(h^a)]$, the relation $h^{a+b} = (h^a)^b$ is proved, with two consequences: $V(h,\alpha)\supset V(h^a,\beta)$; and $a \mapsto h^a$ is continuous for $\|a\| < \gamma(h)$, hence $V(h,\alpha)$ is connected. On the other hand, formula (7) implies that

$(h^a)^b(f) - h^a(f) = \sum_{\substack{n\geq 0 \\ p>0}} h^a_n(f^b_p) \to 0$ as $\|b\| \to 0$: each linear map $h \mapsto h(f)$ is continuous for the topology we have chosen.

b) The restriction of ϕ to the open ball $B[x,\delta(x)]$ is open, continuous (see section 3, Example 1), and 1-1 because any two distinct points in that ball are separated by some function $u \circ p$, $u \in E'$.

c) Let $u \in E'$: given $x \in X$, $u \mapsto \dot{x}(u \circ p) = u[p(x)]$ is a continuous linear form on the Banach space E', identified with $p(x) \in E$; more generally, given $h \in H(X)$, $u \mapsto h(u \circ p)$ is a continuous linear form $\hat{p}(h)$ on E', since the linear map $u \mapsto u \circ p$ is continuous from E' to $A_\delta(X)$ (see section 2, Example 1).

With $f = u \circ p$ we have (see section 2, Example 3) $h^a(u \circ p) = h(u \circ p) + u(a)$ $\forall u \in E'$, hence the relation $\hat{p}(h^a) = \hat{p}(h) + a$ for $\|a\| < \gamma(h)$, which implies:

(i) that $\hat{p}$ is continuous from $H(X)$ to E'', and its restriction to $V[h,\gamma(h)]$ is 1-1 ;

(ii) that $\hat{p}^{-1}(E)$ is open in $H(X)$; since it is closed too by (i), and contains $\dot{X}$, it contains $E(X)$;

(iii) that $\hat{p}$ is an open map from $E(X)$ to E.

Remarks. 1) If any two distinct points in X are separated by $A(X)$ (for instance in the special case when X is an open subset of E), then ϕ is a homeomorphism from X to $\dot{X}$. Be it so or not, the relation $f = \dot{f} \circ \phi$ is a 1-1 correspondence between $f \in A(X)$ and $\dot{f} \in A(\dot{X})$, since $f \circ \phi^{-1}$ is meaningful for $f \in A(X)$; then one may replace X by $\dot{X}$ when constructing an extension of X.

2) Similarly, the remark at the end of section 3 can be made precise as follows: in the situation of Example 2, the set $\dot{Y}_o$ of homomorphisms $f \mapsto \overline{f}(y)$, $y \in Y_o$, is the image of the connected

manifold Y_0 under a local homeomorphism into $H(X)$, or $E(X)$; then again one may replace Y_0 by $\dot{Y}_0$ and consider $E(X)$ as a bigger extension than Y_0, once the following theorem is proved.

Theorem 4. *If X is connected, $E(X)$ is an extension of $\dot{X}$.*

Proof. Given $\dot{f} \in A(\dot{X})$, $\dot{f} \circ \phi \in A(X)$, $\overline{f}(h) = h(\dot{f} \circ \phi)$ is meaningful for $h \in E(X)$ and defines a continuation $\overline{f}$ of $\dot{f}$ to $E(X)$; $\overline{f}$ is continuous by Proposition 3 a) and analytic since (see proof of Proposition 3 a)) $h^{a+tb}(\dot{f} \circ \phi)$ is a holomorphic function of t for $\|a + tb\| < \gamma(h)$.

References

1. H. Cartan and P. Thullen, *Regularitäts- und Konvergenzbereiche*, Math. Ann. 106 (1932), 617-647.

2. G. Coeuré, *Fonctions plurisousharmoniques et fonctions C-analytiques à une infinité de variables*, C. R. Acad. Sci. Paris Sér. A-B 267 (1968), A440-A442.

3. J. M. Exbrayat, A lecture on Alexander's thesis, Séminaire Lelong, 9th year, 1968/69.

4. E. Hille and R. Phillips, *Functional analysis and semigroups*, Amer. Math. Soc. Colloquium Publications Vol. 31, Providence, R. I., 1957.

University of Paris, France and
University of Maryland, College Park, Maryland 20742

ON CERTAIN REPRESENTATIONS OF SEMI-SIMPLE ALGEBRAIC GROUPS

Jun-ichi Igusa

Summary. Suppose that k is an algebraic number field, G a connected semi-simple algebraic group, and ρ a rational representation of G in a vector space X defined over k. The triple (G,X,ρ) or simply ρ is called *admissible* over k if the Weil criterion [9, p. 20] for the convergence of the integral over G_A/G_k of the generalized theta-series

$$\sum_{\xi\in X_k} \Phi(\rho(g)\cdot\xi)$$

is satisfied. (The subscript A denotes the adelization functor relative to k and Φ is an arbitrary Schwartz-Bruhat function on X_A.) It is called *absolutely admissible* if it is admissible over any finite algebraic extension of k. We shall discuss a complete classification of all absolutely admissible representations. (If G is absolutely simple, there are essentially 6 infinite sequences of, and 26 isolated, absolutely admissible triples.) Then, in the general case, we shall discuss the existence and basic properties of what we call the *principal subset* X' of X. Although the adjoint representation of G can never be absolutely admissible, we can say that X' is analogous to the set of regular elements of the Lie algebra of G via its adjoint representation [cf. 4]. The significance of X' is that it permits us to formulate a conjectural Siegel formula in a precise, explicit form. (The restriction to X' corresponds in the classical Siegel case to considering only those terms which contribute to cusp forms.) We shall discuss a weak solution of this conjecture in the general case. Finally, if there exists only one G-invariant, we shall discuss its solution as well as the analytic continuation and the functional equation of the corresponding zeta-function.

1. We shall first explain the admissibility condition using non-adelic language for $k = \mathbb{Q}$. We take a norm $\|x\|$ on $X_{\mathbb{R}}$ and consider the following theta-series:

$$\sum_{\xi \in X_{\mathbb{Z}}} \exp(-\|\rho(g)\cdot\xi\|^2),$$

in which g is a variable point of $G_{\mathbb{R}}$. Then, this series defines a continuous function on $G_{\mathbb{R}}$, and it is invariant with respect to the right multiplication by elements of a certain congruence group of $G_{\mathbb{Z}}$. If we require that this is an L^1-function on the corresponding quotient space for every choice of the norm on $X_{\mathbb{R}}$, we get an analytic condition for ρ to be "small". The admissibility of (G,X,ρ) over $\mathbb{Q}$ more or less expresses this condition in the adelic language.

In the above explanation, we implicitly assumed that $G_{\mathbb{R}}/G_{\mathbb{Z}}$ is very far from being compact, i.e., that G splits over $\mathbb{Q}$. In fact, if $G_{\mathbb{R}}/G_{\mathbb{Z}}$ is compact, we get no condition whatsoever on ρ. Since the Siegel formula is not valid just for any representation, certain restriction on ρ is necessary. If (G,X,ρ) is obtained by twisting (relative to k) from an admissible triple over $\mathbb{Q}$, in which the group is $\mathbb{Q}$-split, then (G,X,ρ) is absolutely admissible in the sense that we have defined. The converse is also true.

Now, the first problem was to classify all absolutely admissible representations. For this purpose, we may assume without losing the generality that G is simply connected. If we have a finite number of triples (G_i,X_i,ρ_i), we define their "sum" as (G,X,ρ), in which $G = G_1 \times G_2 \times \ldots$, $X = X_1 \times X_2 \times \ldots$,

$$\rho(g_1,g_2,\ldots)\cdot(x_1,x_2,\ldots) = (\rho_1(g_1)\cdot x_1 , \rho_2(g_2)\cdot x_2 , \ldots)$$

for every g_i in G_i and x_i in X_i. If the triple (G,X,ρ) is defined over k and is absolutely admissible, we can show that it becomes a sum of absolutely admissible (G_i,X_i,ρ_i) defined over k, in which G_i are k-simple factors of G. The converse is also true. Therefore, we may assume that G is k-simple in (G,X,ρ). Then G is the product of its connected, absolutely simple normal subgroups G_1, G_2, ... Let K denote the smallest field of definition of, say G_1, containing k. Then, there exists an absolutely admissible triple (G_1,X_1,ρ_1) defined over K such that (G,X,ρ) becomes the sum of all conjugates of (G_1,X_1,ρ_1) relative to k. The converse is also true. Therefore, we may assume that G is absolutely simple in (G,X,ρ). Then, the problem decomposes into the following two parts:

First, determine all admissible (G,X,ρ), in which G is a connected, simply connected, simple $\mathbb{Q}$-split group. Second, determine all twistings of such a triple relative to the given algebraic number field k.

In order to settle the first part, we examined each one of the four main types of, and of the five exceptional simple groups. In each case, the criterion is that a certain integral over a cone be convergent. For instance, if the group is SL_{n+1} and u_1, u_2, ..., u_{n+1} denote the weights of the identity mapping of SL_{n+1}, the integral to be examined takes the following form:

$$\int \prod_{\lambda} \sup(1,\exp(-\lambda(u)))\cdot\exp\Big(\sum_{i=1}^{n} (2n-2i+2)u_i\Big)\cdot du_1 \ldots du_n ,$$

in which the lambdas are the weights of ρ each repeated with its multiplicity. Furthermore, the domain of integration is defined by

$$0 < u_1 \leqq u_2 \leqq \ldots \leqq u_n \leqq u_{n+1} .$$

In this and other cases, we had to examine carefully those weights which are unbounded in both ways on the domain of integration. At any rate, this part has been settled, and the result was announced in [3].

We shall pass to the second part of the problem. We recall that there are two well-known methods to construct semi-simple algebraic groups explicitly over k. One is to take the adjoint group of a k-form of the Lie algebra of the given type. Another is to consider a semi-simple algebra A defined over k and take a subgroup defined over k of the group Aut(A) of all invertible linear transformations in A. Since the adjoint representation can never be absolutely admissible, we can not use the first method. If we use the second method, we get a group together with its small representation, the representation space being a suitable subspace of A. For instance, if A is a semi-simple associative algebra with an involution, the subgroup of Aut (A) consisting of those elements which keep the algebra structure (and the involution) invariant is a classical group. By allowing k-irreducible representations to appear with small multiplicities, we get the absolutely admissible representations that have been investigated by Siegel [8] and Weil [9]. On the other hand, if A is a reduced, simple exceptional Jordan algebra, we get a k-form of the simple group of type E_6 as the subgroup of Aut(A) consisting of those elements which keep the norm form invariant. The corresponding representation (with A as its representation space) is the absolutely admissible representation that has been investigated by Mars [5]. We have found (rather unexpectedly) that *all absolutely admissible representations which are not of the Siegel-Weil type are related directly or indirectly to Jordan algebras.* Since the elaboration of this statement will take too much space, we have to refer for the details to our forthcoming papers.

2. Although we have to assume that k is an algebraic number field to define the admissibility, using the "classification theorem", we can define an absolutely admissible triple (G,X,ρ) over an arbitrary field k. We shall denote by Ω a universal domain over k and by $\Omega[X]$ the coordinate ring of X. If we take a k-base of X_k, $\Omega[X]$ becomes isomorphic to the ring of polynomials in $\dim(X)$ letters with coefficients in Ω. At any rate, we can let G operate on $\Omega[X]$ via ρ as a group of degree-preserving automorphisms. We shall denote by $\Omega[X]^G$ the corresponding ring of invariants. Then, first of all, we can show that *$\Omega[X]^G$ is generated over Ω by algebraically independent homogeneous polynomials $f_1, f_2, \ldots, f_N$ with coefficients in k.* Let $I(X)$ denote the affine space Ω^N and $f: X \to I(X)$ the morphism defined over k as

$$f(x) = (f_1(x), f_2(x), \ldots, f_N(x))$$

for every x in X. Then, we can show that *the fiber $f^{-1}(i)$ contains a principal orbit*, i.e., a unique G-orbit of maximal dimension, *for every i in $I(X)$.* We shall denote this G-orbit by $U(i)$ and put

$$X' = \bigcup_{i \in I(X)} U(i).$$

Then X' is a k-open subset of X. This is the *principal subset* that we have mentioned in the Summary. *If x is a point of X', the morphism f is submersive at x* in the sense that the N differentials $df_1, df_2, \ldots, df_N$ are linearly independent at x. In other words, the N hypersurfaces defined by $f_1 - f_1(x) = 0$, $f_2 - f_2(x) = 0, \ldots,$ $f_N - f_N(x) = 0$ are transversal at x. This implies that $\dim(U(i))$ is equal to $\dim(X) - \dim(I(X))$ for every i in $I(X)$. What is more remarkable and important is that *we have*

$$\dim(f^{-1}(i) - U(i)) \leqq \dim(U(i)) - 2$$

for every i. In particular, the boundary $X - X'$ of X' is a k-closed subset of codimension at least equal to 2, and this implies that the principal subset X' is simply connected.

We recall that Kostant proved similar properties for the set of regular elements of a complex semi-simple Lie algebra in [4]. Since no adjoint representation is absolutely admissible, our class and the class of adjoint representations have no representation in common. It is an intriguing problem to find out a group-theoretic characterization of (G,X,ρ) which gives rise to such an excellent fibering of the representation space X. Even a part of the problem, e.g., to find a sufficient condition such that $\Omega[X]^G$ has algebraically independent generators seems quite interesting.

Now, if we consider the stabilizer subgroup H of G at any given point x of the principal subset X', the above-mentioned analogy breaks down completely. In fact, we can show that, in our case, *H is connected, the Levi subgroup exists*, and *it is semi-simple*. In particular, the radical of H is unipotent. Furthermore, *if G is simply connected, H is also simply connected provided that no absolutely simple summand of (G,X,ρ) is the third fundamental representation of Sp_6*.

The existence of X', its properties, and the structure of the stabilizer subgroup of G at any given point of X' may be called the *orbital decomposition theorem*. The proof of this theorem can be reduced by the classification theorem to the case when G is absolutely simple. In this case, there are essentially two new infinite sequences of, and 25 isolated, absolutely admissible representations to be examined. Of these, two isolated absolutely admissible representations, which are quite substantial (if not the most difficult), are examined by Haris in his dissertation (under the assumption of characteristic 0). One of the difficulties comes from the fact that no sufficient condition seems to be known for arbitrarily given points of the representation space to belong to the same G-orbit. Another (less serious) difficulty has been to prove the connectedness of the stabilizer subgroups at various points of the principal subset. If one had a general theorem which guarantees that a specialization (satisfying further conditions if necessary) of a connected subgroup of a group variety is connected, it would have shortened our proof. At any rate, we have made an extensive use of the theory of Jordan algebras. Perhaps the usefulness comes from the fact that the representation space is not just a vector space but it has a certain multiplicative structure. On the other hand, Haris has used the theory of Chevalley groups (in characteristic 0). Although this method is not quite effective, it has the pleasant feature that it is linked with the general theory of algebraic groups.

3. We shall consider a (not necessarily connected) semi-simple algebraic group G and a representation ρ of G in a vector space X defined over an algebraic number field k. Let G_o denote the connected component of the identity in G and ρ_o the restriction of ρ to G_o. We recall that the Weil criterion is the same for (G,X,ρ) and (G_o,X,ρ_o). (This is based on a lemma of Borel to the effect that $G_A/(G_o)_A$ is compact.) We shall assume that (G_o,X,ρ_o) is absolutely admissible and that every G_o-invariant via ρ_o is also a G-invariant via ρ. Then, we can find algebraically independent homogeneous G-invariants f_1, $f_2,\ldots, f_N$ with coefficients in k which generate $\Omega[X]^G$ over Ω. Let $d\mu(g)$ denote the Haar measure on G_A normalized by the condition that G_A/G_k is of measure 1. Similarly, let $|dx|_A$ denote the Haar measure on X_A normalized by the condition that X_A/X_k is of measure 1. Finally, fix a "basic character" χ of the adele group k_A, which is a continuous, non-trivial homomorphism of k_A/k to the multiplicative

group of complex numbers of absolute value 1. Then, the *conjectural Siegel formula* takes the following form

$$\int_{G_A/G_k} \Big(\sum_{\xi\in X_k'} \Phi(\rho(g)\cdot\xi)\Big)\cdot d\mu(g) = \sum_{i^*\in k^N} \int_{X_A} \Phi(x)\chi\Big(\sum_{\alpha=1}^{N} f_\alpha(x)i_\alpha^*\Big)\cdot|dx|_A ,$$

in which Φ is an arbitrary element of the Schwartz-Bruhat space $S(X_A)$. Although the left hand side is convergent (because of our assumption), it is not clear whether the infinite series on the right hand side is meaningful. In fact, if we denote by $I(X)^*$ the dual space of $I(X)$, the major difficulty comes from the fact that the magnitude, especially the L^1-integrability, of the bounded continuous function on $I(X)^*_A = (k_A)^N$ defined by

$$I(X)^*_A \ni i^* \mapsto \int_{X_A} \Phi(x)\chi\Big(\sum_{\alpha=1}^{N} f_\alpha(x)i_\alpha^*\Big)\cdot|dx|_A$$

is not clear except when $f_1, f_2, \ldots, f_N$ are at most quadratic (and except for some other cases that we shall mention later). At any rate, the right hand side is independent of the choice of $f_1, f_2, \ldots, f_N$. Now, if they are at most quadratic, the conjectural Siegel formula has been proved. In such a case, taking Φ from the dense subspace $G(X_A)$ of $S(X_A)$ [cf. 2], we can transform the above formula into a non-adelic form. For instance, if G is the orthogonal group of a definite quadratic form in $2m$ variables with coefficients in $\mathbb{Q}$, we get a classically known identity of Siegel [8, I, p. 394]: On one hand, we have the part of the analytic class invariant consisting of those terms which contribute to cusp forms. It is a certain weighted mean of incomplete class invariants of the form

$$\sum{}' \exp(\pi i\cdot \mathrm{tr}({}^t\xi\sigma\xi z)),$$

in which σ is a positive-definite integer matrix of degree $2m$,

z a point of the Siegel upper-half plane of degree n (for any fixed n satisfying $m > n+1$), and ξ runs over the set of $2m \times n$ integer matrices of rank n. On the other, we have a similarly incomplete Eisenstein-Siegel series of the form

$$\sum{}' \, h(\sigma,\gamma,\delta)\cdot\det(\gamma z+\delta)^{-m},$$

in which γ, δ are integer matrices of degrees n satisfying $\det(\gamma) \neq 0$ and the summation is really with respect to $\gamma^{-1}\delta$ running over the set of all symmetric rational matrices of degree n. The coefficients $h(\sigma,\gamma,\delta)$ are generalized Gaussian sums depending on σ, γ, δ. (The complete formula consists of similar identities, which can be re-written as identities of the above form for smaller n's.)

As we have said, the conjectural Siegel formula has been proved in many cases. However, there are certain cases where the difficulty is still insurmountable. Therefore, we shall discuss a *weak solution* of the problem. Since the boundary of the principal subset X' is a k-closed subset of X of codimension at least equal to 2, X' has "1" as its convergence factor. In other words, if we consider $|dx|_A$ as a sort of product measure on X_A and restrict each factor measure to the corresponding completion of X', we get a measure, say $|dx'|_A$, on X'_A, and $|dx|_A$ becomes its image measure under the continuous injection $X'_A \to X_A$. Therefore, if the conjectural Siegel formula is true, by restricting both sides to X'_A, we get a similar formula in which Φ is an arbitrary continuous function on X'_A with compact support, and X_A, $|dx|_A$ are replaced by X'_A, $|dx'|_A$. It appears that the so-specialized formula can be proved.

Finally, if ρ is absolutely irreducible, and if there exists only one G-invariant, it appears that we can prove the conjectural Siegel formula (in the original form). This is because the non-quadratic invariant that we have to deal with is either a norm form of a simple Jordan algebra or its variant, and by the works of Mars [5] and ours (which is mentioned in [3]), we can handle such an invariant. Also, in this case, let $Z(s,\Phi)$ denote the zeta-function associated with the morphism, in which s is a complex variable and Φ an arbitrary element of $S(X_A)$. For any (absolutely irreducible) polynomial $f(x)$, such a zeta-function has been introduced and examined by Ono [6]. In the present case, $f^{-1}(i)$ is a G-orbit for every $i \neq 0$ and splits into a finite number of G-orbits for $i = 0$. Since the adelization works nicely except for the case when ρ is the third fundamental representation of Sp_6, we can show that $Z(s,\Phi)$ has an analytic continuation to the entire s-plane as a meromorphic function and satisfies a standard functional equation of the form $Z(s,\Phi) = Z(\kappa - s, \Phi^*)$, in which $\kappa \cdot \deg(f) = \dim(X)$ and Φ^* is the Fourier transform of Φ. Also, the poles of $Z(s,\Phi)$ are all of order 1 and they are determined in a simple manner by the modules of the stabilizer subgroups of the G-orbits in $f^{-1}(0)$. It is our understanding that Sato has already discussed such zeta-functions assuming necessary properties of the orbital decomposition. In this connection, we would like to mention the problem to associate a zeta-function to a rather arbitrary polynomial mapping (with coefficients in a number field). If this is possible, perhaps the zeta-function associated with our $f: X \to I(X)$ will have similar properties as those mentioned above.

References

1. H. Braun and M. Koecher, *Jordan-Algebren*, Springer-Verlag, Berlin-Heidelberg-New York, 1966.

2. J. Igusa, *Harmonic analysis and theta-functions*, Acta Math. 120 (1968), 187-222.

3. ———, *Some observations on the Siegel formula*, Rice University Studies 56 (1970).

4. B. Kostant, *Lie group representations on polynomial rings*, Amer. J. Math. 85 (1963), 327-404.

5. J. G. M. Mars, *Les nombres de Tamagawa de certains groupes exceptionnels*, Bull. Soc. Math. France 94 (1966), 97-140.

6. T. Ono, *An integral attached to a hypersurface*, Amer. J. Math. 90 (1968), 1224-1236.

7. I. Satake, *Symplectic representations of algebraic groups satisfying a certain analyticity condition*, Acta Math. 117 (1967), 215-279.

8. C. L. Siegel, *Gesammelte Abhandlungen*, I-III, Springer-Verlag, Berlin-Heidelberg-New York, 1966.

9. A. Weil, *Sur la formule de Siegel dans la théorie des groupes classiques*, Acta Math. 113 (1965), 1-87.

The Johns Hopkins University, Baltimore, Maryland 21218

SOME NUMBER-THEORETICAL RESULTS ON REAL ANALYTIC AUTOMORPHIC FORMS

Tomio Kubota

1. *Introduction*

In some of the previous works of the author, a generalization, in a certain sense, of the theta function was obtained as real analytic automorphic functions which are intimately connected with the reciprocity law of power residues of higher degrees, generalizing the fact that theta functions are related to quadratic reciprocities. But, in constructing such functions, it is necessary to use the reciprocity law itself, and so far no direct way of construction such as theta series is known. To remove this incompleteness, it is necessary to investigate the functions more precisely, and to discover all of their characteristic properties. In the present note, we propose to announce some new results and conjectures which may be of some use for that purpose.

In section 1, we state definitions and a summary of known results, as far as they are necessary in the sequel. In section 2, we explain a relationship that can be expected to exist between generalized theta functions and Gauss sums. In section 3, we shall show that there is some possibility of interpreting the transformation formula of generalized theta functions as a consequence of Poisson's summation formula, although it is still unknown whether such an interpretation is actually meaningful in algebraic or analytic number theory.

We denote by H the three dimensional upper half space, whose points are of the form $u = (z,v)$, $(z \in \mathbb{C}, \; v > 0)$. The space H is the non-hermitian symmetric space $SL(2,\mathbb{C})/SU(2)$, and the operation of $\sigma = \begin{pmatrix} a & b \\ c & d \end{pmatrix} \in SL(2,\mathbb{C})$ on H is given by the linear transformation $\sigma u = (au+b)(cu+d)^{-1}$, where we identify u with the matrix $\begin{pmatrix} z & -v \\ v & \bar{z} \end{pmatrix}$, and any $t \in \mathbb{C}$ with the matrix $\begin{pmatrix} t & \\ & \bar{t} \end{pmatrix}$. By $|u|$ we denote the distance between the point u and $0 = (0,0)$, i.e., $|u|^2 = |z|^2 + v^2$. By u^* we denote the image of u under the transformation determined by $\begin{pmatrix} & -1 \\ 1 & \end{pmatrix}$, i.e., $u^* = (-\bar{z},v)|u|^{-2}$. The group $GL(2,\mathbb{C})$ also operates naturally on H; the operation is given by $\sigma u = [(\det\sigma)^{-\frac{1}{2}}\sigma]u$, $((\det\sigma)^{-\frac{1}{2}}\sigma \in SL(2,\mathbb{C}))$. For a complex number t, we write $tu = (tz, |t|v)$; if $t \neq 0$, then $tu = \begin{pmatrix} t & \\ & 1 \end{pmatrix} u$.

Let F be a totally imaginary algebraic number field of degree $2r$ containing the n-th roots of unity, and let $\mathfrak{o}$ be the ring of integers of F. Furthermore, denote by (a/b) the n-th power residue symbol in F. The principal previous results of the author are, as stated in [1] and [2], that the symbol (a/b) defines a representation χ of degree 1 of a certain subgroup Γ of finite index of $GL(2,\mathfrak{o})$, and that the Eisenstein series containing the character χ give us the way of constructing a space Θ, finite dimensional over $\mathbb{C}$, of real analytic automorphic functions which can be regarded as generalized theta functions.

Explaining the situation more precisely, χ is defined in such a way that $\chi(\sigma) = (c/d)$ or $\chi(\sigma) = 1$ according as $c \neq 0$ or $c = 0$ for $\sigma = \begin{pmatrix} a & b \\ c & d \end{pmatrix}$ in a certain congruence subgroup of Γ, and Ker χ contains no congruence subgroup. The group Γ operates on the product H^r of upper half space similarly to the operation of Hilbert's modular group, and the fundamental domain $\mathcal{D} = \Gamma \backslash H^r$ is of finite volume. So, putting $v(u) = v$ for $u = (z,v) \in H$, and putting $v(u) = \prod v(u_i)$ for $u = (u_1,\ldots,u_r) \in H^r$, we can define an Eisenstein series by

$$E(u,s) = \sum \bar{\chi}(\sigma)v(\sigma u)^s, \qquad (\sigma \in \Gamma_\infty \backslash \Gamma),$$

where Γ_∞ is the group of $\sigma = \begin{pmatrix} a & b \\ c & d \end{pmatrix} \in \Gamma$ with $c = 0$. This series does not converge absolutely unless $\mathrm{Re}\ s > 2$. But it has a meromorphic continuation on the whole s-plane, has a pole of first order at $s = (n+1)/n$, and the residue $\theta(u)$ of $E(u,s)$ at the pole is a square integrable function on $\mathcal{D}$ satisfying $\theta(\sigma u) = \chi(\sigma)\theta(u)$ for $\sigma \in \Gamma$. The series $E(u,s)$ is so to speak the Eisenstein series attached to the cusp ∞ of $\mathcal{D}$, so that a similar series can be defined for each cusp of $\mathcal{D}$, and the space Θ of generalized theta functions is the space of residues at $(n+1)/n$ of all complex linear combinations of such Eisenstein series.

If we put $e(z) = \exp(2\pi\sqrt{-1}(z + \bar{z}))$ for $z \in \mathbb{C}$, and put $e(u) = \prod_i e(z_i)$ for $u = (u_1, \ldots, u_r) \in H^r$ with $u_i = (z_i, v_i) \in H$, then $E(u,s)$ has a Fourier expansion of the form

$$E(u,s) = \sum a_m(v_1, \ldots, v_r, s)e(mu),$$

where m ranges over a lattice in $\mathbb{C}^r$. The functional equation of the Eisenstein series implies a functional equation of each a_m. After expressing the coefficients a_m explicitly by Dirichlet series, we have a zeta function $Z(s)$ whose coefficients are Gauss sums containing the n-th power residue symbol, and a functional equation $\xi(s) = \xi(2-s)$ with

$$\xi(s) = (2\pi)^{-(n-1)rs}\, d_F^{n/2}\, \frac{\Gamma(n(s-1))^r}{\Gamma(s-1)^r}\, Z(s),$$

where d_F is the absolute value of the discriminant of F.

Furthermore, it should be mentioned that the space Θ can be regarded as a space of functions on a topological covering group, called sometimes metaplectic group, of the adèle group of $GL(2,F)$, and has then some meaning in the theory of unitary representation of the metaplectic group. In particular, Θ is mapped into itself by all

Hecke operators of the metaplectic group, which can be defined quite similarly to those of $GL(2,F)$.

2. *Investigations on Hecke operators and Dirichlet series*

From now on, we always treat the case where the basic field is $F = \mathbb{Q}(\sqrt{-3}) = \mathbb{Q}(\rho)$, $\rho = \exp(2\pi\sqrt{-1}/3)$. The space Θ is defined on the upper half space H itself, and there exists in Θ a function $\theta(u)$ which possesses the Fourier expansion

$$(1)\quad \theta(u) = v^{\frac{2}{3}} + \sum_{m} c_m v K_{\frac{1}{3}}(2^2 3^{-\frac{3}{2}}\pi|m|v)e\left(\frac{mu}{3\sqrt{-3}}\right), \qquad (m \in \mathfrak{o},\ \ m \neq 0),$$

where K is a modified Bessel function. The group Γ used in the definition of the space Θ here is the one generated by the principal congruence subgroup $SL(2,\mathfrak{o})_3$ mod 3 of $SL(2,\mathfrak{o})$ and by $\begin{pmatrix} & -1 \\ 1 & \end{pmatrix}$, and the character χ is defined by $\chi(\sigma) = 1$ for $\sigma = \begin{pmatrix} & -1 \\ 1 & \end{pmatrix}$ or for $\sigma = \begin{pmatrix} a & b \\ c & d \end{pmatrix} \in SL(2,\mathfrak{o})_3$ with $c = 0$, and by $\chi(\sigma) = (c/d)$ for $\sigma = \begin{pmatrix} a & b \\ c & d \end{pmatrix} \in SL(2,\mathfrak{o})_3$ with $c \neq 0$; we put here $n = 3$ so that (c/d) is the cubic residue symbol.

We want to get as much information as possible on the coefficients c_m; first we show an application of Hecke operators in the sense of [1]. The result, in its simplest form, is the following

Proposition 1. *Let* $\omega \equiv 1 \pmod 3$ *be a prime number of* F, *and denote by* $T(\omega^3)$ *the Hecke operator defined by*

$$(2)\quad (T(\omega^3)f)(u) = f(\omega^3 u) + \sum_{\substack{c \bmod \omega \\ \omega \nmid c}} f\left(\begin{pmatrix} \omega^2 & c \\ & \omega \end{pmatrix} u\right)(c/\omega)$$

$$+ \sum_{\substack{c \bmod \omega^2 \\ \omega \nmid c}} f\left(\begin{pmatrix} \omega & c \\ & \omega^2 \end{pmatrix} u\right)(c/\omega)^2 + \sum_{c \bmod \omega^3} f\left(\begin{pmatrix} 1 & c \\ & \omega^3 \end{pmatrix} u\right)$$

for a function $f(u)$ on H. On the other hand, denote by $g(\omega)$ the Gauss sum

$$g(\omega) = \sum_{c \bmod \omega} (c/\omega)e(\tfrac{c}{\omega}), \quad \left(= \sum_{c \bmod \omega} (c/\omega)e\left(\frac{c}{3\sqrt{-3}\,\omega}\right)\right),$$

and put $c_1 = 1$, $c_\omega = g(\omega)/|\omega|$, $c_{\omega^2} = 0$, $c_{\omega^{3e}} = |\omega|^e$, $c_{\omega^{1+3e}} = c_\omega|\omega|^e$, *and* $c_{\omega^{2+3e}} = 0$, $(e = 1,2,\ldots)$, *then the Fourier series*

$$f(u) = v^{\frac{2}{3}} + \sum_{m=0}^{\infty} c_{\omega^m}\, vK_{\frac{1}{3}}(2^2 3^{-\frac{3}{2}}\pi|\omega|^m v)e\left(\frac{\omega^m}{3\sqrt{-3}}z\right)$$

has the property $T(\omega^3)f = (|\omega|^2 + |\omega|^4)f$, *i.e., f is an eigenfunction of* $T(\omega^3)$.

Proof. Denote by $T(\omega^3,1)$, $T(\omega^2,\omega)$, $T(\omega,\omega^2)$, resp. $T(1,\omega^3)$ the four operations on f given by the first, second, third, resp. fourth term in the right hand side of the formula (2) defining $T(\omega^3)$. Then, $T(\omega^3) = T(\omega^3,1) + T(\omega^2,\omega) + T(\omega,\omega^2) + T(1,\omega^3)$. On the other hand, putting

$$k(u) = K_{\frac{1}{3}}(2^2 3^{-\frac{3}{2}}\pi v)e\left(\frac{z}{3\sqrt{-3}}\right), \qquad u = (z,v),$$

for the sake of simplicity, we have

$$T(\omega^3,1)\cdot v^{\frac{2}{3}} = |\omega|^2 v^{\frac{2}{3}},$$

$$T(\omega^3,1)\cdot vk(\omega^m u) = |\omega|^3 vk(\omega^{m+3}u), \qquad m \geqq 0,$$

$$T(\omega^2,\omega)\cdot v^{\frac{2}{3}} = 0$$

$$T(\omega^2,\omega)\cdot vk(\omega^m u) = \begin{cases} g(\omega)|\omega|vk(\omega u), & m = 0, \\ 0, & m \geqq 1, \end{cases}$$

$$T(\omega,\omega^2)\cdot v^{\frac{2}{3}} = 0$$

$$T(\omega,\omega^2)\cdot vk(\omega^m u) = \begin{cases} \overline{g(\omega)}|\omega|vk(u), & m = 1, \\ 0, & m \geqq 0,\ m \neq 1, \end{cases}$$

and

$$T(1,\omega^3)\cdot v^{\frac{2}{3}} = |\omega|^4 v^{\frac{2}{3}},$$

$$T(1,\omega^3)\cdot vk(\omega^m u) = \begin{cases} |\omega|^3 vk(\omega^{m-3}u), & m \geqq 3, \\ 0, & 2 \geqq m \geqq 0. \end{cases}$$

Therefore, f being as in the proposition, the coefficient of $vk(\omega^m u)$ in the Fourier expansion of $T(\omega^3)f$ is:

$$\overline{g(\omega)}|\omega|c_\omega + |\omega|^3 c_{\omega^3} = g(\omega)\overline{g(\omega)} + |\omega|^4 = (|\omega|^2 + |\omega|^4)c_1 \quad \text{for} \quad m = 0,$$

$$g(\omega)|\omega|c_1 + |\omega|^3 c_{\omega^4} = (|\omega|^2 + |\omega|^4)c_\omega \quad \text{for} \quad m = 1,$$

$$|\omega|^3 c_{\omega^{m-3}} + |\omega|^3 c_{\omega^{m+3}} = (|\omega|^2 + |\omega|^4)c_{\omega^m} \quad \text{for} \quad m \geqq 2.$$

Furthermore, it is clear that $T(\omega^3)v^{\frac{2}{3}} = (|\omega|^2 + |\omega|^4)v^{\frac{2}{3}}$. This completes the proof.

It follows from the general theory in [1] that the space Θ is decomposed into the sum of one dimensional spaces, each of which is generated by a simultaneous eigenfunction of operators $T(\omega^3)$, ($\omega \equiv 1 \pmod 3$). So proposition 1 may suggest the existence of the fact that the Fourier coefficients of functions in Θ consist of Gauss sums.

To catch the fact more clearly, we now observe the Dirichlet series which is combined with a function in Θ through Mellin transformation.

Proposition 2. *Let* $\theta(u)$ *be as in* (1). *Then the function*

$$\xi(s) = 2^{-1-2s}\, 3^{3s-\frac{3}{2}}\, \pi^{1-2s}\, \Gamma(s - \tfrac{2}{3})\Gamma(s - \tfrac{1}{3}) \sum_m \frac{c_m}{|m|^{2s-1}}$$

is holomorphic on the whole s-plane except two poles of first order at $s = 4/3$ *and* $2/3$*, and satisfies the functional equation* $\xi(s) = \xi(2-s)$.

Proof. The proof is quite similar to the case of the usual zeta function. Denote by $\theta(v)$ the value of $\theta(u)$ at $u = (0,v)$, and consider the Mellin transformation $\int_0^\infty (\theta(v) - v^{\frac{2}{3}})v^{2(s-1)-1}dv$. Then, from $\theta(v) = \theta(1/v)$ and from the formula

$$\int_0^\infty vK_{\frac{1}{3}}(v)v^{2(s-1)-1}dv = 2^{2s-3}\,\Gamma(s-\tfrac{2}{3})\Gamma(s-\tfrac{1}{3}),$$

it follows that

$$\begin{aligned}\xi(s) &= \int_0^\infty (\theta(v) - v^{\frac{2}{3}})v^{2(s-1)-1}dv \\ &= \int_1^\infty (\theta(v) - v^{\frac{2}{3}})(v^{-2(s-1)-1} + v^{2(s-1)-1})dv \\ &\quad + \frac{1}{2(s-1) - \frac{2}{3}} + \frac{1}{-2(s-1) - \frac{2}{3}}.\end{aligned}$$

This completes the proof.

It is now remarkable that there is a strong resemblance between the Dirichlet series in proposition 2 and the Dirichlet series deduced from the Fourier coefficients of the Eisenstein series as was mentioned in section 1. In the present case that $n = 3$, the latter is essentially the series

$$Z(s) = \left(\sum_m \frac{g(m)}{|m|^{2s}}\right)\left(\sum_\alpha \frac{1}{|\alpha|^{6s-4}}\right), \quad (m,\alpha \in \mathfrak{O},\ m \neq 0,\ \alpha \neq 0),$$

and the gamma factor in the functional equation of $Z(s)$ turns out to be $2^{-2s}\,3^{3s/2}\,\pi^{-2s}\,\Gamma(3(s-1))/\Gamma(s-1)$. So if we rewrite $Z(s)$ to have

$$Z(s) = \left(\sum_m \frac{g(m)/|m|}{|m|^{2s-1}}\right)\left(\sum_\alpha \frac{|\alpha|}{|\alpha^3|^{2s-1}}\right),$$

and recall

$$\Gamma(3(s-1))/\Gamma(s-1) = \frac{3^{3(s-1)}}{2\pi\sqrt{3}}\,\Gamma(s-\tfrac{2}{3})\Gamma(s-\tfrac{1}{3}),$$

then propositions 1 and 2 will lead us to the conjecture that the series $\sum c_m/|m|^{2s-1}$ in proposition 2 coincides with $Z(s)$ up to an elementary factor. Thus we again obtain an evidence for the fact that the Fourier coefficients of the functions in Θ are composed of Gauss sums. Moreover, it appears that there probably exists a decomposition of the form $R(u) = \sum g(\nu^3 u)$, $(\nu \in \mathfrak{o})$, with a partial sum $g(u)$ of the Fourier series $R(u) = v^{-\frac{2}{3}}\theta(u)$.

3. *Fourier transformation*

As was mentioned in section 1, the functions in Θ are generalizations which actually reduce to theta functions in the special case with $n = 2$. In this case, the invariance under $\begin{pmatrix} & -1 \\ 1 & \end{pmatrix}$ of the functions in Θ is a consequence of the Poisson summation formula applied essentially to the function $e^{-\pi x^2}$. The aim of this section is to show that there is a general possibility of interpreting the invariance under $\begin{pmatrix} & -1 \\ 1 & \end{pmatrix}$ of the functions in Θ as a consequence of Poisson's formula. To do this, we introduce an integral transformation which we call temporarily in this note Fourier transformation of degree n.

Let $f(u)$ be a function on the upper half space H, such that $f(\zeta u) = f(u)$ for any $(n-1)$-st root ζ of unity. Then, a function $g(u)$ on H is called the Fourier transform of degree n of $f(u)$, and denoted by $F_n f = g$, if

$$(3) \qquad \int_{\mathbb{C}} f(t^{\frac{n}{n-1}}u)\,e(Atw)\,dV(t) = |u|^{-\frac{2(n-1)}{n}}\, g(w^{\frac{n}{n-1}}u^*),$$

where A is a complex number such that $|A| = 1/2$, and $dV(t) = \tfrac{1}{2}|dt \wedge \overline{dt}|$ is the Euclidean volume of $\mathbb{C}$. It is easy to see that the left hand side of (3) is actually expressible in the form of the right hand side with some g. Moreover, g is also invariant under $u \to \zeta u$, and F_n^2 is the identity.

Coming back now to the case that $n = 3$, observe the function $R(u) = v(u)^{-2/3}\theta(u)$, where $\theta(u)$ is as in (1). Then,

$$R(u) = |u|^{-4/3} R(u^*),$$

and the constant term in the Fourier expansion of $R(u)$ is 1. To combine the function $R(u)$ with the transformation in (3), we denote once and for all by F_3 the transformation defined by (3) in which we put $n = 3$ and $A = -\sqrt{-1}/2$, so that the lattice $\ell = 2^{1/2}3^{-1/4}\mathfrak{o}$, $\mathfrak{o}$ being as ever the ring of integers $\mathbb{Z}[\rho]$ of F, is self dual with respect to the additive character $e(At) = \exp(\pi(t-\bar{t}))$ of $\mathbb{C}$. If now we assume the existence of a function $f(u)$ on H such that $F_3 f = f$, and such that $f(u)$ tends to 1 when u tends to 0, then it follows immediately from the ordinary Poisson formula that

$$\sum_{\nu\in\ell} f(\nu^{3/2}u) = |u|^{-4/3} \sum_{\nu\in\ell} f(\nu^{3/2}u^*).$$

Comparing these two formulas with each other, it seems to be natural that $R(u)$ can be expressed as a series of the form $\sum f(\nu^{3/2}u)$, $(\nu\in\ell)$, where f is a function invariant under F_3.

It is, in principle, not very difficult to find a function f such that $R(u) = \sum f(\nu^{3/2}u)$, $(\nu\in\ell)$, when R is known. To explain the situation briefly, we assume that $R(u)$ is invariant under the transformation $u \to \zeta u$ for every 6-th root ζ of unity. This does not cause any substantial restriction in our investigation, because by observing a group Γ' commensurable with our Γ, and a character χ' of Γ' which coincides with the character χ of Γ on $\Gamma\cap\Gamma'$, we can get a function which is invariant under $u \to \zeta u$, and is automorphic with respect to Γ' with the automorphic factor $\chi'(\sigma)$, $(\sigma\in\Gamma')$.

Under the assumption $R(u) = R(\zeta u)$, $(\zeta^6 = 1)$, put

$$f(u) = \frac{1}{6} \sum_{\nu \in \mathfrak{o}} \mu(\nu) R(2^{-\frac{3}{4}} 3^{\frac{3}{8}} \nu^{\frac{3}{2}} u), \tag{4}$$

μ being the Möbius function, then $f(u)$ has all wanted properties, i.e., $\frac{1}{6}\sum f(\nu^{\frac{3}{2}}u) = R(u)$, $(\nu \in \ell)$, and $F_3 f = f$, whenever $F_3 f$ exists. For various reasons, in particular because of those results which are stated in section 2, one can actually expect that $F_3 f$ exists, whereas it is not completely clear that such an f or a transformation such as F_3 is most adequate for the study of the space Θ. The function f in (4) is no longer periodic with respect to z, but $v^{\frac{5}{3}} f$ is still an eigenfunction of the Laplacian.

The same technique of arguments as above works if we replace the transformation F_3 with some other one, for instance with the transformation $f \to g$ defined by

$$\int_{\mathbb{C}} f(t\bar{t}^{-\frac{1}{2}}u) e(Atw) dV(t) = |u|^{-\frac{4}{3}} g(w\bar{w}^{-\frac{1}{2}}u^*).$$

For the case of $n = 2$, the function $f(u) = \exp(-\pi v) e(z/4)$ is invariant under F_2, i.e.,

$$\int_{\mathbb{C}} f(t^2 u) e(tw/2) dV(t) = |u|^{-1} f(w^2 u^*)$$

holds, and this entails all theta formulas and the quadratic reciprocity for a totally imaginary field.

References

1. T. Kubota, *On automorphic functions and the reciprocity law in a number field*, Lectures in Mathematics 2, Kyoto University, 1969.

2. ———, *Some results concerning reciprocity law and real analytic automorphic functions*, Proc. of the Summer Institute on Number Theory held at State University of New York at Stony Brook, 1969.

Nagoya University, Japan

RECENT RESULTS ON ANALYTIC MAPPINGS AND PLURISUBHARMONIC FUNCTIONS IN TOPOLOGICAL LINEAR SPACES

Pierre Lelong

Introduction

In the past four years, results were obtained to extend complex analysis to topological vector spaces, which are not Banach spaces. A good motivation is to solve problems in functional spaces, mostly Fréchet spaces. We give here (see sections 4, 5 of this lecture) examples, concerning the space $A(\Omega)$ of the analytic functions in a domain $\Omega \subset \mathbb{C}^n$, endowed with the topology of compact convergence, and spaces of entire functions. Analytic mappings, plurisubharmonic functions and the notions of polar and negligible sets are introduced through classical problems in functional spaces. We give also a generalization of the classical Banach-Steinhaus theorem for complex vector spaces and polynomial mappings, which leads us to introduce the notion of C-barrelled space. For the proofs of the basic results concerning plurisubharmonic functions, the reader is referred to the book [12] and to my seminar lectures [14] and [15], to the Note [13] and to the theses of Ph. Noverraz [21] and G. Coeuré [4].

1. *Analytic mappings*

1.1. *Mappings* $\mathbb{C} \to F$. As in the classical case $n = 1$, or n finite $\neq 1$, we have to compare many different definitions of the analyticity.

We suppose that E and F are Hausdorff spaces, and F a locally convex space (then $F' \neq \{0\}$). A mapping $f\colon \mathbb{C} \to F$ is called *analytic* in $U \subset \mathbb{C}$ with values in F, if there exists a representation

$$f(x+u) = f(x) + \xi_1 u + \dots + \xi_n u^n + \dots , \qquad u \in \mathbb{C} \tag{1}$$

uniformly convergent in F for $|u| < \rho$, $\rho > 0$, with ξ_n contained in F; a mapping f is called *weak analytic* if for each $x' \in F'$, $x' \circ f$ is an analytic function $\mathbb{C} \to \mathbb{C}$. For a weak analytic mapping f, there exists a convergent Taylor series (1), but the coefficients ξ_n, given by the Cauchy integral, are elements of the weak completion $\tilde{F}$ of F (see Grothendieck [5]). In order to obtain $\xi_n \in F$ it is sufficient to suppose that F has the property (K_c): the convex and closed hull of every compact set in F is a compact set. In the following we assume that F is a quasi-complete space (i.e., each closed and bounded set in F is complete). Then analyticity and weak analyticity for mappings $\mathbb{C} \to F$ coincide.

1.2. *Polynomial mappings.* A mapping $f: E \to F$ is called a homogeneous polynomial mapping of degree m if there exists an m-linear and symmetric mapping $\tilde{f}(x_1,\dots,x_m): E^m \to F$, such that $f(x) = \tilde{f}(x,\dots,x)$; the map $\tilde{f}$ is unique and the isomorphism $\psi: \tilde{f} \to f$ has for inverse

$$\psi^{-1}: f \to \tilde{f}(x_1,\dots,x_m) = \frac{1}{2^m m!} \sum_{\varepsilon_i = \pm 1} \varepsilon_1 \varepsilon_2 \dots \varepsilon_m f(\varepsilon_1 x_1 + \dots + \varepsilon_m x_m).$$

By the isomorphism ψ, the *continuous* m-linear mappings give continuous homogeneous polynomial mappings.

Proposition 1. *We suppose that E and F are locally convex spaces, and denote by $Q^m(E,F)$ the space of the homogeneous polynomial mappings. For $f \in Q^m(E,F)$ the following conditions are equivalent (see [2]):*

(i) *for every continuous semi-norm* q *on* F*, there exists an open set* U_q *in* E*, such that* $q \circ f$ *is bounded in* U_q ;

(ii) *for every continuous semi-norm* q *on* F*, there exists a neighborhood* V_q *of* $0 \quad E$ *such that* $q \circ f$ *is bounded in* V_q ;

(iii) f *is continuous at zero;*

(iv) f *is continuous everywhere.*

Remarks. a) In (i), U_q depends on q. Example: the identity $E \to E$ is continuous if E is a Fréchet space, but the image of a neighborhood of 0 is not a bounded set in E, if E has no bounded neighborhood of 0.

b) The proof (i) $\Rightarrow$ (ii) uses only the property of q to be a plurisubharmonic function, homogeneous of order one, and the following lemma:

Lemma 1. If $L(z)$ is a plurisubharmonic function in E, such that $L(ux) = |u|L(x)$, for every $u \in \mathbb{C}$, then, given a circled set $A \subset E$, and a not necessarily continuous polynomial mapping f, we have

(2) $$\sup_{y \in A} L \circ f(y) \leqq \sup_{y \in A} L \circ f(x_0 + y), \quad \text{for all} \quad x_0 \in E.$$

A set A is called *circled* if $\lambda A = A$, for every $\lambda \in \mathbb{C}$, $|\lambda| \leqq 1$. We consider $g_y(u) = L \circ f(ux + y)$ which is a subharmonic function of $u \in \mathbb{C}$ and has the property: $g_{y'}(u) = g_y(ue^{i\theta})$ if $y' = ye^{-i\theta}$. Then $g(u) = \sup_{y \in A} g_y(u)$ depends only on $|u|$, and is an increasing function of $|u|$; $g(0) \leqq g(1)$ and (2) is proved.

In the following we denote by $P^m(E,F)$ the space of the *continuous* homogeneous polynomial mappings $E \to F$.

1.3. *Analytic mappings* $E \to F$

Definition A. Given an open set $U \subset E$; a mapping $f\colon U \to F$, F quasi-complete and locally convex, is called *analytic* if each $x \in U$ has a neighborhood $x + V \subset U$ such that

$$(3) \qquad f(x+h) = \sum f_n(h), \qquad f_n \in P^n(E,F),$$

with uniform convergence for $h \in V$.

G-analyticity. A mapping $f\colon U \subset E \to F$ is called *G-analytic* if the restriction of f to each affine complex line is an analytic mapping $\mathbb{C} \to F$.

In other terms, f is G-analytic if for every linear mapping $l\colon u \in \mathbb{C} \to z = z_0 + uy \in E$, and for every $x' \in F'$, the function $x' \circ f \circ l(u)$ is analytic $\mathbb{C} \to \mathbb{C}$. We can also consider a disk $\Delta_{x,y}$

$$\Delta_{x,y} = \{z \in E; \quad z = x + uy, \quad u \in \mathbb{C}, \quad |u| \leq 1\},$$

linear image of the unit disk in $\mathbb{C}$, and assume that the mapping f has an analytic restriction on each disk $\Delta_{x,y} \subset U$.

Definition B. A mapping $f\colon U \subset E \to F$ is analytic if *it is G-analytic and continuous.*

Theorem 1. A $\Leftrightarrow$ B.

Sketch of the proof: A $\Rightarrow$ B is obvious. To prove B $\Rightarrow$ A, we remark first that G-analyticity has for consequence

$$f(x+uh) = \sum_0^\infty u^n f_n(h), \qquad u \in \mathbb{C},$$

which converges for $h \in S_x$; S_x is the maximal circled set such that $x + S_x$ is contained in U. The coefficients $f_n(h)$ are homogeneous; they are polynomial mappings because $f_n(h_1 + vh_2)$, $v \in \mathbb{C}$, is a polynomial mapping (of v) of degree $\leqq n$. Therefore $f_n \in \mathcal{Q}^n(E,F)$.

Second step of the proof:

$$f_n(h) = \frac{1}{2\pi} \int_0^{2\pi} f(x + he^{i\theta})e^{-ni\theta} d\theta \in \tilde{F} = F$$

is continuous because f is continuous in x; for each continuous semi-norm q on F, $q \circ f$ is locally bounded; therefore $q \circ f_n(h)$ converges normally, i.e.,

$$\sum_0^\infty \{\sup_{h\in\overline{V}} q \circ f_n(h)\} < \infty$$

for a closed, circled neighborhood $x + \overline{V}$ of x.

The theorem $B \Rightarrow A$ received a beautiful complement from M. A. Zorn for Banach spaces [25]. Zorn's result was extended if E is a Fréchet space by Ph. Noverraz [21]: *if* $f: U \subset E \to F$ *is G-analytic, the set* A *of the points of continuity for* f *is open and closed*; *therefore* $A \neq \emptyset$ *gives* $A = U$ *if* U *is a domain.* A more general formulation was recently given by Bochnak and Siciak [1] and [2]:

Theorem 2. *Given two topological vector spaces* E, F *over* $\mathbb{C}$, *Hausdorff, and* F *locally convex and quasi-complete. For a G-analytic mapping* $f: U \subset E \to F$, *we consider the following properties, where* $\Gamma(F)$ *is the set of the continuous semi-norms on* F:

(a) *for every* $q \in \Gamma(F)$, $q \circ f$ *is locally bounded in* U,

(b) *for every* $q \in \Gamma(F)$ *there exists an open set* $U_q \subset U$ *such that* $q \circ f$ *is bounded in* U_q, *and* $\bigcap U_q \neq \emptyset$,

(c) f *is continuous at some* $x_0 \in U$,

(d) f *is continuous in* U,

(e) f *is analytic in* U.

Then (a) $\Leftrightarrow$ (d) $\Leftrightarrow$ (e).

If moreover E *is a Baire space*, (a), (b), (c), (d), (e) *are equivalent properties.*

As a consequence of the results of M. A. Zorn and Noverraz we have:

Theorem 3 (Generalization of the Hartogs theorem, see Ph. Noverraz [21]). *Let us consider two vector spaces* E_1, E_2, *Hausdorff, which are metric and complete, and a mapping* $f\colon U \to G$ *of a domain* $U \subset E_1 \times E_2$ *to a locally convex, quasi-complete Hausdorff linear vector space* F. *If* $f(x_1,x_2)$, $x_1 \in E_1$, $x_2 \in E_2$ *is separately analytic in* x_1 *and* x_2, *then* f *is an analytic mapping* $E_1 \times E_2 \supset U \to F$.

2. *Plurisubharmonic functions in topological vector spaces*

2.1. The extension to topological vector spaces of the basic properties of the class of plurisubharmonic functions was given by C. O. Kiselman, Ph. Noverraz, G. Coeuré and the author during the past four years. We suppose E Hausdorff and complete.

We recall that a disk $\Delta_{x,y}$ is the set $\Delta_{x,y} = \{z \in E;\ z = x + uy,\ u \in \mathbb{C},\ |u| \leqq 1\}$, i.e., the image of $|u| \leqq 1$ by a linear mapping into E; $\Delta_{x,y}$ is a compact set. If E is a topological vector space, we denote by F_E the filter of the neighborhoods of the origin in E.

Definition 1. A function $x \in G \to f(x) \in \mathbb{R}$, defined in an open set $G \subset E$, with real values $(-\infty \leqq f(x) < \infty)$ is called *plurisubharmonic* if and only if:

a) f is upper semi-continuous, i.e., for every $c \in \mathbb{R}$, the set $f(x) < c$ is open in G.

b) The mean-value inequality holds for the disks $\Delta_{x,y} \subset G$:

$$f(x) \leqq \frac{1}{2\pi} \int_0^{2\pi} f(x + ye^{i\theta})\,d\theta. \tag{1}$$

Remarks. 1) *We denote by $P(G)$ the class of the plurisubharmonic functions in an open set G.*

2) If the condition (1) is replaced by

$$(1)' \qquad f(x) \leqq \int f(x + ye^{i\theta})d\nu(\theta),$$

where ν is a positive Radon-measure, and $\|\nu\| = 1$, we obtain a class $C_\nu \subset P(G)$. If $\nu = \{\frac{1}{2}\delta(0), \frac{1}{2}\delta(\pi)\}$, (1)' becomes

$$(1)'' \qquad f(x) \leqq \tfrac{1}{2}f(x+y) + \tfrac{1}{2}f(x-y)$$

and C_ν is the class of the convex (and continuous) functions in G.

3) If f is an analytic function $E \to \mathbb{C}$, then for $\log|f(x+uy)| = \log \phi_{x,y}(u)$ the mean-value inequality (1) holds. We obtain, as in the finite dimensional case:

Proposition 2. 1) *The convex functions in $G \subset E$ are plurisubharmonic functions.*

2) *If f is an analytic function in G, $\log|f|$, $|f|$, $\chi(\log|f|)$, for a convex and increasing function χ, are plurisubharmonic functions in G.*

3) *If $\{f_i\}$, $1 \leqq i \leqq N$ are plurisubharmonic functions, $g = \sup_i f_i$ is plurisubharmonic, and the same is true for $\sum_1^N c_i f_i$, if $c_i \geqq 0$; $P(G)$ is a convex cone.*

4) *If $f_n \searrow f$ is a decreasing sequence, $f_n \in P(G)$, then $f = \lim f_n \in P(G)$.*

Remark. The constant $-\infty$ will not be considered as a plurisubharmonic function.

2.2. *Transformation by an analytic mapping*. We present now the following result (P. Lelong [13] and [14]):

Theorem 4. *Given an open set* $G \subset E$, *and an analytic mapping* $f = G \to F$, *into a complex locally convex space* F, *with condition* K_c, cf. section 1, (*in particular* F *quasi-complete is appropriate*), *then if* V *is a plurisubharmonic function* $F \to \mathbb{R}$ *defined on* $f(G) + W$, $W \in F_F$, *the function* $V_1 = V \circ f \colon E \to \mathbb{R}$ *is plurisubharmonic in* G.

Proof. The upper semi-continuity of f is obvious. Let us consider a disk $\Delta_{x,y} \subset G$; we have to prove that

$$V \circ f(x) \leqq \frac{1}{2\pi} \int_0^{2\pi} V \circ f(x + ye^{i\theta})d\theta .$$

We consider first for $u \in \mathbb{C}$:

$$f(x + uy) = f(x) + \xi_1 u + \ldots + \xi_q u^q + \ldots \quad .$$

By the hypothesis (K_c), we have $\xi_q \in F$. We write $f_q(u) = = f(x) + \xi_1 u + \ldots + \xi_q u^q$ and consider $q > q_0$ such that $f_q(\Delta_{x,y}) \in f(G) + W$. For such values of q, $V \circ f_q(x + uy)$ is defined for $|u| \leqq 1$. Furthermore $f(\Delta_{x,y})$ is a compact set K in $f(G) + W$ and $M = \sup_{z \in K} V(z)$ is finite. Given $M' > M$, there exists an open neighborhood $U(K)$ of K in F such that $V(z) < M'$ for $z \in U(K)$. We choose $q_1 \geqq q_0$ such that $f_q(x + uy) \in U(K)$ for $q > q_1$ and $\|u\| \leqq 1$. If $q > q_1$ and $|u| \leqq 1$, all the functions $V \circ f_q(x + uy)$ (for fixed x, y) are bounded from above by M'.

We consider the mapping $\Delta_{x,y} \to f_q(\Delta_{x,y})$; the image in F belongs to the finite-dimensional subspace $F_q \subset F$ with basis $\{f(x), \xi_1, \ldots, \xi_q\}$. Let us consider the coordinates in F_q: $\{z_0 = 1,\ z_1 = u, \ldots, z_q = u^q,\ u \in \mathbb{C},\ |u| \leqq 1\}$, and the manifold $S_q = \{f_q(x + uy);\ |u| \leqq 1\}$, image of the disk $\Delta_{x,y}$ by f_q.

The function V is plurisubharmonic on F_q and, by a classical result (cf. [16]), its restriction to S_q is a subharmonic function of u for $|u| \leqq 1$. We obtain using the upper semi-continuity:

$$V \circ f(x) = V \circ f_q(x) \leqq \frac{1}{2\pi} \int_0^{2\pi} V \circ f_q(x + re^{i\theta})d\theta, \quad r < 1,$$

$$V \circ f(x) \leqq \limsup_{r \to 1} \frac{1}{2\pi} \int_0^{2\pi} V \circ f_q(x + re^{i\theta})d\theta \leqq \frac{1}{2\pi} \int_0^{2\pi} V \circ f_q(x + ye^{i\theta})d\theta.$$

By the lemma of Fatou applied to $\phi_q(\theta) = V \circ f_q(x + ye^{i\theta}) \leqq M'$ we obtain

$$V \circ f(x) \leqq \limsup_{q \to \infty} \frac{1}{2\pi} \int_0^{2\pi} V \circ f_q(x + ye^{i\theta})d\theta,$$

$$V \circ f(x) \leqq \frac{1}{2\pi} \int_0^{2\pi} \limsup V \circ f_q(x + ye^{i\theta})d\theta \leqq \frac{1}{2\pi} \int_0^{2\pi} V \circ f(x + ye^{i\theta})d\theta,$$

and theorem 4 is proved.

2.3. *Homogeneous plurisubharmonic functions.* We say that $V \in P(E)$ is homogeneous of order $\sigma > 0$ if for every $u \in \mathbb{C}$

$$V(ux) = u^{\sigma} V(x).$$

As in the finite-dimensional case, the following theorem holds:

Theorem 5. *If* $V \in P(E)$ *is homogeneous of order* $\sigma > 0$, *then* $V(x) \geqq 0$, *and* $\log V$ *is plurisubharmonic in* E.

Proof. $V(x_0) < 0$ gives $V(ux_0) \equiv -\infty$ in u and $V(0) = -\infty$; $V(x_1) \neq -\infty$ gives $V(0) = 0 \neq -\infty$; therefore if $V(x_0) < 0$, we obtain $V(x) \equiv -\infty$ and a contradiction. Therefore $V(x) \geqq 0$ for every $x \in E$.

Now we recall that for a subharmonic function $\phi(u) = V(x+uy) \geqq 0$ $\log \phi(u)$ is subharmonic if and only if for every $\alpha > 0$, $|e^{\alpha u}|\phi(u)$ is subharmonic. We define a linear form l on the space C^2 generated by $\{x, y\}$, with values $l(x) = 0$, $l(y) = \alpha$, $l(x+uy) = \alpha u$. The mapping $f: z \to z\cdot\exp[\sigma^{-1}l(z)]$ is analytic and by theorem 4 the function $V \circ f$ is analytic. Therefore for $z = x+uy$, $u \in \mathbb{C}$,

$$V[z\cdot\exp \frac{1}{\sigma} l(z)] = [\exp l(z)]V(z) = e^{\alpha u}V(x+uy)$$

is analytic, and theorem 5 is proved.

We give an important consequence:

Corollary. If q is a continuous semi-norm on F, and $f: G \subset E \to F$ is an analytic mapping, $\log q \circ f$ *is plurisubharmonic in G.*

Proof. q is plurisubharmonic because it is convex; $q(ux) = |u|q(x)$; therefore $\log q(y) \in P(F)$, by theorem 5; by theorem 4 we obtain $\log q \circ f(x) \in P(U)$.

2.4. *Locally upper-bounded sets of plurisubharmonic functions.* We say that a set f_α, $\alpha \in J$, of functions is *locally upper-bounded* in an open set G if for every $x \in G$ there exists a neighborhood $x + V \subset G$, and $M_x < \infty$, such that $f_\alpha(y) \leqq M_x$ for every α and $y \in x + V$.

For a function g we denote by g^* the least upper-semi-continuous function such that $g^* \geqq g$; $g^*(x) = \limsup_{x' \to x} g(x')$. We write $g^* = \text{u.r.}g$ (upper regularization of g).

The following theorem was given first by C. O. Kiselman [7] for Fréchet spaces, and in the general form in [12].

Theorem 6. If f_α, $\alpha \in J$, *is a locally upper-bounded family of plurisubharmonic functions in* $G \subset E$, *and* $g = \sup f_\alpha$, *then*

$$g^* = \text{u.r.}g$$

defined by $g^*(x) = \limsup_{x' \to x} g(x')$, *is plurisubharmonic in* G.

It is a consequence of the following lemma:

Lemma 2. Given a compact $K \subset E$ *and a Radon-measure* μ *with support in* K, *and an upper-semi-continuous function* $f(x)$ *defined in a neighborhood of* K,

$$h(x) = \int d\mu(a) f(a + x)$$

is an upper-semi-continuous function in a neighborhood of the origin.

Now, if $f_\alpha \in P(G)$,

$$l(x,y) = \int g^*(x + ye^{i\theta})d\theta$$

is an upper-semi-continuous function of (x,y), and we obtain for every disk $\Delta_{x,y} \subset G$:

$$f_\alpha(x) \leqq \frac{1}{2\pi} \int_0^{2\pi} f_\alpha(x + ye^{i\theta})d\theta \leqq \frac{1}{2\pi} \int_0^{2\pi} g^*(x + ye^{i\theta})d\theta \quad \text{for every } \alpha \in J,$$

$$g(x) \leqq \frac{1}{2\pi} \int_0^{2\pi} g^*(x + ye^{i\theta})d\theta. \tag{2}$$

The right side of (2) is an upper-semi-continuous function of x. Therefore we obtain

$$g^*(x) \leqq \frac{1}{2\pi} \int_0^{2\pi} g^*(x + ye^{i\theta})d\theta.$$

and the theorem is proved.

Theorem 7. Let us consider a locally upper-bounded family f_α, $\alpha \in J$, in $G \subset E$, $f_\alpha \in P(G)$, and suppose that J is a directed set with countable cofinal basis A_n for the filter F of the sections. Then $h(x) = \lim_F [\sup_{\alpha \in A_n} f(x)]$ has a plurisubharmonic (see [7] and [12]) *upper regularization.*

Proof. For a given disk $\Delta_{x,y} \subset G$ we consider:

$$h_n(x) = \sup_{\alpha \in A_n} f_\alpha(x) \leqq \frac{1}{2\pi} \int_* h_n(x + ye^{i\theta})d\theta,$$

where $\int_*$ denotes the lower integral (by a more sophisticated argument using the properties of the subharmonic functions of u, $\phi_\alpha(u) = f_\alpha(x + uy)$, the reader can also prove that $h_n(x + uy)$ coincides with a subharmonic function of u except on a set of vanishing capacity in $\mathbb{C}$, and the integral itself does exist).

Then we obtain, using the lemma of Fatou,

$$h(x) = \lim_n h_n(x) \leqq \lim_n \int_* h_n(x + ye^{i\theta})d\theta \leqq \int_* \lim h_n(x + ye^{i\theta})d\theta,$$

$$h(x) \leqq \int_* h(x + ye^{i\theta})d\theta,$$

$$h^*(x) \leqq \int h^*(x + ye^{i\theta})d\theta, \qquad \text{Q.E.D.}$$

Theorem 8. Let us consider a positive measure μ with compact support K in a locally compact space T, and let $f(x,t)$ be a family of plurisubharmonic functions, upper-semi-continuous of $x \times t$ for $x \in G$, $t \in K$. Then:

$$F(x) = \int d\mu(t) f(x,t) \in P(G).$$

It is a consequence of the upper-semi-continuity of F [21].

We have a quite similar result (cf. [12, p. 54]) with different assumptions.

Theorem 9. *Let us consider a locally compact space T, a positive measure μ on T, $\|\mu\| < \infty$, and a domain $G \subset E$; and suppose that E is a metrizable space. We suppose that $f(x,t)$ is a mapping $G \times T \to \mathbb{R}$ such that $f(x + ye^{i\theta}, t)$ is measurable for $d\theta \times \mu$ and that:* (i) *for fixed t, the function $f(x,t)$ is plurisubharmonic in x,* (ii) *for fixed x, $f(x,t)$ is μ integrable,* (iii) *for every compact $K \subset G$, $f(x,t)$ is upper-bounded on $K \times T$. Then $F(x) = \int f(x,t)d\mu(t)$ is plurisubharmonic in G.*

Remark. Definition 1 of the plurisubharmonic class supposes that E is endowed with a topology such that upper-semi-continuous functions are measurable on disks. It is a priori not required for that topology to be a vector space topology. For instance, we will have to consider on E the f-topology: $A \subset E$ is an open set for the f-topology if for every affine subspace $M \subset E$, which is finite-dimensional, $M \cap A$ is open in M: the plurisubharmonic class for the f-topology is the class of functions defined in f-open sets and having plurisubharmonic restrictions to the finite-dimensional subspaces. For such a function f, if we consider on E a linear vector space topology T (which is of course coarser than the f-topology), if f is locally bounded for T, then $f^* = \text{u.r.}f$ (in T) is a T-plurisubharmonic function, i.e., a plurisubharmonic function *on the topological vector space* (E,T).

Given a set $A \subset E$, we denote by A_i the set of the $x \in A$, such that x is an internal point of A, i.e., for every $y \in E$, $y \neq 0$, A contains a disk $\Delta_{x,ry}$, $r > 0$, of center x; if A is f-open, then $A = A_i$.

2.5. *Pseudo-convex f-open sets.* Given an f-open set A, for every $x \in A$, and $y \in E$, $y \neq 0$, we define

$$\rho(x,y) = \sup \{r, \ r > 0, \ \Delta_{x,ry} \subset A\}.$$

Obviously, $\rho(x,y) \geqq 0$, $\rho(x,uy) = |u|^{-1}\rho(x,y)$, $u \in \mathbb{C}$.

Definition 2. An f-open set A is called *pseudo-convex* if $-\log \rho(x,y)$ is a plurisubharmonic function (for the f-topology) of $x \times y$, $x \in A$, $y \in E - \{0\}$.

For $y = 0$, we set $-\log \rho(x,y) = -\infty$ so that $-\log \rho(x,y)$ has a plurisubharmonic continuation to $A \times E$.

For open sets A in $\mathbb{C}^n$ a formally different definition of the pseudo-convexity is often given, requiring only that $\rho(x) = \inf \|x-y\|$, $y \in A$, has the property that $-\log \rho(x)$ is plurisubharmonic. This property has for consequence that $-\log \rho(x,y)$ is plurisubharmonic in $A \times E$ [12], and the two definitions coincide. Then an f-open set $A \subset E$ is pseudo-convex if and only if for each finite-dimensional affine subspace $M \subset E$, $M \cap A$ is open and pseudo-convex

2.6. *Quasi-norms.* Given a linear space E over $\mathbb{C}$, a function $E \ni x \to L(x) \in \mathbb{R}$, $0 \leqq L(x) < +\infty$ will be called a *quasi-norm* on E if

(i) L is homogeneous of degree 1 : $L(ux) = |u| L(x)$,

(ii) L is a plurisubharmonic function in the f-topology; or, equivalently, the restriction of L to the finite-dimensional subspaces are plurisubharmonic functions.

Proposition 3. *L is a quasi-norm on E if and only if there exists a pseudo-convex, f-open and circled set A such that $A = \{x \in E \ ; \ L(x) < 1\}$; L is the gauge of A and $\rho(0,y) = L^{-1}(y)$ if $\rho(x,y)$ is relative to A.*

Proof. If A is f-open, pseudo-convex, circled, $\rho(0,y)$ has the property that $-\log \rho(0,y)$ is f-plurisubharmonic for $y \in E$. Let us consider for given y, $y \neq 0$, the complex line $x = uy$ and the section of A: it is a disk; the gauge of this disk is $l_y(u)$, $u \in \mathbb{C}$, and $l_y[\rho(0,y)] = 1$; then $l_y(u) = \rho^{-1}(0,y)$. If L is the gauge of A, we have $L(x) = l_y(u)$ for $y = ux$;

$$L(x) = l_y(u) = \rho^{-1}(0,x)$$

which proves that $L(x)$ and $\log L(x)$ are plurisubharmonic functions in the f-topology.

Conversely, given a quasi-norm $L(x)$ on E, $A = \{x \in E \; ; \; L(x) < 1\}$ is f-open and has for intersections by every finite-dimensional affine subspace M open and pseudo-convex sets; therefore A is an f-open pseudo-convex set and is obviously circled.

Proposition 4. *Given a finite set $L_1(z),\dots,L_n(z)$ of quasi-norms,* $\sup L_i$, *and* $\sqrt[q]{P_q(L_1,\dots,L_n)}$ *are quasi-norms if P_q is a homogeneous polynomial of degree q with positive coefficients.*

For the proof, it is sufficient to remark that $\log P_q(L_1,\dots,L_n)$ is a plurisubharmonic function of z. The disk property is a consequence of the following: if $p_q(u)$ is the restriction of P_q to the line $z = x + uy$, $|e^{-\sigma u}| p_q(u)$ is a plurisubharmonic function of $u \in \mathbb{C}$ for each given $\sigma \in \mathbb{C}$.

Remarks. 1) If E is not Hausdorff, and $N = \overline{0} \neq 0$, each plurisubharmonic function V defined in $G \subset E$, can be identified with a plurisubharmonic function V_1 defined on $G_1 = p(G)$, where p is the projection $E \to E/N$. For $x \in G$, $y \in G$, $h = x - y \in N$, we have $V(x) = V(y)$ because $x + N$ belongs to every neighborhood of x, and $V(x + uh)$ is an upper-bounded subharmonic function of $u \in \mathbb{C}$.

2) If $E = \prod E_i$, $i \in J$, is a product space, with the product topology, a function $V \in P(G)$, $G \subset E$, has, locally, a representation $V = V_\alpha(x_{i_1},\dots,x_{i_q})$, $x_i \in E_i$, by a plurisubharmonic function defined on $E_I = \prod_{i \in I} E_i$, where I is a finite subset $\{i_1,\dots,i_q\}$ of J. The same is true for analytic functions (see [6]).

3. *Polar sets and negligible sets in topological linear spaces*

3.1. *Polar sets.* We recall some definitions and basic facts (see [12] and [14]): in what follows E is supposed to be Hausdorff and complete.

Definition 3. A set $A \subset E$ is called *polar in a domain* $G \subset E$ if there exists $V \in P(G)$ such that $A \subset A' = \{x \in G\ ;\ V(x) = -\infty\}$; if moreover $V(x) \leqq 0$ in G, A is called *strictly polar in* G.

If A is polar, $\mathring{A} = \emptyset$. More precisely

Proposition 5. *A polar set has no internal points.*

Theorem 10. *Given a sequence* $A_n \subset G$ *of strictly polar sets defined by* $A_n \subset A'_n = \{x \in G\ ;\ V_n(x) = -\infty,\ V_n \in P(G)\}$, *then*

(i) $\bigcup A_n = A$ *is strictly polar in* G *or* $G = \bigcup A'_n$ *and* G *is itself the countable union of strictly polar subsets,*

(ii) *If* E *is a Baire space and if the* A'_n *are closed sets,* A *is a strictly polar subset of* G.

In the following we say that $V \in P(G)$, with values $-\infty \leqq V < \infty$, is continuous if $\exp V$ is continuous: (ii) is true if E is a Baire space and the functions V_n are continuous. For the proof see [14].

To get an example of a domain G which is actually a countable union of strictly polar subsets $A'_n \subset G$, let us consider the Schwartz space $\mathcal{D}(\Omega) = \lim_q \text{ind}\, \mathcal{D}(\Omega_q)$, Ω_q compact in Ω, $\Omega_q \nearrow \Omega$, $\Omega_q \subset \overset{\circ}{\Omega}_{q+1}$. We define:

$$G = \{\phi \in \mathcal{D}(\Omega) \ ; \ \sup_{z\in\Omega} |\phi(z)| \leq 1\}.$$

We suppose that Ω is a domain in $\mathbb{C}$; $\mathcal{D}(\Omega)$ is the space of the C^∞-functions with compact support in G : $\mathcal{D}(\Omega) = \bigcup_q \mathcal{D}(\Omega_q)$; therefore $G = \bigcup_q [G \cap \mathcal{D}(\Omega_q)] = \bigcup_q A_q$.

We consider a measure μ_q with support in $\overset{\circ}{\Omega}_{q+1} - \Omega_q$ and $V_q(\phi) = \log \mu_q(\phi) - \log \|\mu_q\|$. For every $\phi \in \mathcal{D}(\Omega_q)$, we obtain $\mu_q(\phi) = 0$, and $V_q(\phi) = -\infty$; $V_q \not\equiv -\infty$ on G and $V_q \in P(G)$; $V_q(\phi) \leq 0$ if ϕ belongs to G. Then A_q is a strictly polar set in G and G is actually the countable union of the strictly polar sets A_q.

3.2. *Negligible sets.* The following class contains the preceding and probably is larger.

Definition 4. A set $A \subset G$ is called *negligible in a domain* $G \subset E$ if there exists a sequence $V_q \in P(G)$, locally upper-bounded in G, such that

$$A \subset A' = \{x \in G \ ; \ g(x) < g^*(x)\}, \tag{1}$$

and $g = \sup V_q$, $g^* = \text{u.r.}g$.

Theorem 11 (see [15]). *A polar set in G is a negligible set in G.*

The following situation allows us to state that a negligible set $A \subset G$ is a polar set:

Theorem 12 (see [15]). a) *If in* (1), $g^*(x) \equiv 0$ *or if g^* is a pluriharmonic function in G* (i.e., g^* *and* $-g^*$ *plurisubharmonic in* G), *and if there exists* $\xi \in G$, *such that* $g(\xi) = g^*(\xi)$, *then A is a strictly polar set.*

b) *If moreover E is a Baire space, and the V_q are continuous, ξ does exist and A is meager and strictly polar in G.*

3.3 *Conic sets.* If G is all the space E and A is a conic set (i.e., $\lambda A = A$ for every $\lambda \in \mathbb{C}$), then A is said to be polar (resp. negligible) if there exists a neighborhood U of the origin in E such that $A \cap U$ is polar (resp. negligible) in U.

4. *Applications to algebras of holomorphic functions*

We use now the preceding notions for problems in classical functional spaces and applications to problems of "exceptional sets".

I. We consider first the Fréchet space $A(\Omega)$ of the holomorphic functions in $\Omega \subset \mathbb{C}^n$, and denote by $\eta \subset A(\Omega)$ the subset of the functions which are holomorphic in $\Omega' \supset \Omega$, $\Omega' \neq \Omega$. The topology of $A(\Omega)$ will be defined by the semi-norms $p_s(f) = \sup_{z \in K_s} |f(z)|$, $K_s \subset \mathring{K}_{s+1}$, $K_s \nearrow \Omega$.

Theorem 13. *There exist only two possibilities:* $\eta = A(\Omega)$ *or* $\eta = \bigcup F_{p,m}$ *and $F_{p,m}$ is a closed, convex and negligible set such that $\mathring{F}_{p,m} = \emptyset$; $F_p = \bigcup_m F_{p,m}$ is a convex cone, and $\mathring{F}_p = \emptyset$; η is meager and is the countable union of closed, convex and negligible sets in $A(\Omega)$.*

For the proof see [15]; it is a consequence of the following property: if we consider an open ball B_p in $\mathbb{C}^n$ with center $z_p \in \Omega$, and radius $r_p > 0$, such that B_p contains exterior points and boundary points of Ω, the set $F_{p,m}$ of the functions which are holomorphic in $\Omega \cup B_p$ and for which $|f(z)| \leqq m$ in B_p, is a closed subset in $A(\Omega)$; $\mathring{F}_{p,m} = \emptyset$. Let us consider $D_y^q f(z_p) = [\frac{\partial^q}{\partial u^q} f(z_p + uy)]_{u=0}$, $u \in \mathbb{C}$, $y \neq 0$, $y \in \mathbb{C}^n$, which are the derivatives of f of order q, for given z_p. Then the functions $D_y^q f(z_p)$ can be considered as functions

$$f \times y \to \mathbb{C}, \quad f \in A(\Omega); \quad y \in \mathbb{C}^n.$$

The functions $V_q(z,y,f) = \log \frac{1}{q!}|D_y^q f(z)|$ are continuous in $z \times f$. The functions

$$V'_q(z,f) = \sup_{\|y\|=1} \frac{1}{q} \log[\frac{1}{q!} D_y^q f(z)]$$

are plurisubharmonic functions locally (relative to f) upper-bounded. If the ball $\|z - z_0\| < 2\rho$ is contained in $K_s \subset \Omega$, then we obtain, using the Cauchy integral

$$\frac{1}{q!} |D_y^q f(z) - D_y^q f_0(z_0)| \leq \rho^{-q}[p_s(f - f_0) + \frac{q+1}{\rho} \|z - z_0\| p_s(f_0)].$$

Then if the distance of z_p to $C\Omega$ is $\delta_p < r_p$, we have:

(i) $V'_q(f,z_p) \leq -\log r_p + q^{-1}\log m$, if $f \in F_{p,m}$, $-\log r_p < -\log \delta_p$,

(ii) $\limsup_q V'_q(f,z_p) = -\log \delta_p$, if $f \notin F_{p,m}$.

Therefore, there exists q' and $\alpha > 0$, ($\alpha < -\log \delta_p + \log r_p$), such that $g_{q'}(f) = \sup_{q \geq q'} V'_q(z_p,f)$ has the property:

$$g_{q'}(f) < -\log \delta_p - \alpha, \quad \text{if } f \in F_{p,m},$$

$$g^*_{q'}(f) \geq -\log \delta_p$$

($g^*_{q'}(f)$ is the upper regularization *relative to* f). Then

$$F_{p,m} \subset \{f \in A(\Omega) \; ; \; g_{q'}(f) < g^*_{q'}(f)\}$$

is a negligible set; moreover, $F_{p,m}$ is convex and closed; for $F_p = \bigcup F_{p,m}$ we have by obvious topological considerations $\mathring{F}_p = \emptyset$ if $\eta \neq A(\Omega)$ and the theorem is proved.

We give two other results (see [15]).

Theorem 14. *Given a bounded set in B, there exists* $U_B(f) \in P[A(\Omega)]$ *such that* $U_B(f) = -\infty$ *if* $f \in B \cap \eta$.

Theorem 15. *Given a Banach space M of functions* $f \in A(\Omega)$, *if the topology on M is finer than the topology induced by the $A(\Omega)$-topology, then the set* $\eta_1 \subset M$ *of the continuable functions in M coincides with M or is a strictly polar and meager cone in M.*

In other words: if Ω is a domain of holomorphy, η_1 is a strictly polar cone in M, if M contains a function having Ω as holomorphy domain.

Theorem 14 and Theorem 15 are obtained as consequences of Theorem 12, (see P. Lelong [15]).

II. In a natural way it is possible to give very similar results for Fréchet spaces of entire functions. Given a plurisubharmonic function $L(z)$, homogeneous of order $\rho > 0$, we consider the space $E_{\rho,L}$ of the entire functions f in $\mathbb{C}^n$ such that

$$\Lambda_f = \text{u.r.}\lambda_f = \text{u.r.}(\limsup_{t \to +\infty} t^{-\rho} \log|f(tz)|) \leqq L(z).$$

$E_{\rho,L}$ becomes a Fréchet space with the help of the semi-norms

$$p_q(f) = \sup_{z \in \mathbb{C}^n} |f(z)e^{-\phi_q(z)}|,$$

where $\phi_q > \phi_{q-1}$ is a decreasing sequence of homogeneous *continuous* plurisubharmonic functions such that $\lim \phi_q = L$ and $\phi_q(z) - \phi_{q-1}(z) \geqq a_q > 0$ for $\|z\| = 1$ (it is easy to verify that the topology defined on $E_{\rho,L}$ in this way does not depend on the sequence ϕ_q). The following result was obtained in [15]:

Theorem 16. *In* $E_{\rho,L}$ *the set η of the entire functions f which have a regularized indicatrix* $\Lambda_f \neq L$ *is contained in a countable union of closed convex cones* F_q, $\mathring{F}_q = \emptyset$, *which are negligible.*

A result similar to Theorem 15 states that in a Banach space M contained in $E_{\rho,L}$, η is all of M, or is a strictly polar cone [15].

5. *C-barrelled spaces*

5.1. We give now applications to the continuous polynomial mappings (see section 1). We recall that a topological vector space E over $\mathbb{R}$ is called *barrelled* if every set which is convex, closed, and absorbing is a neighborhood of the origin. In other words E over $\mathbb{R}$ is barrelled if, given a set $p_\iota(x)$ of continuous semi-norms such that $\pi(x) = \sup_\iota p_\iota(x)$ is finite for every $x \in E$, there exists a continuous majorant $\pi^*(x)$ of $\pi(x)$; then $\pi^*(x)$ is a continuous semi-norm for E. An equivalent form is: if the set $\{p_\iota(x)\}$ is bounded for each $x \in E$, then there exists a neighborhood U of the origin and a number $M \in \mathbb{R}$ such that $p_\iota(x) \leqq M$ for $x \in U$ and all ι.

Definition 5. A topological vector space E over $\mathbb{C}$ with topology T will be called a *C-barrelled* space if, given a set of 1-homogeneous plurisubharmonic functions $L_\iota(x)$ such that $\pi(x) = \sup_\iota L_\iota(x)$ is bounded for every $x \in E$, there exists for $\pi(x)$ a plurisubharmonic majorant in E.

The functions $L_\iota(x)$ are then locally upper-bounded and $\pi^*(x) = \text{u.r.}\pi(x)$ is a plurisubharmonic majorant for $\pi(x)$; $\pi^*(0) = 0$ because $\pi^*(ux) = |u|\pi(x)$. Then:

Proposition 6. *The topological vector space (E,T) is C-barrelled if and only if for each set $L_\iota(x)$ of semi-continuous quasi-norms which is upper-bounded at each $x \in E$, there exists a neighborhood V of the origin in (E,T) and $M > 0$ such that $L_\iota(x) \leqq M$ for $x \in V$ and for every ι.*

5.2. *First extension of the Banach-Steinhaus theorem.* The classical Banach-Steinhaus theorem states that if $H \subset L(E,F)$ is a set of linear and continuous mappings $E \to F$, bounded at each $x \in E$, then H is equicontinuous if one of the following assumptions is true:

(i) E is barrelled and F is locally convex,

(ii) E is a Baire space.

We give now the following generalization of (i):

Theorem 17. *Let us consider a set* $P_\iota(x)$, $\iota \in J$, *of continuous, homogeneous polynomial mappings* $E \to F$. *We suppose* P_ι *of degree* n_ι ; E *and* F *are topological vector spaces over* $\mathbb{C}$. *Moreover we suppose that* E *is a C-barrelled space and for the topology of* F, *there exists a basis* $\{V_\alpha ; \alpha \in A\}$ *of circled neighborhoods of* $0 \in F$, *such that the gauge* $q_\alpha(y)$ *of* V_α *is plurisubharmonic (for example,* q_α *can be a fundamental set of continuous semi-norms in* F). *We suppose that, for every* $x \in E$, *there exists a bounded set* $B_x \subset F$, *and a finite complex number* σ_x *such that*

$$(1) \qquad P_\iota(x) \subset \sigma_x^{n_\iota} B_x \quad \text{for all } \iota \in J.$$

Then:

(i) *For each* q_α, *there exists a neighborhood* $U_\alpha \in F_E$ *of the origin in* E *with gauge* $p_\alpha(x)$ *such that we have*

$$q_\alpha \circ P_\iota(x) \leqq [p_\alpha(x)]^{n_\iota}$$

for every $\iota \in J$; $p_\alpha(x)$ *is a quasi-norm in* E *and is continuous at the origin.*

(ii) *For every* $V_\alpha \in F_F$, *in* F, *pseudo-convex and circled,*

$$(2) \qquad U'_\alpha = \bigcap_\iota P_\iota^{-1}(V_\alpha)$$

contains a pseudo-convex and circled neighborhood of $0 \in E$.

Proof. For (i): from (1) we obtain, if $q_\alpha(y) \leqq c_{\alpha,x}$ for $y \in B_x$:

$$q_\alpha \circ P_\iota(x) \leqq c_{\alpha,x} \cdot \sigma_x^{n_\iota}.$$

We can suppose $c_{\alpha,x} \geqq 1$, and write $c'_{\alpha,x} = c_{\alpha,x} \cdot \sigma_x$. Then

(3) $$q_\alpha \circ P_\iota(x) \leqq [c'_{\alpha,x}]^{n_\iota}.$$

We obtain: $L_{\alpha,\iota}(x) = [q_\alpha \circ P_\iota(x)]^{1/n_\iota} \leqq c'_{\alpha,x}$, and $L_{\alpha,\iota}(x)$ is a set of 1-homogeneous plurisubharmonic functions on E such that $\pi_\alpha(x) = \sup_\iota L_{\alpha,\iota}(x)$ is upper-bounded for every $x \in E$. Then, using the property of E to be C-barrelled, we can assert that $\pi_\alpha(x)$ has a plurisubharmonic majorant

$$\pi^*_\alpha(x) = \text{u.r.}\pi_\alpha(x)$$

and $L_{\alpha,\iota}(x) \leqq \pi^*_\alpha(x)$ for every $x \in E$, $\iota \in J$.

We define $U_\alpha = \{x \in E \ ; \ \pi^*_\alpha(x) < 1\}$, and $p_\alpha(x) = \pi^*_\alpha(x)$. Then by (3), we obtain

(4) $$q_\alpha \circ P_\iota(x) \leqq [p_\alpha(x)]^{n_\iota}$$

and (i) is proved because $\pi^*_\alpha(x)$ is a 1-homogeneous plurisubharmonic function in E.

To prove (ii), we remark that U'_α defined by (2) is given by the conditions $q_\alpha \circ P_\iota(x) < 1$ for every $\iota \in J$; using (4) we see that U'_α contains the set $p_\alpha(x) < 1$ which is a circled, pseudo-convex open neighborhood of $0 \in E$ because $p_\alpha(x) = \pi^*_\alpha(x)$ is 1-homogeneous and plurisubharmonic and $\pi^*_\alpha(0) = 0$.

6. *Generalization of the Banach-Steinhaus theorem in Baire spaces*

Now we give a complement to the theorem of Banach-Steinhaus, if E is a Baire space. As for Theorem 17, we consider continuous and homogeneous polynomial mappings $P_\iota(x) : E \to F$, degree $P_\iota = n_\iota$; moreover, we suppose that F_F has a basis $\{V_\alpha\}$ such that the gauge $q_\alpha(y)$ of V_α is plurisubharmonic, continuous and homogeneous.

Theorem 18. *Given a plurisubharmonic quasi-norm q_α on F, and a family of homogeneous continuous polynomial mappings $P_\iota(x) : E \to F$, $(\iota \in J)$, if E is a Baire space, there exist only two possibilities:*

(i) *The functions*

$$f_{\alpha,\iota}(x) = \frac{1}{n_\iota} \log q_\alpha \circ P_\iota(x)$$

have a plurisubharmonic majorant and there exists a quasi-norm $p_\alpha(x)$ on E, which is plurisubharmonic in E, such that

$$q_\alpha \circ P_\iota(x) \leqq [p_\alpha(x)]^{n_\iota},$$

or:

(ii) *The set $A \subset E$ of points x such that the $f_{\alpha,\iota}(x)$ are upper-bounded for $\iota \in J$, is a conic, strictly polar and meager set in E.*

Proof. Let us suppose that there exists an open set U_α such that $f_{\alpha,\iota}(x) \leqq M$ for $x \in U_\alpha$, and every $\iota \in J$. We consider a point $x_0 \in U_\alpha$, and a circled neighborhood W of $0 \in E$, such that $x_0 + W \subset U_\alpha$. Then $f_{\alpha,\iota}(x) \leqq M_\alpha$, for every $\iota \in J$, and $x \in W$ (see Lemma 1, section 1), and the functions

$$L_{\alpha,\iota}(x) = [q_\alpha \circ P_\iota(x)]^{1/n_\iota}$$

are upper-bounded in W : $\pi_\alpha(x) = \sup_\iota L_{\alpha,\iota}(x)$ has an upper regularization $\pi^*_\alpha(x)$ which is a plurisubharmonic function and $\pi^*_\alpha(0) = 0$. We define $p_\alpha(x) = \pi^*_\alpha(x)$ and (i) is proved.

As a consequence, if (i) is not true, the set A of the points $x \in E$ in which $f_{\alpha,\iota}(x)$ is not bounded for $\iota \in J$, has no interior point. Let us consider a neighborhood U of the origin in E. There exists a sequence $P_{\iota_p}(x)$, denoted by $P_p(x)$, of the polynomial mappings, P_p of degree n_p, such that

$$\sup_{x \in U} \left[\frac{1}{n_p} \log q \circ P_p(x)\right] = \sigma_p,$$

$\sigma_p > 1$, and $\sigma_p < \sigma_{p+1}$, $\sigma_p \to +\infty$. Then

$$g_p(x) = \frac{1}{n_p \sigma_p} \log q_\alpha \circ P_p(x)$$

is a plurisubharmonic function on E. We have

$$g_p(x) \leqq 1, \quad x \in U.$$

Moreover

$$g_p(\lambda x) = \frac{1}{\sigma_p} \log|\lambda| + g_p(x).$$

As a consequence we obtain for $g(x) = \lim_p \sup g_p(x)$:

$$g(x) \leqq 1, \quad x \in E,$$

$$g^*(x) = \text{u.r.}g(x) \leqq 1,$$

$g^*(x)$ is a plurisubharmonic function in E; consequently $g^*(x) \equiv a$, for some constant $a \leqq 1$. We will prove that $a = 1$. The functions

$$g_p'(x) = \sup_s g_{p+s}(x)$$

satisfy

$$g_p'(\lambda x) \leqq g_p'(x) + \frac{1}{\sigma_p} \log|\lambda|,$$

and the same is true for $g_p'^*$. As a consequence $\gamma(x) = \lim g_p'^*(x)$

is a plurisubharmonic function which is a constant $a' \leqq 1$. By the assumptions we have $a' = 1$. Each set

$$\eta_{p,m} = \{x \in E \; ; \; g_p'^*(x) - g_p'(x) \geqq \frac{1}{m}\}$$

is closed and $\mathring{\eta}_{p,m} = \emptyset$. Therefore the set of the $x \in E$ such that $g^*(x) < \gamma(x) - \frac{1}{m}$ is meager, and we have $a = 1$, $g^*(x) \equiv 1$. If in x the set of values $\frac{1}{n_\iota} \log q_\alpha \circ P_\iota(x)$ is upper-bounded, the sequence

$$\frac{1}{n_p} \log q_\alpha \circ P_p(x)$$

is upper-bounded and $g(x) = 0$. Therefore x belongs to the set $A' = \{x \in E \; ; \; g(x) < g^*(x)\}$. There exists $\xi \in E$ such that $g(\xi) = g^*(\xi) = 1$ because the sets $\eta_{p,m}$ are closed and $\mathring{\eta}_{p,m} = \emptyset$, and $\bigcup_p \eta_{p,m}$ is meager in the Baire space E. Then by Theorem 12, of section 3, the set $A \subset E$ of the points x, such that $\frac{1}{n_\iota} \log q_\alpha \circ P_\iota(x)$ is upper-bounded, is contained in the cone A', which is meager and strictly polar, and Theorem 18 is proved.

Corollary. Let us suppose now that F has a metrizable topology defined by $\{V_m\}$, circled basis of F_F, and q_m gauge of V_m is plurisubharmonic and continuous. Then, given the assumptions of Theorem 18, *there are only two possibilities:*

(i) *for each q_m, there exists $p_m(x)$, plurisubharmonic and* 1-*homogeneous, such that*

$$q_m \circ P_\iota(x) \leqq [p_m(x)]^{n_\iota},$$

or:

(ii) *for each $x \in E$, $x \notin A$, there exists q_m such that* $\limsup_\iota \frac{1}{n_\iota} \log q_m \circ P_\iota(x) = +\infty$ *and A is a strictly polar cone and a meager set.*

Proof. If (i) is not true, there exists $m \in \mathbb{N}$, such that

$$\limsup_{\iota} \frac{1}{n_\iota} \log q_m \circ P_\iota(x) = +\infty$$

except on a set A_m which has the given properties; $A = \bigcup A_m$ has the same properties and the corollary is proved.

References

1. J. Bochnak and J. Siciak, *Fonctions analytiques dans les espaces vectoriels topologiques réels et complexes*, C. R. Acad. Sci. Paris Sér. A-B 270 (1970), A643-A645.

2. ————, *Analytic functions in topological vector spaces*, Studia Math., to appear.

3. N. Bourbaki, *Eléments de mathématique*, Livre V, *Espaces vectoriels topologiques*, Actualités Sci. Indust., No. 1189, 1229, Hermann, Paris, 1966, 1955.

4. G. Coeuré, *Fonctions plurisousharmoniques sur les espaces vectoriels topologiques et applications à l'étude des fonctions analytiques*, Ann. Inst. Fourier (Grenoble), to appear.

5. A. Grothendieck, *Sur certains espaces de fonctions holomorphes*, I. J. Reine Angew. Math. 192 (1953), 35-64.

6. A. Hirschowitz, *Bornologie des espaces de fonctions analytiques en dimension infinie*, Séminaire Lelong, 1970.

7. C. O. Kiselman, *On entire functions of exponential type and indicators of analytic functionals*, Acta Math. 117 (1967), 1-35.

8. P. Lelong, *Les fonctions plurisousharmoniques*, Ann. Sci. École Norm. Sup. (3) 62 (1945), 301-338.

9. ————, *On a problem of M. A. Zorn*, Proc. Amer. Math. Soc. 2 (1951), 12-19.

10. ————, *Fonctions entières de type exponentiel dans* $\mathbb{C}^n$, Ann. Inst. Fourier (Grenoble) 16 (1966), fasc. 2, 269-318.

11. ————, *Noncontinuous indicators for entire functions of* $n \geqq 2$ *variables and of finite order*, Proc. Sympos. Pure Math. 11 (1968), 285-297, Amer. Math. Soc., Providence, R. I.

12. ————, *Fonctionnelles analytiques et fonctions entières (n variables)*, Cours d'été Montréal, 1967, Les Presses de l'Université de Montréal, 1968.

13. ————, *Fonctions plurisousharmoniques et ensembles polaires dans les espaces vectoriels topologiques*, C. R. Acad. Sci. Paris Sér. A-B 267 (1968), A916-A918.

14. P. Lelong, *Fonctions plurisousharmoniques dans les espaces topologiques*, Séminaire P. Lelong (Analyse), Année 1967-1968, Lecture Notes in Mathematics, Vol. 71, Springer-Verlag, Berlin-Heidelberg-New York, 1968, exposé No. 17.

15. ————, *Fonctions plurisousharmoniques et ensembles polaires sur une algèbre de fonctions holomorphes*, Séminaire P. Lelong (Analyse), Année 1968-1969, Lecture Notes in Mathematics, Vol. 116, Springer-Verlag, Berlin-Heidelberg-New York, 1970, exposé No. 1.

16. ————, *Fonctions plurisousharmoniques et formes différentielles positives*, Gordon & Breach, Paris-London-New York, 1968.

17. ————, *Fonctions et applications de type exponentiel dans les espaces vectoriels topologiques*, C. R. Acad. Sci. Paris Sér. A-B 269 (1969), A420-A422.

18. A. Martineau, *Les supports des fonctionnelles analytiques*, Séminaire P. Lelong (Analyse), Année 1968-1969, Lecture Notes in Mathematics, Vol. 116, Springer-Verlag, Berlin-Heidelberg-New York, 1970.

19. P. Mazet, *Nullstellensatz en géométrie analytique banachique*, Séminaire Lelong, 1970.

20. L. Nachbin, *Topology on spaces of holomorphic mappings*, Ergebnisse der Mathematik und ihrer Grenzgebiete, Band 47, Springer-Verlag, Berlin-Heidelberg-New York, 1969.

21. Ph. Noverraz, *Fonctions plurisousharmoniques et analytiques dans les espaces vectoriels topologiques complexes*, Ann. Inst. Fourier (Grenoble) 19 (1969), fasc. 2, 419-493.

22. J.-P. Ramis, *Sous-ensembles analytiques d'une variété analytique banachique*, Ergebnisse der Mathematik und ihrer Grenzgebiete, Band 53, Springer-Verlag, Berlin-Heidelberg-New York, 1970.

23. C. E. Rickart, *Analytic functions of an infinite number of complex variables*, Duke Math. J. 36 (1969), 581-597.

24.. F. Trèves, *Locally convex spaces and linear partial differential equations*, Die Grundlehren der mathematischen Wissenschaften in Einzeldarstellungen, Vol. 146, Springer-Verlag, Berlin-Heidelberg-New York, 1967.

25. M. A. Zorn, *Characterization of analytic functions in Banach spaces*, Ann. of Math. (2) 46 (1945), 585-593.

University of Paris, France

REMARKS ON THE KOBAYASHI METRIC

H. L. Royden[1]

1. *Introduction*

Let M be a complex analytic manifold of dimension n, and let Δ_R denote the disk $\{z: |z| < R\}$ in the complex plane. We define a function d_M^* from $M \times M$ into $[0,\infty]$ by setting

$$d_M^*(p,q) = \inf \tfrac{1}{2} \log \frac{R+1}{R-1}$$

where the infimum is taken over all real $R > 1$ for which there is an analytic map $\phi\colon \Delta_R \to M$ with $\phi(0) = p$, $\phi(1) = q$, and where $d^* = \infty$ if there is no such mapping. Clearly, $d^*(p,q) = d^*(q,p)$, and if $f\colon M \to N$ is holomorphic, then

$$d_N^*(f(p),f(q)) \leqq d_M^*(p,q).$$

In particular, if $N \subset M$, $d_M^* \leqq d_N^*$.

The function d^* may not satisfy the triangle inequality, but following Kobayashi [9], we introduce the pseudometric d_M by setting

$$d_M(p,q) = \inf \sum_{i=0}^{k} d_M^*(p_i,p_{i+1})$$

where the infimum is taken over all finite sequences $\langle p_0,p_1,\ldots,p_k\rangle$ with $p_0 = p$, $p_k = q$. Then d_M is a pseudometric for M, which we call the Kobayashi (pseudo-) metric. Note that $d_M(p,q) \leqq d_M^*(p,q)$, and the value of d_M is always finite. We shall call d_M^* the unreduced Kobayashi distance (even though it need not satisfy the triangle

[1] This work was partially supported by NSF Grant GP 11911.

inequality), and reserve the symbols d_M and d_M^* for these functions.

If, for a fixed value of the integer k, we set

$$d_M^{(k)}(p,q) = \inf \sum_{i=0}^{k} d_M^*(p_i,p_{i+1})$$

with the infimum taken over all sequences $\langle p_0,\ldots,p_k\rangle$ of length k with $p_0 = p$, $p_k = q$, then

$$d_M^{(k+1)}(p,q) \leqq d_M^{(k)}(p,q)$$

and

$$d_M(p,q) = \lim_{k\to\infty} d_M^{(k)}(p,q).$$

If $\Delta = \Delta_r$ is a disk in $\mathbb{C}$, then d_Δ and d_Δ^* are the same and are equal to the Poincaré distance for Δ; i.e.,

$$d_\Delta^*(z,\zeta) = d(z,\zeta) = \tfrac{1}{2} \log \frac{|r^2-z\bar{\zeta}| + r|z-\zeta|}{|r^2-z\bar{\zeta}| - r|z-\zeta|}.$$

If D is the polydisk $\Delta_{r_1} \times \Delta_{r_2} \times \ldots \times \Delta_{r_n}$ in $\mathbb{C}^n$, then

$$d_D(a,\zeta) = d_D^*(z,\zeta) = \max_i \tfrac{1}{2} \log \frac{|r_i^2-z_i\bar{\zeta}_i| + r_i|z_i-\zeta_i|}{|r_i^2-z_i\bar{\zeta}_i| - r_i|z_i-\zeta_i|}.$$

Thus d_D is equivalent to the Euclidean metric for D. The equivalence is uniform on each compact subset of D but not on D.

Since any point of a manifold M is contained in a coordinate polydisk D and $d_D \geqq d_M$, we see that the topology induced on M by d_M is weaker than the standard topology for M. We shall show (Theorem 2) that if d_M is a metric, then it induces the standard topology on M.

2. *The infinitesimal metric*

The purpose of this section is to introduce a form F_M which is a differential metric in the sense of Grauert and Reckziegel [4] and to show that the Kobayashi metric d_M is the integrated form of F_M.

If $\langle x,\xi\rangle$ is an element of the tangent bundle of M, we define

$$F_M = \inf \frac{1}{R} ,$$

where R ranges over all positive real numbers for which there is an analytic map $\phi\colon \Delta_R \to M$ with $\phi(0) = x$ and $\phi'(0) = \xi$. For the disk Δ_R in $\mathbb{C}$, we have

$$F_{\Delta_R}(z,dz) = \frac{R\,dz}{R^2 - |z|^2} ;$$

i.e., the form F agrees with the differential form of the Poincaré metric. For the polydisk $D = \Delta_{r_1} \times \ldots \times \Delta_{r_n}$, we have

$$F_D(x,\xi) = \max_i \frac{r_i|\xi_i|}{r_i^2 - |x_i|^2} .$$

The following proposition is an immediate consequence of the definition of F_M:

Proposition 1. *If* $f\colon N \to M$ *is a holomorphic map, then*

$$F_M(f(x), f^*\xi) \leqq F_N(x,\xi).$$

In particular, if $N \subset M$ *then*

$$F_M(x,\xi) \leqq F_N(x,\xi).$$

If $\langle x,y;\xi,\eta\rangle$ *is an element of the tangent bundle of* $M \times N$ *with* $\langle x,\xi\rangle$ *and* $\langle y,\eta\rangle$ *elements of the tangent bundles of* M *and* N, *then*

$$F_{M\times N}(x,y;\xi,\eta) = \max(F_M(x,\xi), F_N(y,\eta)).$$

Proposition 2. *The function F_M is non-negative, and we have*

$$F_M(x,\alpha\xi) = |\alpha|\, F_M(x,\xi).$$

If K is a compact set contained in a coordinate polydisk, there is a constant C_K such that

$$F_M(x,\xi) \leqq C_K \|\xi\|$$

for all $x \in K$ *where* $\|\xi\| = \max |\xi_i|$.

Proof. The first statement follows from the fact that if $\psi(z) = \phi(\alpha z)$, then $\psi'(0) = \alpha\phi'(0)$, and ψ maps the disk of radius $|\alpha|^{-1}R$ into M. The second statement follows from the fact that for the coordinate polydisk D, we have

$$F_M(x,\xi) \leqq F_D(x,\xi) = \max_i \frac{r_i|\xi_i|}{r_i^2 - |x_i|^2} \leqq C_K\|\xi\| ,$$

since $|x_i|$ is bounded away from r_i on K.

In order to establish regularity properties of F, we need the following lemma:

Lemma 1. *Let ϕ be an analytic map of the disk Δ_R into M with* $\phi'(0) \neq 0$. *Then for each* $r < R$, *there is a* $\lambda > 0$ *and a holomorphic map f of the polydisk* $\Delta_r \times \Delta_\lambda \times \ldots \times \Delta_\lambda$ *into M such that f is a biholomorphic map on some neighborhood of the origin, and* $f|\Delta_r \times 0 \times \ldots \times 0 = \phi|\Delta_r$.

The proof of this lemma in the general case is quite complicated, and we consider here only the case when $M \subset \mathbb{C}^n$. In this case, choose coordinates in $\mathbb{C}^n$ so that $\phi'(0) = \langle 1,0,\ldots,0\rangle$, and define a map $g\colon \Delta_R \times \mathbb{C}^{n-1} \to \mathbb{C}^n$ by setting

$$g^1(t_1,\ldots,t_n) = \phi^1(t_1) \quad \text{and}$$

$$g^k(t_1,\ldots,t_n) = \phi^k(t_1) + t_k, \quad 2 \leq k \leq n.$$

Then g is holomorphic on $\Delta_R \times \mathbb{C}^{n-1}$ and biholomorphic on a neighborhood of the origin. Moreover, $g|\Delta_R \times 0 \times \ldots \times 0 = \phi$. Let $0 = g^{-1}[M]$, and set $D_\lambda = \Delta_r \times \Delta_\lambda \times \ldots \times \Delta_\lambda$. Then $\{\overline{D}_\lambda\}$ is a decreasing family of compact sets whose intersection is contained in 0. Hence $\overline{D}_\lambda \subset 0$ for some λ. Take $f = g|D_\lambda$.

The following lemma is an immediate consequence of Lemma 1 and the definition of F_M:

Lemma 2. *Let* $\varepsilon > 0$ *and an element* $\langle x,\xi\rangle$ *of the tangent bundle of* M *be given. Let* τ_1 *be that element of the tangent space of* $\mathbb{C}^n$ *whose coordinates are* $(1,0,\ldots,0)$. *Then there is a holomorphic map* f *of a polydisk* D *into* M *which is biholomorphic on a neighborhood of* 0 *and for which* $f(0) = x$, $f^*\tau_1 = \xi$, *and*

$$F_D(0,\tau_1) < F_M(x,\xi) + \varepsilon.$$

Proposition 3. *The function* F_M *is upper semi-continuous on the tangent bundle of* M.

Proof. Choose $\varepsilon > 0$ and an element $\langle x,\xi\rangle$ of the tangent bundle of M. Let $f\colon D \to M$ be the holomorphic mapping satisfying the conditions of Lemma 2. Since F_D is continuous, there is a neighborhood U of $\langle 0,\tau_1\rangle$ in D such that $F_D(t,\tau) < F_D(0,\tau_1) + \varepsilon$ for $\langle t,\tau\rangle \in D$. Since f is biholomorphic on a neighborhood of 0, $\langle f,f^*\rangle$ maps U onto a set containing a neighborhood V of $\langle x,\xi\rangle$. If $\langle y,\eta\rangle \in V$, then $y = f(t)$, $\eta = f^*(\tau)$ for some $\langle t,\tau\rangle \in U$, and we have

$$\begin{aligned} F_M(y,\eta) &= F_M(f(t),f^*(\tau)) \\ &\leq F_D(t,\tau) \\ &< F_D(0,\tau_1) + \varepsilon \\ &< F_M(x,\xi) + 2\varepsilon. \end{aligned}$$

Thus F_M is upper semicontinuous.

Let γ be a differentiable curve in M, and let $x\colon [a,b] \to M$ be a parametrization of γ. Since $\langle x,\dot{x}\rangle$ is a continuous map into the tangent bundle of M, and F_M is upper semicontinuous, the function $F_M(x,\dot{x})$ is upper semicontinuous on $[a,b]$. Hence it is bounded and measurable, and we may define the length of γ by

$$l(\gamma) = \int_\gamma F_M = \int_a^b F_M(x(t),\dot{x}(t))dt.$$

Since F_M is homogeneous with respect to ξ, this integral is independent of the parametrization chosen. The following theorem tells us that d_M is the integrated form of F_M:

Theorem 1. *The Kobayashi metric d_M is the integrated form of F_M; that is,*

$$d_M(p,q) = \inf \int_\gamma F_M$$

where the infimum is taken over all differentiable curves γ joining p to q.

Proof. Define $\bar{d}(p,q)$ to be $\inf \int_\gamma F_M$ over all differentiable curves γ joining p to q. From the upper bound on F_M given in Proposition 2, it follows that this is the same as the infimum taken over all piece-wise differentiable curves joining p to q. Thus $\bar{d}$ is a pseudometric, and in order to show that $\bar{d} \leq d_M$, it will suffice to show that $\overline{d}(p,q) \leqq d_M^*(p,q)$. Let ϕ be an analytic map of Δ_R into M with $\phi(0) = p$, $\phi(1) = q$, and

$$\tfrac{1}{2} \log \frac{R+1}{R-1} < d_M^*(p,q) + \varepsilon .$$

Let γ be the image of $[0,1]$ under ϕ. Then

$$\begin{aligned} \overline{d}(p,q) &\leqq \int_\gamma F_M \\ &\leqq \int_0^1 F_{\Delta_R} = \tfrac{1}{2} \log \frac{R+1}{R-1} \\ &\leqq d_M^*(p,q) + \varepsilon . \end{aligned}$$

Thus $\overline{d} \leqq d_M$.

To prove the inequality in the opposite direction, let γ be a differentiable curve joining p to q such that $\int_\gamma F_M < \overline{d}(p,q) + \varepsilon$, and let $x(t)$ be a parametrization of γ with $0 \leqq t \leqq 1$. Since $F_M(x(t),\dot{x}(t))$ is upper semicontinuous on $[0,1]$, it is the limit of a decreasing sequence of continuous functions on $[0,1]$. Thus by the Lebesgue convergence theorem, there is a continuous function h on $[0,1]$ such that $h(t) > F(x(t),\dot{x}(t))$ and

$$\int_0^1 h(t)dt < \overline{d}(p,q) + \varepsilon .$$

Since h is continuous, it is Riemann integrable, and there is a $\delta > 0$ such that if $0 = t_0 < t_1 < \ldots < t_k = 1$ is any subdivision of $[0,1]$ with $t_{i+1} - t_i < \delta$, and $s_1,\ldots,s_k$ are points of $[0,1]$ with $|s_i - t_i| < \delta$, then

$$\sum_{i=1}^{k} h(s_i)(t_i - t_{i-1}) < \overline{d}(p,q) + \varepsilon .$$

For each $s \in [0,1]$, there is a map f of $D = \Delta_R \times \Delta_\lambda \times \ldots \times \Delta_\lambda$

into M such that $\frac{1}{R} < h(s)$, $f(0) = x(s)$, $f'(0) = \dot{x}(s)$ and such that f is biholomorphic in a neighborhood of the origin. Thus there is an open interval I_s containing s so that on I_s we have $x = f \circ y$ where y is a differentiable map of I_s into D with $y(s) = (0,0,\ldots,0)$ and $y'(s) = (1,0,\ldots,0)$.

Since $y(t) = (t-s,0,\ldots,0) + O(|t-s|^2)$, it follows from the explicit expression for d_D on a polydisk that

$$d_D(y(t),y(u)) \leqq (1+\varepsilon)\,\frac{|t-u|}{R} < (1+\varepsilon)h(s)|t-u|$$

for all t and u in some sufficiently small interval I_s' containing s. We may suppose that the length of each I_s' is less than δ. The Kobayashi metric decreases under holomorphic maps, and so

$$d_M(x(t),x(u)) \leqq (1+\varepsilon)h(s)|t-u|$$

for $t, u \in I_s'$.

By the Lebesgue covering lemma ([7], p. 154), there is an $\eta > 0$ such that if $t, u \in [0,1]$ and $|t-u| < \eta$, then there is an s with $t, u \in I_s'$. Let $0 = t_0 < t_1 < \ldots < t_k = 1$ be a subdivision of $[0,1]$ with $t_i - t_{i-1} < \eta$, and choose s_i so that $t_i, t_{i-1} \in I_{s_i}'$. Then $|t_i - s_i| < \delta$ and $t_i - t_{i-1} < \delta$ since the length of I_{s_i}' is less than δ. Hence

$$\begin{aligned} d_M(p,q) = d_M(x(0),x(1)) &\leqq \sum_{i=1}^{k} d_M(x(t_i),x(t_{i-1})) \\ &\leqq (1+\varepsilon)\sum_{i=1}^{k} h(s_i)(t_i - t_{i-1}) \\ &< (1+\varepsilon)(\bar{d}(p,q) + \varepsilon). \end{aligned}$$

Since ε was arbitrary, we have

$$d_M(p,q) \leqq \bar{d}_M(p,q).$$

3. *Hyperbolic manifolds*

We say that a manifold M is hyperbolic at a point x if there is a neighborhood U of x and a positive constant c such that $F_M(y,\eta) \geqq c\|\eta\|$ for all $y \in U$. We say that M is hyperbolic if it is hyperbolic at each point. The manifold M is said to be Kobayashi hyperbolic if d_M is a metric. We shall show that hyperbolicity in our sense is equivalent to Kobayashi hyperbolicity.

In a slight modification of a definition given by Wu [10], we say a complex manifold M with a metric d compatible with its topology is tight if the family of holomorphic mappings of the disk Δ into M is equicontinuous. A family F of mappings of a topological space X into a topological space Y is said to be an even family if, given $x \in X$, $y \in Y$ and a neighborhood U of y, there is a neighborhood V of x and a neighborhood W of y such that for every $f \in F$, we have $f[V] \subset U$ whenever $f(x) \in W$. Equicontinuity of F with respect to any metric inducing the topology of Y implies that F is an even family (Kelley [7], p. 237).

Theorem 2. *Let M be a complex manifold. Then the following statements are equivalent:*

i) *The family $A(\Delta,M)$ of holomorphic maps of the disk Δ into M is equicontinuous for some metric d inducing the topology of M i.e., (M,d) is tight.*

ii) *The family $A(\Delta,M)$ is an even family.*

iii) *M is hyperbolic.*

iv) *d_M is a metric; i.e., M is Kobayashi hyperbolic.*

v) *The Kobayashi metric d_M induces the usual topology of M.*

Proof. (i) => (ii) by Kelley ([7], p. 237). To show that (ii) => (iii), let D be a coordinate polydisk about a point x. Since the maps of the unit disk Δ into M are an even family, there is a disk Δ_δ about 0 and a neighborhood W of x such that if

$\phi(0) = y \in W$, then $\phi[\Delta_\delta] \subset D$. If ϕ maps Δ_R into M with $\phi(0) = y \in W$, then $\phi[\Delta_{\delta R}] \subset D$. Hence for $y \in W$, we have $\delta F_D(y,\eta) \leqq F_M(y,\eta)$. We may suppose that $\overline{W}$ is a compact subset of D. Then for $y \in W$, we have

$$F_M(y,\eta) \geqq \delta F_D(y,\eta) \geqq c\|\eta\|$$

foe some positive constant c, and so M is hyperbolic at x.

Theorem 1 implies that (iii) => (iv).

To show that (iv) => (v), let ρ be any metric inducing the standard topology on M. The topology induced by d_M is weaker than the usual one, and to show they are equivalent, we need only show that for each neighborhood D of a point x, there is an $\varepsilon > 0$ such that $d_M(x,y) < \varepsilon$ implies $y \in D$. We can take D to be a coordinate polydisk $\|z\| < a$ with $x = 0$. Let $S = \{z \in D : \|z\| = a/2\}$. Then S is compact. Since $d_M(x,y)$ is continuous as a function of y in the usual topology, there is a $y_0 \in S$ such that

$$\inf_{y\ S} d_M(x,y) = d_M(x,y_0).$$

By (iv), $d_M(x,y_0) = \varepsilon > 0$. If now u is not in the disk $D' = \{z \in D : \|z\| \leqq a/2\}$, then each differentiable curve γ joining x to u must pass through a point $y \in S$. Thus the length of γ must be greater than the length of the part from x to y which is at least ε. Hence $d_M(x,u) \geqq \varepsilon$, and $\{v : d_M(x,v) < \varepsilon\} \subset D'$. Thus d_M induces the usual topology.

Since a holomorphic map is distance decreasing for the Kobayashi metric, (v) => (i).

Corollary. If M is hyperbolic, the family $A(N,M)$ of holomorphic mappings of N into M is an even family.

Proposition 4. *Let M be a hyperbolic manifold of dimension m and N a manifold of dimension $n \leqq m$. If there is a holomorphic map $f: N \to M$ whose Jacobian has rank n everywhere, then N is hyperbolic. In particular, every subdomain of a hyperbolic manifold is hyperbolic, and every bounded domain in $\mathbb{C}^n$ is hyperbolic.*

The product $M \times N$ of two hyperbolic manifolds is hyperbolic.

Proof. This follows directly from Proposition 1 and the definition of hyperbolicity.

4. *Tautness and completeness*

Modifying a defintition of Wu [10] slightly, we say a complex manifold M is taut if the family $A(\Delta, M)$ of holomorphic maps of a disk into M is a normal family.

Proposition 5. *If M is taut, it is hyperbolic, and F_M is continuous on the tangent bundle.*

Proposition 6. *For a complex manifold M, the following statements are equivalent*

i) *M is taut.*

ii) *For each integer $k \geqq 2$, positive real number r and point $x \in M$, the set $\{y : d_M^{(k)}(x,y) \leqq r\}$ is compact.*

iii) *For each $r > 0$ and each $x \in M$, the set $\{y : d^{(2)}(x,y) \leqq r\}$ is compact.*

A manifold M is called a complete hyperbolic manifold if it is hyperbolic and d_M is a complete metric. Since d_M is an inner metric (i.e., one that comes from arc length), the Hopf-Rinow-Myers theorem can be formulated as follows:

Proposition 7. *For a hyperbolic manifold M, the following are equivalent:*

i) *M is complete.*

ii) *For each $x \in M$ and each $r > 0$, the set $\{y : d_M(x,y) \leqq r\}$ is compact.*

iii) *Each isometry of $[0,a)$ into M can be extended to an isometry of $[0,a]$.*

Corollary. *A complete hyperbolic manifold is taut.*

Proposition 8. *If $\tilde{M}$ is a covering manifold of M, then $\tilde{M}$ is complete hyperbolic if and only if M is.*

The product of two complete hyperbolic manifolds is complete hyperbolic.

If M and N are complete hyperbolic submanifolds of a manifold V, then $M \cap N$ is complete hyperbolic.

Lemma 2. *Let U and V be subdomains of a hyperbolic manifold M. For $x \in U$, define*

$$d^*(x) = d_V^*(x, V \sim U) = \inf_{y \in V \sim U} d_V^*(x,y).$$

Then

$$F_V(x,\xi) \leqq F_{U \cap V}(x,\xi) \leqq \operatorname{ctgh} d^*(x) \cdot F_V(x,\xi).$$

Theorem 3. *Let M be a hyperbolic manifold and V a subdomain such that $\overline{V}$ is complete with respect to d_M. Then V will be a complete hyperbolic manifold if and only if each boundary point of V (i.e., each point of $\overline{V} \sim V$) has a neighborhood U for which $U \cap V$ is a complete hyperbolic manifold.*

Observe that $\overline{V}$ will be complete if $\overline{V}$ is compact or if M is complete. Thus if V is a bounded domain in $\mathbb{C}^n$, we always have $\overline{V}$

complete with respect to the Kobayashi metric of a large polydisk containing $\overline{V}$.

Theorem 4. *Let V be a domain on a hyperbolic manifold M, and suppose that* $\overline{V}$ *is complete with respect to* d_M*. Then V is a complete hyperbolic manifold if for every boundary point x of V, there is a coordinate polydisk D containing x and an analytic function f in D with* $f(x) = 0$ *and* $f \neq 0$ *in* $D \cap V$.

References

1. T. J. Barth, *Taut and tight complex manifolds*, Proc. Amer. Math. Soc. 24 (1970), 429-431.

2. C. J. Earle and R. S. Hamilton, *A fixed point theorem for holomorphic mappings*, Proc. of the Summer Institute on Global Analysis at Berkeley, California, 1968.

3. D. A. Eisenmann, *Hyperbolic mappings into tight manifolds*, Bull. Amer. Math. Soc. 76 (1970), 46-48.

4. H. Grauert and H. Reckziegel, *Hermitesche Metriken und normale Familien holomorpher Abildungen*, Math. Z. 89 (1965), 108-125.

5. W. Kaup, *Reelle Transformationsgruppen und invariante Metriken auf komplexen Räumen*, Invent. Math. 3 (1967), 43-70.

6. ———, *Hyperbolische komplexe Räume*, Ann. Inst. Fourier (Grenoble) 18 (1968), 303-330.

7. J. L. Kelley, *General topology*, Van Nostrand, Princeton, N. J., 1955.

8. P. Kiernan, *On the relations between taut, tight and hyperbolic manifolds*, Bull. Amer. Math. Soc. 76 (1970), 49-51.

9. S. Kobayashi, *Invariant distances on complex manifolds and holomorphic mappings*, J. Math. Soc. Japan 19 (1967), 460-480.

10. H. Wu, *Normal families of holomorphic mappings*, Acta Math. 119 (1967), 193-233.

Stanford University, Stanford, California 94305

$\bar{\partial}$- COHOMOLOGY SPACES ASSOCIATED WITH A CERTAIN GROUP EXTENSION

Ichiro Satake

The purpose of this paper is to give a new aspect on unitary representations of a semi-direct product of Lie groups of a certain type. We shall be concerned with unitary representations obtained on the (square-integrable) $\bar{\partial}$- cohomology spaces attached to a pair (G,K) of a Lie group G and a compact subgroup K such that the homogeneous space G/K has a structure of a hermitian manifold. (More precisely, we assume the conditions (A1-2) in section 1.) Part I is a summary of basic facts concerning such $\bar{\partial}$- cohomology spaces. In Part II, we shall consider the case where G is a semi-direct product $G = G_2G_1$ and $K = K_2 \times K_1$, (G_i,K_i) $(i = 1,2)$ being two pairs satisfying the above conditions. (More precisely, we assume the conditions (B1-2) in section 5.) We will obtain, under certain additional conditions ((C1-4) in sections 7 and 9), a relationship between the $\bar{\partial}$- cohomology spaces for (G,K) and for (G_1,K_1) and (G_2,K_2). As a typical example, where all these conditions are satisfied, we shall consider in section 10 the case where (G_1,K_1) is any pair satisfying the conditions (A1-2), G_2 is a nilpotent group associated with a symplectic space (V,A) (V : a real vector space, A : a non-degenerate alternating bilinear form on $V \times V$) (see section 3), and where the semi-direct product $G = G_2G_1$ is defined by a symplectic representation $\rho: G_1 \to Sp(V,A)$ satisfying a condition (B2)'. A group extension of this type appears in the classical theory of theta-functions (see [13]), and also in a recent theory of Kuga's fibre varieties. In this special case, our result yields a complete description of the irreducible unitary representations of G of discrete series in terms of the corresponding irreducible (projective) unitary representations of G_1.

I. $\bar{\partial}$-*cohomology spaces attached to a homogeneous hermitian manifold*

1. Let G be a connected unimodular Lie group and K a compact subgroup of G. We denote by $\mathfrak{g}$ and $\mathfrak{k}$ the Lie algebras of G and K, respectively. Then there exists a subspace $\mathfrak{m}$ of $\mathfrak{g}$ such that one has

(1) $\mathfrak{g} = \mathfrak{k} + \mathfrak{m}$ (direct sum), $\mathrm{ad}(k)\mathfrak{m} \subset \mathfrak{m}$ for all $k \in K$.

We fix such a subspace $\mathfrak{m}$ once and for all. The complexifications of $\mathfrak{g}$, $\mathfrak{k}$, $\mathfrak{m}$, ... will be denoted by $\mathfrak{g}_{\mathbb{C}}$, $\mathfrak{k}_{\mathbb{C}}$, $\mathfrak{m}_{\mathbb{C}}$, ... We assume that there is given a complex structure J on $\mathfrak{m}$ satisfying the following conditions.

(A1) J is K-invariant (i.e., J commutes with all $\mathrm{ad}(k)|\mathfrak{m}$, $k \in K$) and the $(\pm i)$-eigenspace $\mathfrak{m}_{\pm}$ of J in $\mathfrak{m}_{\mathbb{C}}$ is a complex subalgebra of $\mathfrak{g}_{\mathbb{C}}$. (From the first condition, one has $[\mathfrak{k}_{\mathbb{C}}, \mathfrak{m}_{\pm}] \subset \mathfrak{m}_{\pm}$, so that $\mathfrak{k}_{\mathbb{C}} + \mathfrak{m}_{\pm}$ is also a complex subalgebra of $\mathfrak{g}_{\mathbb{C}}$).

(A2)[1] There exists a *complex form* $G_{\mathbb{C}}$ of G (i.e., a connected complex Lie group with the Lie algebra $\mathfrak{g}_{\mathbb{C}}$, containing G as a closed subgroup corresponding to $\mathfrak{g}$) such that, if one denotes by $K_{\mathbb{C}}$, $M_{\pm}$ the complex analytic subgroups of $G_{\mathbb{C}}$ corresponding to $\mathfrak{k}_{\mathbb{C}}$, $\mathfrak{m}_{\pm}$, respectively, then $K_{\mathbb{C}}$, $M_{\pm}$, $K_{\mathbb{C}} M_{\pm}$ are all closed and one has

(2) $$G \cap K_{\mathbb{C}} M_{-} = K, \quad K_{\mathbb{C}} \cap M_{-} = \{e\}.$$

---

[1] For simplicity, this condition is stated in a slightly stronger form than we need. Note that the condition (A2) (in the present form) implies that K is connected.

Under these conditions, it is clear that $GK_{\mathbb{C}}M_-$ is an open subset of $G_{\mathbb{C}}$ and one has a natural injection

$$(3) \qquad G/K \approx GK_{\mathbb{C}}M_-/K_{\mathbb{C}}M_- \subset G_{\mathbb{C}}/K_{\mathbb{C}}M_- ,$$

by which the homogeneous space $\mathcal{D} = G/K$ is identified with an open G-orbit in the complex homogeneous space $G_{\mathbb{C}}/K_{\mathbb{C}}M_-$. Since the isotropy subgroup K is compact, one can also define a G-invariant hermitian metric on $\mathcal{D}$. Thus $\mathcal{D}$ has a structure of a homogeneous hermitian manifold.

Now, let τ be an irreducible representation of K in a finite dimensional complex vector space F. Consider the direct product $G \times F$ and define a left (resp. right) action of G (resp. K) on $G \times F$ as follows:

$$(4) \qquad G \times F \ni (g,x) \begin{cases} \mapsto (g'g,x) & \text{for } g' \in G, \\ \mapsto (gk,\tau(k)^{-1}x) & \text{for } k \in K. \end{cases}$$

Then the quotient space $E_\tau = (G \times F)/K$ has a natural structure of a vector bundle over $\mathcal{D} = G/K$ with fibre F. Since the action of K on $G \times F$ can naturally be extended to a holomorphic action of $K_{\mathbb{C}}M_-$ on $G_{\mathbb{C}} \times F$ and since E_τ can also be identified with an open subbundle of $(G_{\mathbb{C}} \times F)/K_{\mathbb{C}}M_-$, E_τ has a natural structure of a holomorphic vector bundle. Since there exists a K-invariant hermitian metric on F, E_τ has also a structure of a hermitian vector bundle. In the following, we fix the invariant hermitian metrics on $\mathcal{D}$ and on F once and for all.

For a non-negative integer q, one denotes by $C^q(E_\tau)$ the space of all E_τ-valued C^∞-forms of type $(0,q)$ on $\mathcal{D}$, and by $C^q_c(E_\tau)$ the subspace of $C^q(E_\tau)$ formed of all forms with compact support. We define a standard L_2-norm in $C^q_c(E_\tau)$ by the formula

$$\|\omega\|^2 = \int_{\mathcal{D}} {}^t\omega \wedge *\omega\# .$$

(For the meaning of these notations, see [1], [5a].) The completion $L_2^q(E_\tau)$ of $C_c^q(E_\tau)$ with respect to this norm may be identified with the Hilbert space of all (equivalence-classes of) E_τ-valued square-integrable forms of type $(0,q)$ on $\mathcal{D}$. One defines the operators d'' and δ'':

$$C_c^q(E_\tau) \underset{\delta''}{\overset{d''}{\rightleftarrows}} C_c^{q+1}(E_\tau)$$

in the usual manner, and then defines the operators $\bar\partial$ and θ:

$$L_2^q(E_\tau) \underset{\theta}{\overset{\bar\partial}{\rightleftarrows}} L_2^{q+1}(E_\tau)$$

as the *formal adjoints* of δ'' and d'', respectively. (Namely, for $\omega \in L_2^q(E_\tau)$, $\bar\partial\omega$ is defined if and only if there exists an element $\omega' \in L_2^{q+1}(E_\tau)$ such that one has $(\omega, \delta''\eta) = (\omega', \eta)$ for all $\eta \in C_c^{q+1}(E_\tau)$ and, when that is so, one puts $\omega' = \bar\partial\omega$. The definition of θ is similar.) Then $\bar\partial$ (resp. θ) is a closed operator, defined almost everywhere in $L_2^q(E_\tau)$ (res. $L_2^{q+1}(E_\tau)$), which extends d'' (resp. δ''). Moreover, by virtue of the completeness of $\mathcal{D}$, $\bar\partial$ and θ are mutually adjoint (cf. [1], Prop. 5). A form $\omega \in L_2^q(E_\tau)$ is called *harmonic* if both $\bar\partial\omega$ and $\theta\omega$ are defined and $= 0$; one denotes by $H^q(E_\tau)$ the closed subspace of $L_2^q(E_\tau)$ formed of all harmonic forms. Then it is clear that the q-th $\bar\partial$-cohomology space (i.e., Ker $\bar\partial \ominus$ (closure of Im $\bar\partial$)) coincides with $H^q(E_\tau)$.

2. Following Matsushima-Murakami ([5]), we shall now replace the spaces $C^q(E_\tau)$, ... by the corresponding spaces $C^q(G,K,\tau)$,..., to be defined below, which are more convenient for our purpose. In the following, an element $X \in \mathfrak{g}_{\mathbf{C}}$ will always be viewed as a left

invariant (complex) vector field on G, which assigns to each $g \in G$ a (complex) tangent vector X_g at g to G. By definition, $C^q(G,K,\tau)$ (or, more precisely, $C^q(G,K,J,\tau)$) is the space of all F-valued C^∞-forms ω of degree q on G satisfying the following conditions:

$$(5) \quad \begin{cases} i(X)\omega = 0 & \text{for all } X \in \mathfrak{k}_{\mathbb{C}} + \mathfrak{m}_+ , \\ \omega(gk;X) = \tau(k)^{-1}\omega(g;\mathrm{ad}(k)X) & \text{for } g \in G, \ k \in K, \end{cases}$$

where in the second equality the symbol X stands for a q-tuple of elements taken from $\mathfrak{g}_{\mathbb{C}}$. We denote by π (resp. $\tilde{\pi}$) the canonical projection $G \to \mathcal{D}$ (resp. $G \times F \to E_\tau$) as well as the induced linear maps of the tangent spaces. Then it is easy to see that one has a canonical isomorphism

$$C^q(E_\tau) \cong C^q(G,K,\tau)$$

given by the correspondence $\tilde{\omega} \leftrightarrow \omega$, $\tilde{\omega} \in C^q(E_\tau)$, $\omega \in C^q(G,K,\tau)$ defined by

$$(6) \quad \tilde{\omega}(\pi(g);\pi(X_g)) = \tilde{\pi}(g,\omega(g;X))$$

([5a], [6a]). If we denote by $C^q_c(G,K,\tau)$ the subspace of $C^q(G,K,\tau)$ formed of all forms with compact support, then the above isomorphism induces an isomorphism $C^q_c(E_\tau) \cong C^q_c(G,K,\tau)$. Using this isomorphism, one transports the norm $\| \ \|$ and the operators d'', δ'' from $C^q_c(E_\tau)$ to $C^q_c(G,K,\tau)$, and then defines the completion $L^q_2(G,K,\tau)$ of $C^q_c(G,K,\tau)$, which is isomorphic to $L^q_2(E_\tau)$, and the natural extensions $\bar{\partial}$ and θ of the operators d'' and δ'' to $L^q_2(G,K,\tau)$.

Now let $(X_1,\ldots,X_m)$ be a basis of $\mathfrak{m}_+$ over $\mathbb{C}$ such that $(X_\alpha + \bar{X}_\alpha)_{1 \leq \alpha \leq m}$ forms an orthonormal basis of $\mathfrak{m}$ with respect to the (fixed) hermitian metric, and put $X_{\bar{\alpha}} = \bar{X}_\alpha$ $(1 \leq \alpha \leq m)$. Let A_q be the set of all q-tuples taken from the set of indices $\{1,\ldots,m\}$, and for $(\alpha) = (\alpha_1,\ldots,\alpha_q) \in A_q$ and for $\omega \in L_2^q(G,K,\tau)$ put

$$\omega_{(\bar{\alpha})}(g) = \omega(g;X_{(\bar{\alpha})}) = \omega(g;X_{\bar{\alpha}_1},\ldots,X_{\bar{\alpha}_q}) \quad (g \in G).$$

Then one sees at once that for $\omega \in L_2^q(G,K,\tau)$ the L_2-norm is given by

$$\|\omega\|^2 = c \sum_{\alpha \in A_q} \int_G \|\omega_{(\bar{\alpha})}(g)\|_F^2 \, dg \, , \tag{7}$$

where $\| \ \|_F$ denotes the (fixed) hermitian norm on F, dg is a biinvariant Haar measure on G, and c is a positive constant. When the Haar measure dg is normalized in such a way that one has $dg = \dot{dz}\cdot dk$, $\int_K dk = 1$, and $\dot{dz}$ is the standard volume element defined by the (fixed) hermitian metric on $\mathcal{D}$, then one has $c = 2^q/q!$. On the other hand, for $\omega \in C_c^q(G,K,\tau)$ one has[2]

$$
\begin{aligned}
(d''\omega)_{(\bar{\alpha})} &= \sum_{i=1}^{q+1} (-1)^{i+1} X_{\bar{\alpha}_i}\, \omega_{(\ldots\hat{\bar{\alpha}}_i\ldots)} \;+ \\
&\quad + \sum_{i<j} \sum_{\alpha'} (-1)^{i+j}\, \bar{c}_{\alpha_i\alpha_j}^{\alpha'}\, \omega_{\bar{\alpha}'(\ldots\hat{\bar{\alpha}}_i\ldots\hat{\bar{\alpha}}_j\ldots)}, \\
(\delta''\omega)_{(\bar{\alpha})} &= -2\sum_{\alpha'=1}^{m} X_{\alpha'}\,\omega_{\bar{\alpha}'(\bar{\alpha})} \;+ \\
&\quad + \sum_{i=1}^{q-1} \sum_{\alpha',\alpha''=1}^{m} (-1)^i\, c_{\alpha'\alpha''}^{\alpha_i}\, \omega_{\bar{\alpha}'\bar{\alpha}''(\ldots\hat{\bar{\alpha}}_i\ldots)},
\end{aligned}
\tag{8}
$$

[2] These formulas are proved in [5a], [6a] for the case where (G,K) is a hermitian symmetric pair (in which case, the subalgebra $\mathfrak{m}_\pm$ is abelian). However, the same idea can also be applied to our general case.

where $c^{\alpha}_{\alpha'\alpha''}$ is the structure constant defined by $[X_{\alpha'}, X_{\alpha''}] = \sum c^{\alpha}_{\alpha'\alpha''} X_{\alpha}$ $(1 \leqq \alpha', \alpha'' \leqq m)$. One denotes by $H^q(G,K,\tau)$ the closed subspace of $L^q_2(G,K,\tau)$ formed of all harmonic forms. Then one has a canonical isomorphism $H^q(E_\tau) \cong H^q(G,K,\tau)$.

3. From the (left) action of G on $E_\tau = (G\times F)/K$, induced from that on $G \times F$, one obtains in a natural manner an action of G on the spaces $C^q(E_\tau), \ldots$; the action of G on $C^q_c(E_\tau)$, preserving the norm $\| \ \|$, can naturally be extended to that on $L^q_2(E_\tau)$. The corresponding action of G on $L^q_2(G,K,\tau)$ is given by

(9) $$(T_{g'}\omega)_{(\bar{\alpha})}(g) = \omega_{(\bar{\alpha})}(g'^{-1}g) \quad \text{for } \omega \in L^q_2(G,K,\tau).$$

Moreover, the unitary operators T_g $(g \in G)$, commuting with the operators $\bar{\partial}$ and θ, leave the space of harmonic forms $H^q(G,K,\tau)$ invariant. Thus, one obtains (continuous) unitary representations of G on $L^q_2(E_\tau) \cong L^q_2(G,K,\tau)$ and on $H^q(E_\tau) \cong H^q(G,K,\tau)$, which we call the *regular representations* of G on these spaces.

It is a basic problem in the theory of unitary representations to determine when the space $H^q(E_\tau)$ is $\neq \{0\}$ and irreducible, in which case one obtains an irreducible unitary representation of G of discrete series. As was suggested by Langlands and Kostant, one may further ask if all irreducible unitary representations of G of discrete series are obtained in this manner. For $q = 0$, the irreducibility of $H^0(E_\tau)$ is known in general (Kobayashi [4]). In the case where (G,K) is a hermitian symmetric pair, i.e., G is a connected semisimple Lie group of hermitian type and K is a maximal compact subgroup, these problems are almost settled by Okamoto-Ozeki [8] and Narasimhan-Okamoto [7]. Their results have been extended by Schmid [10] to the case where G is a connected semisimple Lie group with a compact Cartan subgroup K.

Let us give another example, in which the group G is nilpotent. Let V be a $2n$-dimensional real vector space supplied with a non-degenerate alternating bilinear form A. Let $\tilde{V}$ be a central extension of V by the 1-dimensional torus T (the multiplicative group of complex numbers of absolute value 1) defined by the factor set $\varepsilon(\frac{1}{2}A(u,v))$ $(u,v \in V)$, where one puts $\varepsilon(\xi) = e^{2\pi i \xi}$ for $\xi \in \mathbb{R}$. The group $\tilde{V}$ may be viewed as a group formed of all pairs (ξ,u), $\xi \in T$, $u \in V$, endowed with the multiplication given by

$$(\xi,u)(\eta,v) = (\xi\eta\cdot\varepsilon(\tfrac{1}{2}A(u,v)),\ u+v). \tag{10}$$

The Lie algebra $\tilde{\mathfrak{v}}$ of $\tilde{V}$ may also be viewed as a Lie algebra formed of all pairs $(2\pi i\xi,u)$, $\xi \in \mathbb{R}$, $u \in V$, with the bracket product:

$$[(2\pi i\xi,u),(2\pi i\eta,v)] = (2\pi i A(u,v),0).$$

The Lie group $\tilde{V}$ has a natural complexification $\tilde{V}_{\mathbb{C}}$ formed of all pairs (ξ,u) with $\xi \in \mathbb{C}^{\times}$, $u \in V_{\mathbb{C}}$, with the same multiplication as above. Therefore, if one puts $G = \tilde{V}$, $K = T$, $\mathfrak{m} = \{(0,u) \mid u \in V\} \subset \tilde{\mathfrak{v}}$ and takes any complex structure J on $\mathfrak{m}$ induced by a complex structure I on V such that the $(\pm i)$-eigen space $W_{\pm}$ of I in $V_{\mathbb{C}}$ is totally isotropic, then these data satisfy the conditions (A1), (A2). Note that this condition on I is equivalent to saying that the bilinear form $A(x,Iy)$ $(x,y \in V)$ is symmetric; when that is so, the signature of the corresponding quadratic form $A(x,Ix)$ is of the form $(2r,2n-2r)$. When the irreducible representation τ of T is given by

$$(l):\qquad \xi \mapsto \xi^{l} \quad \text{for } \xi \in T,$$

one denotes the corresponding (holomorphic) line bundle by $E_{(l)}$, and the corresponding spaces of square-integrable forms and harmonic forms by $L_2^q(E_{(l)}) \cong L_2^q(\tilde{V},T,l)$ and $H^q(E_{(l)}) \cong H^q(\tilde{V},T,l)$, respectively.

Then one obtains the following result.

Proposition 1. *Let* $l \neq 0$ *and let* $(2r, 2n-2r)$ *be the signature of the quadratic form* $lA(x, Ix)$ $(x \in V)$. *Then* $H^q(\tilde{V}, T, l)$ *is* $\neq \{0\}$ *and irreducible for* $q = n - r$ *and* $= \{0\}$ *otherwise.*

When $r = n$, this may be regarded as a special case of the vanishing theorem of Andreotti-Vesentini [1]. A direct proof for the general case was given by Okamoto (unpublished). Here we shall sketch a proof for the case where $A(x, Ix)$ is positive-definite, for the general case can easily be reduced to that case.

4. For $u \in V_{\mathbb{C}}$, one writes $u = u_+ + u_-$ with $u_\pm \in W_\pm$. Then the map $u \mapsto u_+$ gives a $\mathbb{C}$-linear isomorphism $(V, I) \cong W_+$, so that one may identify the homogeneous space $\mathcal{D} = \tilde{V}/T$ with W_+ by the correspondence $\pi((\xi, u)) \leftrightarrow u_+$. The action of $\tilde{V}$ on W_+ is then given by $\tilde{u}(w) = w + u_+$. For $\tilde{u} = (\xi, u)$ and $w, w' \in W_+$, put

$$(11) \qquad \begin{cases} \eta_0(\tilde{u}, w) = \xi \cdot \varepsilon(A(u_-, w + \tfrac{1}{2}u_+)), \\ \kappa_0(w', w) = \varepsilon(A(\bar{w}, w')). \end{cases}$$

Then η_0 and κ_0 have the properties of a canonical automorphy factor and the associated kernel function for $(\tilde{V}, W)$ (cf. [6b], [9]); in particular, one has

$$(12) \qquad \begin{aligned} \eta_0(\tilde{u}\tilde{u}', w) &= \eta_0(\tilde{u}, \tilde{u}'(w))\eta_0(\tilde{u}', w), \\ \kappa_0(\tilde{u}(w'), \tilde{u}(w)) &= \eta_0(\tilde{u}, w')\kappa_0(w', w)\overline{\eta_0(\tilde{u}, w)} \end{aligned}$$

for $\tilde{u}, \tilde{u}' \in \tilde{V}$, $w, w' \in W_+$. It follows that the automorphy factor $\eta_0(\tilde{u}, w)^l$ holomorphically trivializes the holomorphic line bundle $E_{(l)}$ over W_+; namely one has a holomorphic isomorphism $E_{(l)} \cong W_+ \times \mathbb{C}$

given by the correspondence

$$\tilde{\pi}((\xi,u),\zeta) \leftrightarrow (u_+, \eta_o(\tilde{u},0)^l \zeta). \tag{13}$$

Therefore the correspondence $f \leftrightarrow h$ $(f \in L_2^0(\tilde{V},T,l))$ defined by the relation

$$h(u_+) = \eta_o(\tilde{u},0)^l f(\tilde{u}) \tag{14}$$

gives an isomorphism of $L_2^0(\tilde{V},T,l)$ onto the Hilbert space $L_2(W_+)^{(l)}$ formed of all $\mathbb{C}$-valued measurable functions h on W_+ such that

$$\|h\|^2 = \int_{W_+} |h(w)|^2 \kappa_o(w,w)^{-l} \, dw < \infty, \tag{15}$$

where $dw = 2^{-n} \prod dw_i d\bar{w}_i$, (w_i) denoting an orthonormal coordinate of $w \in W_+$ with respect to the positive-definite hermitian form $2iA(\bar{w},w)$. Under this isomorphism the subspace $H^0(\tilde{V},T,l)$ corresponds to the subspace $\mathcal{F}(W_+)^{(l)}$ of $L_2(W_+)^{(l)}$, called a *Fock space*, formed of all holomorphic functions in $L_2(W_+)^{(l)}$ if $l > 0$. When $l < 0$, one clearly has $H^0(\tilde{V},T,l) = \{0\}$; in this case, one defines $\mathcal{F}(W_+)^{(l)}$ as the subspace of $L_2(W_+)^{(l)}$ formed of all functions h such that $\kappa_o(w,w)^{-l}\overline{h(w)}$ is holomorphic. The action of $\tilde{V}$ on $L_2(W_+)^{(l)}$ and $\mathcal{F}(W_+)^{(l)}$ is given by

$$(T_{\tilde{u}}h)(w) = \eta_o(\tilde{u}^{-1},w)^{-l} h(\tilde{u}^{-1}(w)). \tag{16}$$

The unitary representation of $\tilde{V}$ on $\mathcal{F}(W_+)^{(l)}$ $(l \neq 0)$ thus obtained is known to be non-trivial and irreducible (see [2], [9]). Moreover it is well known (Stone- von Neumann) that all irreducible representations of $\tilde{V}$ of discrete series are obtained in this way.

Now we suppose $l > 0$. We denote by $L_2^q(W_+)^{(l)}$ (resp. $\mathcal{F}^q(W_+)^{(l)}$) the space of all $\mathbb{C}$-valued measurable (resp. holomorphic) forms ω of type $(0,q)$ (resp. $(q,0)$) on W_+ which is square-integrable in the sense that $\omega(w,\overline{Y})$ (resp. $\omega(w,Y)$) belongs to $L_2(W_+)^{(l)}$ for any q-tuple Y taken from the Lie algebra of W_+. Then by a correspondence similar to (14) one has a canonical isomorphism

$$L_2^q(\tilde{V},T,l) \cong L_2^q(W_+)^{(l)}, \tag{17}$$

which preserves the norm $\|\ \|$ and the operators $\overline{\partial}$ and θ.

Now one obtains the following lemmas, which can be proved without much difficulty.

Lemma 1. *For any* $h \in \mathcal{F}(W_+)^{(l)}$ *and* $\zeta \in \mathcal{F}^q(W_+)^{(l)}$ *the integral*

$$\omega(w,\overline{Y}) = \int_{W_+} h(w')\overline{\zeta}(w-w', Y)\kappa_0(w-w', w')^l \, dw' \cdot l^{\frac{n}{2}} \tag{18}$$

is absolutely convergent and defines an element ω *of* $L_2^q(W_+)^{(l)}$. *By the correspondence* $\omega \leftrightarrow h \otimes \overline{\zeta}$, *one has an isomorphism of Hilbert spaces*

$$L_2^q(W_+)^{(l)} \cong \mathcal{F}(W_+)^{(l)} \otimes \overline{\mathcal{F}^q(W_+)^{(l)}}. \tag{19}$$

Lemma 2. *Let* (ψ_ν) *be an orthonormal basis of* $\mathcal{F}(W_+)^{(l)}$, *and let* $\omega \in L_2^q(W_+)^{(l)}$, $\omega \leftrightarrow \sum_{\nu=1}^{\infty} \psi_\nu \otimes \overline{\zeta}_\nu$, $\zeta_\nu \in \mathcal{F}^q(W_+)^{(l)}$, *in the sense of Lemma* 1. *Then* $\overline{\partial}\omega$ *is defined if and only if all* $d\zeta_\nu$ $(\nu = 1,2,\ldots)$ *are defined and one has* $\sum \|d\zeta_\nu\|^2 < \infty$. *When that is so, one has* $\overline{\partial}\omega \leftrightarrow \sum \psi_\nu \otimes \overline{d\zeta_\nu}$. *One has also a similar statement for* $\theta\omega$.

Thus the determination of $\overline{\partial}$-cohomology of $L_2^q(W_+)^{(l)}$ is reduced to that of the usual d-cohomology of $\mathcal{F}^q(W_+)^{(l)}$. But, if one puts $\Delta = d\delta + \delta d$, then an easy computation shows that

$$(\Delta\zeta)_{(\beta)} = 2\pi l\Big(\sum_{k=1}^{n} w_k \frac{\partial}{\partial w_k} + q\Big)\zeta_{(\beta)} \quad \text{for} \quad \zeta \in \mathcal{F}^q(W_+)^{(l)},$$

whence follows immediately that, if ζ is harmonic, one has $\zeta = 0$ except for the case $q = 0$, in which case ζ is a constant. Thus one has

$$(20) \qquad H^q(\tilde{V},T,l) \begin{cases} = \{0\} & \text{for } q > 0, \\ \cong \mathcal{F}(W_+)^{(l)} & \text{for } q = 0, \end{cases}$$

which proves our assertion for $l > 0$. The case $l < 0$ then follows immediately by Serre's duality theorem.

II. *Unitary representations of a semi-direct product*

5. Let G_i be a connected unimodular Lie group and K_i a compact subgroup of G_i, for $i = 1, 2$. We suppose that, for each pair (G_i, K_i), one has a decomposition $\mathfrak{g}_i = \mathfrak{k}_i + \mathfrak{m}_i$ of the corresponding Lie algebras satisfying (1) and a complex structure J_i on $\mathfrak{m}_i$ satisfying the conditions (A1-2) with respect to a complex form $G_{i\mathbb{C}}$ of G_i. We further assume that there is given a continuous homomorphism ρ of G_1 into the automorphism group $\mathrm{Aut}(G_2)$ of G_2 satisfying the following conditions:

(B1) For every $g_1 \in G_1$, $\rho(g_1)$ leaves K_2 elementwise invariant and leaves $\mathfrak{m}_2$ invariant (as a whole). Moreover, $\rho(g_1)$ leaves the Haar measure of G_2 invariant.

(B2) ρ can be extended to a holomorphic homomorphism of $G_{1\mathbb{C}}$ into $\mathrm{Aut}(G_{2\mathbb{C}})$, and $\rho(K_{1\mathbb{C}} M_1^+)$ leaves $\mathfrak{m}_2^+$ invariant, where $\mathfrak{m}_i^{\pm}$ denotes the $(\pm\sqrt{-1})$-eigensubspace of J_i in $\mathfrak{m}_{i\mathbb{C}}$, for $i = 1, 2$, and $M_i^{\pm}$ is the corresponding complex analytic subgroup of $G_{i\mathbb{C}}$.

Under these conditions, one constructs a semi-direct product $G = G_2 \cdot G_1$ with the multiplication defined by

$$(21)\qquad (g_2g_1)(g_2'g_1') = (g_2\cdot\rho(g_1)g_2')(g_1g_1') \qquad (g_i,\ g_i'\in G_i).$$

The Lie algebra $\mathfrak{g}$ of G is then a semi-direct sum of the Lie algebras $\mathfrak{g}_1$ and $\mathfrak{g}_2$. It is clear that $K = K_2\cdot K_1$ $(\cong K_1\times K_2)$ is a compact subgroup of G and one has a decomposition $\mathfrak{g} = \mathfrak{k} + \mathfrak{m}$ satisfying (1) with $\mathfrak{k} = \mathfrak{k}_1 + \mathfrak{k}_2$, $\mathfrak{m} = \mathfrak{m}_1 + \mathfrak{m}_2$. Moreover, one has a complex structure $J = J_1 \oplus J_2$ on $\mathfrak{m}$ satisfying the conditions (A1-2) with respect to the complex form $G_{\mathbb{C}}$ of G, given by the semi-direct product of $G_{1\mathbb{C}}$ and $G_{2\mathbb{C}}$. One denotes by π_i the canonical projection $G_i \to D_i = G_i/K_i$.

Let τ_i be an irreducible representation of K_i on a complex vector space F_i $(i = 1, 2)$ and let $\tau = \tau_1\otimes\tau_2$ be the irreducible representation of $K = K_1\times K_2$ on the tensor product $F = F_1\otimes F_2$ given by

$$(22)\qquad \tau(k_1k_2) = \tau_1(k_1)\otimes\tau_2(k_2) \qquad \text{for } k_i\in K_i.$$

Our purpose here is to study the relation between $H^q = H^q(G,K,J,\tau)$ and $H_i^q = H^q(G_i,K_i,J_i,\tau_i)$ $(i = 1, 2)$.

Let $m_i = \frac{1}{2}\dim\mathfrak{m}_i$ and, for a non-negative integer q, let A_q (resp. B_q) be the set of all q-tuples taken from the set of indices $\{1,\dots,m_1\}$ (resp. $\{1,\dots,m_2\}$). We fix once and for all an orthonormal basis $(X_1,\dots,X_{m_1})$ (resp. $(Y_1,\dots,Y_{m_2})$) of $\mathfrak{m}_1^+$ (resp. $\mathfrak{m}_2^+$) in the sense as explained in section 2. Let (q_1,q_2) be a pair of non-negative integers such that $q_1 + q_2 = q$. For $\omega\in L^q = L_2^q(G,K,J,\tau)$ and $(\alpha)\in A_{q_1}$, $(\beta)\in B_{q_2}$, one puts

$$\omega_{(\bar\alpha)(\bar\beta)}(g) = \omega(g\,;X_{\bar\alpha_1},\dots,X_{\bar\alpha_{q_1}},Y_{\bar\beta_1},\dots,Y_{\bar\beta_{q_2}}).$$

A form $\omega\in L^q$ is called *of subtype* (q_1,q_2) if $\omega_{(\bar\alpha')(\bar\beta')} = 0$ for all $(\alpha')\in A_{q_1'}$, $(\beta')\in B_{q_2'}$ such that $(q_1',q_2')\neq(q_1,q_2)$. We denote by $L^{(q_1q_2)} = L_2^{(q_1q_2)}(G,K,J,\tau)$ the subspace of L^q formed of all forms of subtype (q_1,q_2). Then it is clear that $L^{(q_1q_2)}$ is a closed G-invariant subspace of L^q and one has

$$(23)\qquad L^q = \sum_{q_1+q_2=q} L^{(q_1q_2)} \quad \text{(direct sum)}.$$

For $\omega \in C_c^{(q_1q_2)} = L^{(q_1q_2)} \cap C_c(G,K,J,\tau)$, one has

$$d''\omega = d_1''\omega + d_2''\omega,$$

$$\delta''\omega = \delta_1''\omega + \delta_2''\omega$$

with $d_1''\omega \in C_c^{(q_1+1,q_2)}$, $d_2''\omega \in C_c^{(q_1,q_2+1)}$, $\delta_1''\omega \in C_c^{(q_1-1,q_2)}$, $\delta_2''\omega \in C_c^{(q_1,q_2-1)}$. Then, from (8) one obtains

$$(24)\quad \begin{cases} (d_1''\omega)_{(\bar\alpha)(\bar\beta)} = \sum\limits_{i=1}^{q_1+1} (-1)^{i+1} X_{\bar\alpha_i}\, \omega_{(\dots\hat{\bar\alpha}_i\dots)(\bar\beta)} + \\ \quad + \sum\limits_{i<i'} \sum\limits_{\alpha'} (-1)^{i+i'}\, \bar c^{\alpha'}_{\alpha_i\alpha_{i'}}\, \omega_{\bar\alpha'(\dots\hat{\bar\alpha}_i\dots\hat{\bar\alpha}_{i'}\dots)(\bar\beta)} \\ \quad + \sum\limits_{i,j} \sum\limits_{\beta'} (-1)^{i+j+1}\, \overline{\rho(X_{\alpha_i})}^{\beta'}_{\beta_j}\, \omega_{(\dots\hat{\bar\alpha}_i\dots)\bar\beta'(\dots\hat{\bar\beta}_j\dots)}, \\ (d_2''\omega)_{(\bar\alpha)(\bar\beta)} = \sum\limits_{i=1}^{q_2+1} (-1)^{q_1+j+1}\, Y_{\bar\beta_j}\, \omega_{(\bar\alpha)(\dots\hat{\bar\beta}_j\dots)} + \\ \quad + \sum\limits_{j<j'} \sum\limits_{\beta'} (-1)^{q_1+j+j'}\, \bar c'^{\beta'}_{\beta_j\beta_{j'}}\, \omega_{(\bar\alpha)\bar\beta'(\dots\hat{\bar\beta}_j\dots\hat{\bar\beta}_{j'}\dots)}, \end{cases}$$

where one puts $[X_{\alpha'}, X_{\alpha''}] = \sum\limits_{\alpha} c^{\alpha}_{\alpha'\alpha''} X_\alpha$, $[Y_{\beta'}, Y_{\beta''}] = \sum\limits_{\beta} c'^{\beta}_{\beta'\beta''} Y_\beta$

and $[X,Y_{\beta'}] = \rho(X)Y_{\beta'} = \sum_{\beta} \rho(X)^{\beta}_{\beta'} Y_{\beta}$ for $X \in \mathfrak{m}_1^+$. One also has quite similar formulas for δ_1'' and δ_2''. One defines the operators $\bar{\partial}_1$, $\bar{\partial}_2$, θ_1, θ_2 on $L^{(q_1 q_2)}$ as a formal adjoint of δ_1'', δ_2'', d_1'', d_2'', respectively. Then they are densely defined closed operators, and it is clear that, for $\omega \in L^{(q_1 q_2)}$, $\bar{\partial}\omega$ (resp. $\theta\omega$) is defined if and only if both $\bar{\partial}_1\omega$ and $\bar{\partial}_2\omega$ (resp. $\theta_1\omega$ and $\theta_2\omega$) are defined and when that is so one has $\bar{\partial}\omega = \bar{\partial}_1\omega + \bar{\partial}_2\omega$ (resp. $\theta\omega = \theta_1\omega + \theta_2\omega$). In particular, $\omega \in L^{(q_1 q_2)}$ is harmonic if and only if all $\bar{\partial}_1\omega$, $\bar{\partial}_2\omega$, $\theta_1\omega$, $\theta_2\omega$ are defined and $= 0$. We denote by $H^{(q_1 q_2)} = L^{(q_1 q_2)} \cap H^q$ the space of all harmonic forms of subtype (q_1, q_2). Then it is clear that

$$H^q \supset \sum_{q_1+q_2=q} H^{(q_1 q_2)} \quad \text{(direct sum)}. \tag{25}$$

In general, it is not known whether or not one has the equality sign here[3].

---

[3] When one has a discrete subgroup Γ of G, one can define the cohomology space $H^q(\Gamma\backslash G, K, \tau)$ in a similar manner as in section 2. When the quotient space $\Gamma\backslash G$ is compact, an equality similar to (25) has been proved by Kuga in a certain special case; Shahshahani [11] also obtained a similar equality in a somewhat different context. It should be noted that the usual way of proving such an equality breaks down, for, in general, the Laplacian $\Box'' = d''\delta'' + \delta''d''$ does *not* commute with the projection operator $L^q \to L^{(q_1 q_2)}$.

6. Now let $\omega \in L^{(q_1 q_2)}$. Then one has

$$
(26)\quad \begin{cases}
i(X)\omega = 0 \qquad \text{for all } X \in \mathfrak{k}_{\mathbb{C}} + \mathfrak{m}_+ , \\[2ex]
\omega_{(\bar{\alpha})(\bar{\beta})}(g_2 g_1 k_1) = \\
\qquad (\tau_1(k_1) \otimes 1_{F_2})^{-1} \omega(g_2 g_1 ; \operatorname{ad}(k_1) X_{(\bar{\alpha})}, \operatorname{ad}(k_1) Y_{(\bar{\beta})}), \\[2ex]
\omega_{(\bar{\alpha})(\bar{\beta})}(g_2 k_2 g_1) = \\
\qquad (1_{F_1} \otimes \tau_2(k_2))^{-1} \omega(g_2 g_1 ; X_{(\bar{\alpha})}, \operatorname{ad}(k_2) Y_{(\bar{\beta})}), \\[2ex]
\|\omega\|^2 = c' \displaystyle\sum_{\substack{(\alpha) \in A_q \\ (\beta) \in B_q}} \int_{G_2 \times G_1} \|\omega_{(\bar{\alpha})(\bar{\beta})}(g_2 g_1)\|^2_{F_1 \otimes F_2} \, dg_2 dg_1 < \infty,
\end{cases}
$$

where $g_i \in G_i$, $k_i \in K_i$. If the Haar measure dg_i of G_i $(i = 1, 2)$ is normalized as explained in section 2, then one has $c' = 2^{q_1+q_2}/q_1!q_2!$. It follows that for all $(\alpha) \in A_{q_1}$ and for almost all $g_1 \in G_1$ (in the sense of measure-theory) one can define an F-valued square-integrable form $\omega_2^{g_1,(\alpha)}$ of degree q_2 on G_2 by putting

$$
(27)\qquad \omega_2^{g_1,(\alpha)}(g_2 ; Y) = \omega(g_2 g_1 ; X_{(\bar{\alpha})}, \operatorname{ad}(g_1)^{-1} Y).
$$

For each $g_1 \in G_1$, the transform $(\rho(g_1)|\mathfrak{m}_2) J_2 (\rho(g_1)|\mathfrak{m}_2)^{-1}$ of the complex structure J_2 is also a complex structure on $\mathfrak{m}_2$ satisfying the conditions (A1-2). Since this complex structure depends only on $z = \pi_1(g_1)$, we denote it by J_z. Then, from (26) it is clear that $\omega_2^{g_1,(\alpha)}$ may be viewed as an element of $F_1 \otimes L_2^{q_2}(G_2, K_2, J_z, \tau_2)$ in an obvious sense and one has

$$(28) \qquad \|\omega\|^2 = c_1 \sum_{(\alpha)\in A_{q_1}} \int_{G_1} \|\omega_2^{g_1,(\alpha)}\|^2 \, dg_1 < \infty ,$$

where $\| \ \|$ on the right side denotes the product norm in $F_1 \otimes L_2^{q_2}(G_2,K_2,J_z,\tau_2)$ and $c_1 = 2^{q_1}/q_1!$ if the Haar measures are normalized. Moreover, in view of (8), (24), it is easy to see that $\bar\partial_2\omega$ is defined if and only if $\bar\partial(\omega_2^{g_1,(\alpha)})$ is defined for all $(\alpha)\in A_{q_1}$ and almost all $g_1 \in G_1$ and one has

$$\int_{G_1} \|\bar\partial(\omega_2^{g_1,(\alpha)})\|^2 \, dg_1 < \infty ;$$

and, when that is so, one has

$$(29) \qquad (\bar\partial_2\omega)^{g_1,(\alpha)} = (-1)^{q_1} \,\bar\partial(\omega^{g_1,(\alpha)}).$$

One has also a similar statement for $\theta_2\omega$. It follows, in particular, that one has $\bar\partial_2\omega = \theta_2\omega = 0$ if and only if $\omega^{g_1,(\alpha)}$ is harmonic for all (α) and almost all g_1. Therefore, if one puts

$$L_0^{(q_1q_2)} = \{\omega\in L^{(q_1q_2)} \mid \bar\partial_2\omega = \theta_2\omega = 0\}, \quad \text{and}$$

$$H_z^{q_2} = H^{q_2}(G_2,K_2,J_z,\tau_2),$$

then for $\omega\in L^{(q_1q_2)}$ one has

$$(30) \quad \omega\in L_0^{(q_1q_2)} \Longleftrightarrow \omega_2^{g_1,(\alpha)} \in F_1 \otimes H^{q_2}_{\pi_1(g_1)} \quad \text{for all } (\alpha)\in A_{q_1} \text{ and almost all } g_1\in G_1.$$

7. We denote by $T_2^z(g_2)$ $(g_2 \in G_2)$ the *regular* representation of G_2 on the space $H_z^{q_2}$, i.e., for $\zeta \in H_z^{q_2}$ and $g_2, g_2' \in G_2$, one has

(31) $$(T_2^z(g_2)\zeta)(g_2' ; Y) = \zeta(g_2^{-1}g_2' ; Y).$$

Also, for $g_1 \in G_1$, one puts

(32) $$(T_2^z(g_1)\zeta)(g_2 ; Y) = \zeta(g_1^{-1}g_2 g_1 ; \mathrm{ad}(g_1)^{-1}Y).$$

Then it is clear that $T_2^z(g_1)\zeta \in H_{g_1(z)}^{q_2}$ and the map $T_2^z(g_1): H_z^{q_2} \to H_{g_1(z)}^{q_2}$ is an isomorphism of Hilbert spaces. From the definitions, one also has

(33) $$T_2^z(g_1)T_2^z(g_2)T_2^z(g_1)^{-1} = T_2^{g_1(z)}(g_1 g_2 g_1^{-1})$$

for all $g_1 \in G_1$, $g_2 \in G_2$ and $z \in \mathcal{D}_1$. We denote by $z_o = \pi_1(e)$ the origin of $\mathcal{D}_1 = G_1/K_1$ and for $g_2 \in G_2$ put $T_2(g_2) = T_2^{z_o}(g_2)$.

Now we assume that

(C1) The unitary representation $(H_{z_o}^{q_2}, T_2)$ of G_2 is non-trivial and irreducible.

Then the representation T_2 is of discrete series. Therefore, a continuous deformation of it given by $(H_{g_1(z_o)}^{q_2}, T_2^{g_1(z_o)})$ $(g_1 \in G_1)$ should be trivial. In other words, for every $z \in \mathcal{D}_1$, there exists a unitary isomorphism U_{z,z_o} from $H_{z_o}^{q_2}$ onto $H_z^{q_2}$. Put $U_{z',z} = U_{z',z_o} \circ U_{z,z_o}^{-1}$ for $z, z' \in \mathcal{D}_1$; then one has

(34) $$U_{z'',z'} \circ U_{z',z} = U_{z'',z} \qquad \text{for } z, z', z'' \in \mathcal{D}_1.$$

From (33) and from the irreducibility of $H_{z_o}^{q_2}$, one has

(35) $$U_{g_1 g_1'(z_o), g_1(z_o)} \circ T_2^{z_o}(g_1) = \alpha(g_1, g_1') T_2^{g_1'(z_o)}(g_1) \circ U_{g_1'(z_o), z_o}$$

for $g_1, g_1' \in G_1$

with $\alpha(g_1, g_1') \in T$. Clearly α is a T-valued 2-cocycle on G_1 satisfying the condition

(36) $$\alpha(g_1, k_1) = 1 \qquad \text{for } g_1 \in G_1, \ k_1 \in K_1,$$

and the cohomology class of α is uniquely determined, independently of the choice of the isomorphisms U_{z,z_o}. It follows from (35) that, if one puts

(37) $$\dot{T}_2(g_1) = U_{z_o, g_1(z_o)} \circ T_2^{z_o}(g_1) \qquad \text{for } g_1 \in G_1,$$

then one has

(38) $$\dot{T}_2(g_1) \circ \dot{T}_2(g_1') = \alpha(g_1, g_1') \dot{T}_2(g_1 g_1') \qquad \text{for } g_1, g_1' \in G_1,$$

i.e., $\dot{T}_2$ is a projective unitary representation of G_1 in $H_{z_o}^{q_2}$ with the factor set α. In the following, we assume that the system of isomorphisms $\{U_{z,z_o}\}$ is so chosen that the following *continuity condition* is satisfied.

(C2) The projective representation $\dot{T}_2$ is continuous in the sense that, for every $\zeta \in H_{z_o}^{q_2}$, the map $g_1 \mapsto \dot{T}_2(g_1)\zeta$ is a continuous map from G_1 into $H_{z_o}^{q_2}$.

Then the factor set α is also continuous. By putting $\dot{T}_2(g_2 g_1) = T_2(g_2)\dot{T}_2(g_1)$ for $g_1 \in G_1$, $g_2 \in G_2$, one obtains a (continuous) projective unitary representation of the semi-direct product $G = G_2 \cdot G_1$.

8. Now take an orthonormal basis (ζ_ν) of $H_{z_o}^{q_2}$. Then, for every $z \in \mathcal{D}_1$, $(U_{z,z_o}\zeta_\nu)$ is an orthonormal basis of $H_z^{q_2}$. By (26), (30) and (C1), one sees that for every $\omega \in L_0^{(q_1 q_2)}$ there exists a sequence (ω^ν) of elements in $L_2^{q_1}(G_1,K_1,\tau_1)$ such that one has

$$(39) \quad \omega_{(\bar{\alpha})(\bar{\beta})}(g_2 g_1) = \sum_{\nu=1}^{\infty} \omega^\nu(g_1; X_{(\bar{\alpha})})(U_{g_1(z_o),z_o}\zeta_\nu)(g_2; \mathrm{ad}(g_1)Y_{(\bar{\beta})})$$

and (if the Haar measures are normalized) one has

$$(40) \quad \|\omega\|^2 = \sum_{\nu=1}^{\infty} \|\omega^\nu\|^2.$$

Thus the correspondence $\omega \leftrightarrow \sum_{\nu=1}^{\infty} \omega^\nu \otimes \zeta_\nu$ gives an isomorphism

$$(41) \quad L_0^{(q_1 q_2)} \cong L_2^{q_1}(G_1,K_1,\tau_1) \otimes H_{z_o}^{q_2}.$$

Let $T_1(g_1)$ be the regular representation of G_1 on $L_2^{q_1}(G_1,K_1,\tau_1)$ and put

$$(T_1'(g_1)\omega_1)(g_1';X) = \alpha(g_1,g_1^{-1}g_1')^{-1}\omega_1(g_1^{-1}g_1'; X)$$

for $\omega_1 \in L_2^{q_1}(G_1,K_1,\tau_1)$ and $g_1, g_1' \in G_1$. Then it is easy to see that T_1' is a continuous projective unitary representation of G_1 on $L_2^{q_1}(G_1,K_1,\tau_1)$ with the factor set α^{-1} and that, if $T(g)$ denotes the regular representation of G on $L_2^q(G,K,\tau)$, one has under the isomorphism (41)

$$(42) \quad T(g_2 g_1) \mid L_0^{(q_1 q_2)} \leftrightarrow T_1'(g_1) \otimes \dot{T}_2(g_2 g_1)$$

for all $g_1 \in G_1$, $g_2 \in G_2$.

9. Now to proceed further, we first need the following assumption, which enables us to replace the projective unitary representations T_1', $\dot{T}_2$ of G_1 by ordinary ones.

(C3)[4] There exist a finite sheeted covering group (G_1^o, λ) of G_1 and a continuous map $\chi\colon G_1^o \to T$ such that one has

$$(43) \qquad \alpha(\lambda(g^o), \lambda(g^{o'})) = \chi(g^o)\chi(g^{o'})\chi(g^o g^{o'})^{-1}$$

for all $g^o, g^{o'} \in G_1^o$.

The Lie algebra of G_1^o may be identified with the Lie algebra $\mathfrak{g}_1$ of G_1. Put[5] $K_1^o = \lambda^{-1}(K_1)$. Then it follows from (43) and (36) that one has

$$(44) \qquad \chi(g^o k^o) = \chi(g^o)\chi(k^o) \qquad \text{for } g^o \in G_1^o,\ k^o \in K_1^o.$$

In particular, the restriction of χ on K_1^o is a (unitary) character of K_1^o. Define an irreducible representation τ_1' of K_1^o by

$$(45) \qquad \tau_1'(k^o) = \chi(k^o)\tau_1(\lambda(k^o)) \qquad \text{for } k^o \in K_1^o.$$

Then it is clear that the correspondence $\omega_1^o \leftrightarrow \omega_1$ defined by the relation

$$(46) \qquad \omega_1^o(g^o; X) = \chi(g^o)^{-1}\omega_1(\lambda(g^o); X) \qquad (g^o \in G_1^o)$$

---

[4] If there exist a positive integer r and a continuous map $\beta\colon G_1 \to T$ such that one has $\alpha(g,g')^r = \beta(g)\beta(g')\beta(gg')^{-1}$ for all $g, g' \in G_1$, then one can easily construct an r-sheeted (cyclic) covering group (G_1^o, λ) and a continuous map χ satisfying the condition (C3).

[5] In general, the pair (G_1^o, K_1^o), together with the data $\mathfrak{m}_1$, J_1, ..., does not satisfy the condition (A2); in particular, K_1^o may not be connected. However, one can easily see that, after a slight modification, our results in sections 1 and 2 remain true.

gives an isomorphism

$$L_2^{q_1}(G_1^o, K_1^o, \tau_1') \cong L_2^{q_1}(G_1, K_1, \tau_1). \tag{47}$$

If one denotes by $T_1^o(g_1^o)$ the regular representation of G_1^o on $L_2^{q_1}(G_1^o, K_1^o, \tau_1')$, then under the above isomorphism one has

$$T_1^o(g_1^o) \leftrightarrow \chi(g_1^o) T_1'(\lambda(g_1^o)) \qquad \text{for } g_1^o \in G_1^o. \tag{48}$$

On the other hand, if one puts

$$T_2^o(g_1^o) = \chi(g_1^o)^{-1}\, \dot{T}_2(\lambda(g_1^o)), \tag{49}$$

then T_2^o is an (ordinary) unitary representation of G_1^o on $H_{z_o}^{q_2}$. Thus, under the *amended isomorphism*

$$L_0^{(q_1 q_2)} \cong L_2^{q_1}(G_1^o, K_1^o, \tau_1') \otimes H_{z_o}^{q_2} \tag{41'}$$

given by the correspondence $\omega \leftrightarrow \sum \omega^{\nu_o} \otimes \chi \cdot \zeta_\nu$, one has

$$T(g_2 \cdot \lambda(g_1^o)) \mid L_0^{(q_1 q_2)} \leftrightarrow T_1^o(g_1^o) \otimes T_2(g_2) T_2^o(g_1^o) \tag{42'}$$

for all $g_1^o \in G_1^o$, $g_2 \in G_2$.

Next, we strengthen the condition (C2) to a certain differentiability condition. In general, let $\{\zeta^{g_1}\}$ $(g_1 \in G_1)$ be a family of forms on G_2 such that, for each $g_1 \in G_1$ one has $\zeta^{g_1} \in H_z^{q_2}$ $(z = \pi_1(g_1))$. Such a family $\{\zeta^{g_1}\}$ is called *differentiable* if the assignment $g_1 \mapsto \zeta^{g_1}(g_2 ; \mathrm{ad}(g_1) Y_{(\bar{\beta})})$ is C^∞ for all $(\beta) \in B_{q_2}$ and all $g_2 \in G_2$. A family $\{\zeta^{g_1}\}$ is called *harmonic* if it is differentiable and if the following conditions are satisfied:

$$
(50)\quad
\begin{cases}
\bar{X}(\zeta^{g_1}(g_2\,;\,\mathrm{ad}(g_1)Y_{(\bar{\beta})})) + \\
\quad + \sum_{j=1}^{q_2} \sum_{\beta'=1}^{m_2} (-1)^j\, \overline{\rho(X)}^{\beta'}_{\beta_j}\, \zeta^{g_1}(g_2\,;\,\mathrm{ad}(g_1)Y_{\bar{\beta}'(\cdots\hat{\beta}_j\cdots)}) = 0, \\
\\
X(\zeta^{g_1}(g_2\,;\,\mathrm{ad}(g_1)Y_{(\bar{\beta})})) - \\
\quad - \sum_{j=1}^{q_2} \sum_{\beta'=1}^{m_2} (-1)^j\, \rho(X)^{\beta_j}_{\beta'}\, \zeta^{g_1}(g_2\,;\,\mathrm{ad}(g_1)Y_{\bar{\beta}'(\cdots\hat{\beta}_j\cdots)}) = 0
\end{cases}
$$

for all $X \in \mathfrak{m}_1^+$ and for all $(\beta) \in B_{q_2}$, $g_2 \in G_2$. Similar notions are also defined on the group G_1°.

Now we put

$$
(51)\qquad \zeta_\nu^{g_1^\circ} = \chi(g_1^\circ)(U_{z,z_\circ}\zeta_\nu) \qquad (z = \pi_1(\lambda(g_1^\circ)))
$$

and assume that

(C4) The families $\{\zeta_\nu^{g_1}\}$ $(g_1 \in G_1)$ for $\nu = 1,2,\ldots$ are harmonic.

Then, by (8), (24), one can easily show that, for $\omega \in L_0^{(q_1q_2)}$, $\bar{\partial}_1\omega$ is defined if and only if $\bar{\partial}(\omega^{\nu\circ})$ is defined for all ν and one has $\sum \|\partial\omega^{\nu\circ}\|^2 < \infty$; and, when that is so, one has $(\bar{\partial}_1\omega)^{\nu\circ} = \bar{\partial}(\omega^{\nu\circ})$. One also has a similar statement for $\theta_1\omega$. In particular, ω is harmonic if and only if all $\omega^{\nu\circ}$'s are harmonic. Summing up, one obtains the following theorem.

Theorem 1. *For* $i = 1,2$, *let* G_i *be a connected Lie group,* K_i *a compact subgroup of* G_i, J_i *a complex structure on* $\mathfrak{m}_i$ *satisfying the conditions* (A1-2), *and* τ_i *an irreducible representation of* K_i. *Let* $G = {}^{`}G_2G_1$ *be a semi-direct product of* G_1 *and* G_2 *defined by a continuous homomorphism* $\rho\colon G_1 \to \mathrm{Aut}(G_2)$ *satisfying the conditions* (B1-2). *Furthermore, suppose that the conditions* (C1-4) *are satisfied for* q_2 ; *let* $\lambda\colon G_1^{\circ} \to G_1$ *be a covering homomorphism and* $\chi\colon G_1^{\circ} \to T$ *a continuous map as stated in* (C3) *and define an irreducible representation* τ_1' *of* $K_1^{\circ} = \lambda^{-1}(K_1)$ *by* (45). *Then, for any non-negative integer* q_1, *one has an isomorphism of Hilbert spaces*

$$(52) \qquad H^{(q_1q_2)}(G_2\cdot G_1\,,\,K_1\times K_2\,,\,J_1\oplus J_2\,,\,\tau_1\otimes\tau_2)$$
$$\cong H^{q_1}(G_1^{\circ},K_1^{\circ},J_1,\tau_1')\otimes H^{q_2}(G_2,K_2,J_2,\tau_2).$$

Moreover, if one denotes the regular representations of G, G_1° *and* G_2 *on these spaces by* T, T_1° *and* T_2, *respectively, and define a unitary representation* T_2° *of* G_1° *by* (32), (37), (49), *then under the above isomorphism one has*

$$(53) \qquad T(g_2\cdot\lambda(g_1^{\circ})) \leftrightarrow T_1^{\circ}(g_1^{\circ}) \otimes T_2(g_2)T_2^{\circ}(g_1^{\circ})$$

for all $g_1^{\circ}\in G_1^{\circ}$, $g_2\in G_2$.

It follows that, under the same assumptions, the unitary representation T of G on $H^{(q_1q_2)}$ is irreducible, if and only if the unitary representation T_1° of G_1° on $H^{q_1}(G_1^{\circ},K_1^{\circ},J_1,\tau_1')$, or what amounts to the same, the projective unitary representation T_1' of G_1 on $H^{q_1}(G_1,K_1,J_1,\tau_1)$, is irreducible.

10. Finally, as an example, let us consider the case $G_2 = \tilde{V}$, $K_2 = T$ defined in sections 3 and 4. For simplicity, we assume here that the complex structure I on V satisfies the Riemann condition, i.e., the quadratic form $A(x,Ix)$ is positive-definite. Let $G_1' = Sp(V,A)$ be the corresponding symplectic group and K_1' a maximal compact subgroup of G_1' formed of all $g_1' \in G_1'$ which commutes with I. Then, as is well known, the pair (G_1', K_1') is a hermitian symmetric pair, so that all conditions in section 1 are satisfied. To be more precise, let $\mathfrak{g}_1' = \mathfrak{k}_1' + \mathfrak{m}_1'$ be a Cartan decomposition of the Lie algebra $\mathfrak{g}_1'$ of G_1' corresponding to K_1'. Then I belongs to the center of $\mathfrak{k}_1'$ and so $J_1' = \frac{1}{2}\mathrm{ad}(I)|\mathfrak{m}_1'$ is a K_1'-invariant complex structure on $\mathfrak{m}_1'$, of which the $(\pm i)$-eigenspace $\mathfrak{m}_1'^{\pm}$ in $\mathfrak{m}_{1\mathbb{C}}'$ is an abelian subalgebra of $\mathfrak{g}_{1\mathbb{C}}'$. Call $K_{1\mathbb{C}}'$, $M_1'^{\pm}$ the complex analytic subgroups of $G_{1\mathbb{C}}' = Sp(V_{\mathbb{C}}, A)$ corresponding to $\mathfrak{k}_{1\mathbb{C}}'$, $\mathfrak{m}_1'^{\pm}$, respectively. Then, clearly the condition (A2) is also satisfied. Now, every element $g_1' \in G_1'$ defines an automorphism $(\xi, u) \mapsto (\xi, g_1'(u))$ of $G_2 = \tilde{V}$, leaving $K_2 = T$ elementwise invariant and leaving $\mathfrak{m}_2 = \{(0,u) \mid u \in V\}$ invariant. Conversely, it can readily be seen that all automorphisms of $\tilde{V}$ satisfying this condition are obtained in this way. It is also easy to see that an element $g_1' \in G_1'$ belongs to $K_{1\mathbb{C}}' M_1'^{+}$ if and only if one has $g_1'(W_+) \subset W_+$, $W_\pm$ denoting the $(\pm i)$-eigenspace of I in $V_{\mathbb{C}}$. Thus for the pairs (G_1', K_1') and $(\tilde{V}, T)$ (together with other data described above) and for the natural injection $G_1' \subset \mathrm{Aut}(\tilde{V})$, the conditions (B1-2) are satisfied. More generally, for any pair (G_1, K_1) with $\mathfrak{m}_1$, J_1, ... satisfying (A1-2) and $(G_2, K_2) = (\tilde{V}, T)$, the conditions (B1-2) may be paraphrased as follows:

(B1)' ρ is a symplectic representation of G_1 into $G_1' = Sp(V,A)$.

(B2)' ρ can be extended to a holomorphic representation of $G_{1\mathbb{C}}$ into $G_{1\mathbb{C}}' = Sp(V_{\mathbb{C}}, A)$, and one has $\rho(K_{1\mathbb{C}} M_1^{+}) \subset K_{1\mathbb{C}}' M_1'^{+}$.

It follows from these conditions that ρ induces in a natural manner a holomorphic mapping of the homogeneous space $\mathcal{D}_1 = G_1/K_1$ into the *Siegel space* $\mathcal{D}_1' = G_1'/K_1'$.

Now we assume $l > 0$. Then, as we have seen in section 4, one has $H^{q_2}(\tilde{V},T,l) = \{0\}$ for $q_2 > 0$ and $H^0(\tilde{V},T,l) \cong \mathcal{F}(W_+)^{(l)}$. Hence the condition (C1) is satisfied for $q_2 = 0$. We shall show that, in the situation described above, the conditions (C2-4) are also satisfied. To do that, we may (hence will) assume that $G_1 = G_1'$, $\rho = \mathrm{id.}$, for the proofs in the general case can easily be reduced to this special case. We shall make use of a canonical automorphy factor and the associated kernel function for $G = \tilde{V}\cdot G_1'$ and $\mathcal{D} = W_+ \times \mathcal{D}_1'$ defined as follows (cf. [9])[6]. Namely, for $\tilde{g} = \tilde{u}g \in G$, $\tilde{u} = (\xi,u) \in \tilde{V}$, $g \in G_1'$, $\tilde{z} = (w,z)$, $\tilde{z}' = (w',z') \in \mathcal{D}$, one puts

$$(54)\quad \begin{cases} \eta(\tilde{g},\tilde{z}) = \xi\cdot\varepsilon(\tfrac{1}{2}A(u,u_{g(z)}) + A(u,J_1 w) + \tfrac{1}{2}A(gw,J_1 w)) \\ \kappa(\tilde{z}',\tilde{z}) = \varepsilon(\tfrac{1}{2}A(\bar{z}K_1^{-1}w', w') + A(\bar{w}, K_1^{-1}w') + \tfrac{1}{2}A(\bar{w}, K_1^{-1}z'\bar{w})), \end{cases}$$

where $J_1 = J_1(g,z) = {}^t(cz+d)^{-1}$, $K_1 = K_1(z',z) = 1 - z'\bar{z}$ are the canonical automorphy factor and the associated kernel function for $G_1' = Sp(V,A)$ and $\mathcal{D}_1' = G_1'/K_1'$; as usual, an element z of the Siegel space $\mathcal{D}_1'$ is viewed as a $\mathbb{C}$-linear map from W_- into W_+ satisfying the Siegel condition, and for $u = u_+ + u_- \in V$ one puts $u_z = u_+ - zu_-$. We put further

$$(55)\quad \begin{cases} \eta_z(\tilde{g},w) = \eta(\tilde{g},(w,z)), \\ \kappa_z(w',w) = \kappa((w',z),(w,z)). \end{cases}$$

[6] All formulas in [9] are given in terms of the upper half-space realization of the Siegel space D'. It is easy to translate them into the ones given here by a Cayley transformation.

One denotes by $\mathcal{F}(W_+,z)^{(l)}$ the Hilbert space formed of all holomorphic functions h on W_+ such that

$$(56) \qquad \|h\|^2 = \int_{W_+} |h(w)|^2 \kappa_z(w,w)^{-l} d\!\!\!{}^{-}_z w < \infty,$$

where $d\!\!\!{}^{-}_z w = \det(K_1(z,z))^{-1} d\!\!\!{}^{-} w$. The action of $\tilde{V}$ on $\mathcal{F}(W_+,z)^{(l)}$ is defined by

$$(57) \qquad h(w) \mapsto \eta_z(\tilde{u}^{-1},w)^{-l} h(w-u_z).$$

Then it is immediate that the correspondence $f \leftrightarrow h$ defined by

$$(58) \qquad f(\tilde{u}) = \eta_z(\tilde{u},0)^{-l} h(u_z)$$

gives a unitary isomorphism (of the representation-spaces of $\tilde{V}$)

$$(59) \qquad H^0(\tilde{V},T,J_z,l) \cong \mathcal{F}(W_+,z)^{(l)}.$$

On the other hand, for $z, z' \in \mathcal{D}_1'$, a unitary isomorphism $\mathbb{U}_{z',z}: \mathcal{F}(W_+,z)^{(l)} \to \mathcal{F}(W_+,z')^{(l)}$ is given explicitly as follows:

$$(60) \qquad (\mathbb{U}_{z',z} h)(w') = \gamma_{z',z} \int_{W_+} h(w) \kappa(\tilde{z}',\tilde{z})^l \kappa(\tilde{z},\tilde{z})^{-l} d\!\!\!{}^{-}_z w \cdot l^n$$

with

$$\gamma_{z',z} = \det(K_1(z',z))^{-\frac{1}{2}} \cdot \det(K_1(z,z))^{\frac{1}{4}} \cdot \det(K_1(z',z'))^{\frac{1}{4}},$$

where the branch of the analytic function $\det(K_1(z',z))^{\frac{1}{2}}$ is so chosen that it is > 0 for $z = z'$. (Cf. [9]. The integral on the right side is absolutely convergent for all $h \in \mathcal{F}(W_+,z)$.) One then has the relation (34); in particular, $\mathbb{U}_{z,z} = \mathrm{id}$. Note also that

$$\gamma_{z,z_0} = \det(K_1(z,z))^{\frac{1}{4}} = |\det(J_1(g,z_0))|^{\frac{1}{2}} \qquad (z = \pi_1'(g)).$$

Therefore one defines a unitary isomorphism $U_{z,z_0} : H^0(\tilde{V},T,J_{z_0},l) \to H^0(\tilde{V},T,J_z,l)$ by

$$(61)\quad (U_{z,z_0}f)(\tilde{u}) = \eta_z(\tilde{u},0)^{-l}(\mathbb{U}_{z,z_0}h)(u_z)$$

$$= \det(K_1(z,z))^{\frac{1}{2}}\eta_z(\tilde{u},0)^{-l}\int_{W_+} h(w')\kappa(\tilde{z},\tilde{z}')^{l}\kappa(\tilde{z}',\tilde{z}')^{-l}dw'\cdot l^n$$

where $f\in H^0(\tilde{V},T,J_{z_0},l)$ and $h\in\mathcal{F}(W_+)^{(l)} = \mathcal{F}(W_+,z_0)^{(l)}$ are related by (58) (with $z = z_0$), and one writes $\tilde{z} = (w,z)$, $\tilde{z}' = (w',z_0)$. It is then clear that the condition (C2) is satisfied.

As is well known, there exists a (unique) two-sheeted covering group $(G_1'^{\circ},\lambda')$ of $G_1' = Sp(V,A)$, which uniformizes the function $\det(J_1(g,z_0))^{\frac{1}{2}}$. Thus one has a continuous function χ' on $G_1'^{\circ}$ such that one has

$$(62)\quad \chi'(g^{\circ})^2 = \det(J_1(\lambda'(g^{\circ}),z_0))/|\det(J_1(\lambda'(g^{\circ}),z_0)|\quad \text{for } g^{\circ}\in G_1'^{\circ}.$$

Then one has the relation (43) (see [9]), and the condition (C3) is satisfied.

Now, let (ψ_ν) be an orthonormal basis of $\mathcal{F}(W_+)^{(l)}$ corresponding to an orthonormal basis (ζ_ν) of $H^0(\tilde{V},T,J_{z_0},l)$ under the isomorphism (59). Then, from (61), one has for $g^{\circ}\in G_1'^{\circ}$

$$\zeta_\nu^{g^{\circ}}(\tilde{u}) = \chi'(g^{\circ})(U_{g(z_0),z_0}\zeta_\nu)(\tilde{u})$$

$$= \Phi(\tilde{u},g^{\circ})\Psi_\nu(w,z)$$

with

$$\Phi(\tilde{u},g^{\circ}) = \det(J_1(\lambda'(g^{\circ}),z_0))^{\frac{1}{2}}\eta(\tilde{u}\lambda'(g^{\circ}),(0,z_0)),$$

$$\Psi_\nu(w,z) = \int_{W_+}\psi_\nu(w')\kappa((w,z),(w',z_0))^{l}\kappa_{z_0}(w',w')^{l}dw'\cdot l^n.$$

Since Ψ_ν is holomorphic in $(w,z)\in W_+ \times \mathcal{D}_1'$, one has $X\Psi_\nu = 0$ for all $X\in \mathfrak{m}_1'^{-}$. On the other hand, the two-valued function $\Phi_{\tilde{u}}(g) = \Phi(\tilde{u},\lambda'^{-1}(g))$ on G_1' may be extended (locally) to a holomorphic function in a neighbourhood of g in G_1' and one has

$$\Phi_{\tilde{u}}(g\exp(\xi X)) = \Phi_{\tilde{u}}(g) \qquad \text{for all} \quad X\in \mathfrak{p}_{1\mathbb{C}}' + \mathfrak{m}_1'^{-}, \quad \xi\in\mathbb{R}.$$

Hence one has $X\Phi_{\tilde{u}} = 0$. Thus the condition (C4) is also satisfied. The case $l < 0$ can easily be reduced to the case $l > 0$ considered above.

In the general case, where one has a symplectic representation $\rho\colon G_1 \to G_1' = Sp(V,A)$ satisfying (B2)$'$, one can construct an (at most two-sheeted) covering group (G_1°,λ) of G_1 such that there exists a continuous homomorphism $\rho^{\circ}\colon G_1^{\circ} \to G_1'^{\circ}$ with $\lambda'\circ\rho^{\circ} = \rho\circ\lambda$. Then, the covering group (G_1°,λ) together with the map $\chi = \chi'\circ\rho^{\circ}$ satisfies the condition (C3), and all the requirements of Theorem 1 are satisfied. Thus, one obtains the following result.

Theorem 2. *Let G_1 be a connected Lie group, K_1 a compact subgroup of G_1, J_1 a complex structure on $\mathfrak{m}_1$ satisfying the conditions* (A1-2), *and τ_1 an irreducible representation of K_1. Let $\rho\colon G_1 \to Sp(V,A)$ be a symplectic representation of G_1 satisfying the condition* (B2)$'$, *and let $G = \tilde{V}\cdot G_1$ be the semi-direct product of G_1 and $\tilde{V}$ defined by ρ. Then, for $K = K_1 \times T$, $J = J_1 \oplus J_2$, $\tau = \tau_1 \otimes (l)$ with $l \neq 0$, and for any non-negative integer q_1, the space $H^{(q_1 q_2)}(G,K,J,\tau)$ is isomorphic to the tensor product of $H^{q_1}(G_1^{\circ},K_1^{\circ},J_1,\tau_1')$ and $\mathcal{F}(W_+)^{(l)}$ if $l > 0$ and $q_2 = 0$ or if $l < 0$ and $q_2 = n$, and is $= \{0\}$ otherwise, where (G_1°,λ) is an (at most two-sheeted) covering group of G_1, $K_1^{\circ} = \lambda^{-1}(K_1)$, and $\tau_1' = \chi\cdot(\tau_1\circ\lambda)$, $\chi = \chi'\circ\rho^{\circ}$ are as defined above. Moreover, under this isomorphism, one has a similar relation between unitary representations as described in Theorem* 1.

In particular, for $l > 0$, $q_2 = 0$, or $l < 0$, $q_2 = n$, the space $H^{(q_1 q_2)}(G,K,J,\tau)$ is $\neq \{0\}$ and irreducible if and only if the corresponding space $H^{q_1}(G_1^{\circ},K_1^{\circ},J_1,\tau_1')$ is. On the other hand, it can easily be shown that any irreducible unitary representation U of G of discrete series, such that $U(\xi) = \xi^l$ $(l \neq 0)$ for $\xi \in T$, can be obtained as the tensor product of an irreducible unitary representation U_1° of G_1° of discrete series and the unitary representation of the semi-direct product $\tilde{V} \cdot G_1^{\circ}$ on the Fock space $\mathcal{F}(W_+)^{(l)}$ constructed above[7]. Thus, in this case, the problem of determining all such irreducible unitary representations of G, along with their realization on the $\bar{\partial}$-cohomology spaces, is completely reduced to the corresponding problem for the group G_1°.

References

1. A. Andreotti and E. Vesentini, *Carleman estimates for the Laplace-Beltrami equations on complex manifolds*, Inst. Hautes Études Sci. Publ. Math. 25 (1965), 81-130.

2. V. Bargman, *On a Hilbert space of analytic functions and associated integral transform*, I, Comm. Pure Appl. Math. 14 (1961) 187-214.

3. Harish-Chandra, (a) *Representations of semisimple Lie groups*, IV, Amer. J. Math. 77 (1955), 743-777; V, ibid. 78 (1956), 1-41; VI, ibid. 564-628. (b) *Discrete series for semisimple Lie groups*, I, Acta Math. 115 (1965), 241-318; II, ibid. 116 (1966), 1-111.

4. S. Kobayashi, *Irreducibility of certain unitary representations*, J. Math. Soc. Japan 20 (1968), 638-642.

5. Y. Matsushima and S. Murakami, (a) *On vector bundle valued harmonic forms and automorphic forms on symmetric Riemannian manifolds*, Ann. of Math. (2) 78 (1963), 365-416. (b) *On certain cohomology groups attached to hermitian symmetric spaces*, Osaka J. Math. 2 (1965), 1-35; II, ibid. 5 (1968), 223-241.

6. S. Murakami, (a) *Cohomology groups of vector-valued forms on symmetric spaces*, Lecture notes at Univ. of Chicago, Summer 1966. (b) *Facteurs d'automorphie associés à un espace hermitien symétrique*, Geometry of homogeneous bounded domains, Centro Int. Mat. Estivo, 3° Ciclo, Urbino, 1967.

---

[7] For a more precise description of this latter representation, see [9].

7. M. S. Narasimhan and K. Okamoto, *An analogue of the Borel-Weil-Bott theorem for symmetric pairs of non-compact type*, Ann. of Math. (2) 91 (1970), 486-511.

8. K. Okamoto and H. Ozeki, *On square-integrable $\bar{\partial}$-cohomology spaces attached to Hermitian symmetric spaces*, Osaka J. Math. 4 (1967), 95-110.

9. I. Satake, (a) *On unitary representations of a certain group extension* (in Japanese), Sûgaku 21 (1969), 241-253. (b) *Fock representations and theta-functions*, Proceedings of Conference on Riemann surface theory, Stony Brook, 1969.

10. W, Schmid, *The conjecture of Langlands*, Ann. of Math., to appear.

11. M. Shahshahani, *Discontinuous subgroups of extensions of semi-simple Lie groups*, Dissertation, University of California at Berkeley, 1970.

12. D. Shale, *Linear symmetries of free boson fields*, Trans. Amer. Math. Soc. 103 (1962), 149-167.

13. A. Weil, *Sur certains groupes d'opérateurs unitaires*, Acta Math. 111 (1964), 143-211.

University of California, Berkeley, California 94720

CLASS FIELDS OVER REAL QUADRATIC FIELDS IN THE THEORY OF MODULAR FUNCTIONS

Goro Shimura

1. *Introduction*

It is well-known that elliptic modular functions or forms are closely connected with imaginary quadratic fields in various ways. For example, as Hecke showed, the L-function of such a field with a Grössen-character can be associated, through Mellin transformation, with a few cusp forms belonging to a certain congruence subgroup. Further, the class fields over an imaginary quadratic field can be generated by special values of elliptic or elliptic modular functions. Compared with this, the connection of real quadratic fields with modular functions is not so close, although we know at least the following two facts: (i) the L-function of a real quadratic field with a character of finite order corresponds to holomorphic modular forms of weight 1 (i.e., "Dimension -1") under Mellin transformation (Hecke [3]); (ii) the L-function of a real quadratic field with a Grössen-character corresponds to non-holomorphic automorphic forms under Mellin transformation (Maass [5]). However, no theory analogous to complex multiplication has been known for real quadratic fields (see Note 1 at the end of the paper).

The purpose of this lecture is to show that the arithmetic of a real quadratic field k is closely connected with certain *holomorphic* modular forms of weight $w \geqq 2$, and also that some class fields over k can be generated by points of finite order on certain abelian varieties obtained from such modular forms. These do not quite parallel the known results for real or imaginary fields, mentioned above, but still there is some similarity.

To be more explicit, we fix a positive integer $N > 1$, and a non-trivial quadratic character χ of $(\mathbb{Z}/N\mathbb{Z})^{\times}$ (see Note 2), and put

$$\Gamma_0(N) = \{ \begin{pmatrix} a & b \\ c & d \end{pmatrix} \in SL_2(\mathbb{Z}) \mid c \equiv 0 \mod (N)\},$$

$$\Gamma_\chi = \{ \begin{pmatrix} a & b \\ c & d \end{pmatrix} \in \Gamma_0(N) \mid \chi(a) = 1\}.$$

For a positive integer $w > 1$, let $S_{w,\chi}$ denote the vector space of all cusp forms g satisfying

$$g\left(\frac{az+b}{cz+d}\right)\cdot(cz+d)^{-w} = \chi(d)g(z) \quad \text{for all} \quad \begin{pmatrix} a & b \\ c & d \end{pmatrix} \in \Gamma_0(N).$$

For an obvious reason, we assume

$$(1.1) \qquad \chi(-1) = (-1)^w.$$

After Hecke, we define a $\mathbb{C}$-linear endomorphism $T(n)$ of $S_{w,\chi}$, for a positive integer n, by

$$(1.2) \quad g|T(n) = n^{w-1} \sum_{\substack{a>0 \\ ad=n}} \sum_{b=0}^{d-1} \chi(a)g((az+b)/d)d^{-w} \qquad (g \in S_{w,\chi}),$$

and a $\mathbb{C}$-linear automorphism H of $S_{w,\chi}$ by

$$(1.3) \qquad g|H = g(-1/Nz)(N^{\frac{1}{2}}z)^{-w}.$$

Hecke proved that the operators $T(n)$, for all n, form a commutative algebra $\mathcal{R}_{\mathbb{C}}$ over $\mathbb{C}$, whose rank is exactly the dimension of $S_{w,\chi}$ over $\mathbb{C}$. Further one has

$$(1.4) \qquad T(n)H = \chi(n)HT(n) \quad \textit{if } n \textit{ is prime to } N.$$

Hereafter we make the following assumption (on N and χ), though this is not absolutely necessary:

(1.5) *$R_{\mathbb{C}}$ is generated only by the $T(n)$ for all n prime to N.*

This is satisfied, for instance, if χ is a primitive character modulo N (Hecke [4, Satz 24]). Under (1.5), $R_{\mathbb{C}}$ is semi-simple (but the converse is false). Taking $\mathbb{Q}$ instead of $\mathbb{C}$, we obtain

Proposition 1. *Let R denote the algebra generated by the $T(n)$ over $\mathbb{Q}$ for all n. Then $R_{\mathbb{C}} = R \otimes_{\mathbb{Q}} \mathbb{C}$. Moreover, under (1.5), R is a direct sum of fields, each of which is a totally imaginary quadratic extension of a totally real algebraic number field.*

Let $R = K_1 \oplus \ldots \oplus K_r$ be the direct sum decomposition of R with fields $K_1, \ldots, K_r$. Let K_ν be a subfield of $\mathbb{C}$ isomorphic to K_ν, and e_ν the identity element of K_ν. Then we can find an element

$$f_\nu(z) = \sum_{n=1}^{\infty} a_{\nu n} e^{2\pi i n z}$$

of $e_\nu \cdot S_{\omega,\chi}$ with the following properties:

(1.6) *K_ν is generated by the numbers $a_{\nu n}$ over $\mathbb{Q}$;*

(1.7) *The functions $f_{\nu\sigma}(z) = \sum_{n=1}^{\infty} a_{\nu n}^{\sigma} e^{2\pi i n z}$, for all the isomorphisms σ of K_ν into $\mathbb{C}$, form a basis of $e_\nu \cdot S_{\omega,\chi}$ over $\mathbb{C}$;*

(1.8) $f_{\nu\sigma} | T(n) = a_{\nu n}^{\sigma} \cdot f_{\nu\sigma}$.

The fields K_ν are unique up to conjugacy over $\mathbb{Q}$.

By the Hecke theory, the Dirichlet series $L(s, f_{\nu\sigma}) = \sum_{n=1}^{\infty} a_{\nu n}^{\sigma} n^{-s}$ associated with $f_{\nu\sigma}$ has the Euler-product

$$L(s, f_{\nu\sigma}) = \prod_p (1 - a_{\nu p}^{\sigma} p^{-s} + \chi(p) p^{\omega-1-2s})^{-1}. \tag{1.9}$$

(In (1.2) and (1.9), we put $\chi(n) = 0$ if n is not prime to N.) If ρ denotes the restriction of the complex conjugation to K_ν, we obtain

$$a_{\nu n}^{\sigma} = \chi(n) a_{\nu n}^{\sigma\rho} \quad \textit{if } n \textit{ is prime to } N. \tag{1.10}$$

Let K'_ν denote the maximal (totally) real subfield of K_ν, which is meaningful in view of Proposition 1. Then $a_{\nu n} \in K'$ if $\chi(n) = 1$.

Let us now assume that w is even, so that $\chi(-1) = 1$ by (1.1). (We shall make a brief discussion about the case of odd w in §4.) Then the character χ corresponds to a unique real quadratic field k. For example, $k = \mathbb{Q}(\sqrt{N})$ if $\chi(a) = \left(\frac{a}{N}\right)$, N is square-free, and $N \equiv 1 \bmod (4)$. Now our main interest lies in the following relation between the arithmetic of k and the Fourier coefficients $a_{\nu p}$ for rational primes p not dividing N.

(I) *There is a prime ideal $\mathfrak{l}_\nu$ in K_ν, ramified over K'_ν, with the following properties:*

(I_a) $a_{\nu p} \equiv 0 \bmod \mathfrak{l}_\nu$ *if* $\chi(p) = -1$.

(I_b) *If* $\chi(p) = 1$ *and* $p = N_{k/\mathbb{Q}}(\alpha)$ *with a totally positive algebraic integer* α *of* k, *then*

$$\mathrm{Tr}_{k/\mathbb{Q}}(\alpha^{w-1}) \equiv a_{\nu p} \mod \mathfrak{l}_\nu \cap K'_\nu .$$

(I_c) *If u is the fundamental unit of k, then*

$$\mathrm{Tr}_{k/\mathbb{Q}}(u^{w-1}) \equiv 0 \mod N(\mathfrak{l}_\nu).$$

This is neither a theorem nor a conjecture, but merely a somewhat vague statement of what we find empirically from the numerical values of $a_{\nu p}$, and what we can prove for a few smaller N and w. Because of lack of data, it is safe to restrict the validity of (I) to the case where k is of class number one. Probably it is necessary to modify the statement in the case of larger class numbers.

Actually the existence of $\mathfrak{l}_\nu$ satisfying (I_a) can easily be shown if we assume that $a_{\nu p}$, for at least one p, generates a prime ideal $\mathfrak{l}_\nu$ ramified over K'_ν. In fact, (I_a) then follows from (1.10). It is the properties (I_b, I_c) which are really striking, and by which the relatively trivial (I_a) can acquire a new significance.

As a simple example, let us consider the case $N = 29$, $w = 2$. Then $S_{w,\chi}$, with $\chi(a) = \left(\frac{a}{29}\right)$, is of dimension 2. As Hecke showed (Werke, 903-905), there are two cusp forms $\sum_n a_n e^{2\pi i n z}$ and $\sum_n a_n^\rho e^{2\pi i n z}$ with a_n in $K = \mathbf{Q}(\sqrt{-5})$; further $a_p \equiv 0 \bmod (\sqrt{-5})$ if $\chi(p) = -1$. Now the fundamental unit u of $k = \mathbf{Q}(\sqrt{29})$ is $(5+\sqrt{29})/2$, so that $\mathrm{Tr}_{k/\mathbf{Q}}(u) = 5$. Thus ($I_a$) and ($I_c$) are true with $\mathfrak{P}_\nu = (\sqrt{-5})$. (In this case, R itself is isomorphic to K.) We can actually prove (I_b) in this case. More examples will be discussed later.

(II) *If* $w = 2$, *the maximal ray class field over* k *modulo* $N(\mathfrak{P}_\nu)P_\infty$ *can be generated over* k *by the coordinates of certain points of finite order on the Jacobian variety of the modular function field for* Γ_χ, *where* P_∞ *denotes the product of the two archimedean primes of* k. *Moreover, the above relations* (I_a, I_b) *describe the reciprocity law in the class field.*

Again this is neither a theorem nor a conjecture. We can prove the statement under certain conditions on N and $N(\mathfrak{P}_\nu)$. It is at least true for the above special case $N = 29$, $w = 2$.

2. *Preliminary considerations about class fields over a real quadratic field*

A real quadratic field k has "less" class fields than imaginary quadratic fields, since it has a unit of infinite order. To explain this in a quantitative way, let $\mathfrak{o}_k$ denote the ring of algebraic integers in k, $\mathfrak{m}$ an integral ideal in k, h the class number of k, and u the fundamental unit of k. Further let P_∞ be as in (II), and $F_\mathfrak{m}$ the maximal ray class field over k modulo $\mathfrak{m}P_\infty$. Then one has

$$[F_\mathfrak{m} : k] = 2h\cdot\phi(\mathfrak{m})/\mu ,$$

where $\phi(\mathfrak{m})$ denotes the number of invertible elements in the ring

$\mathfrak{v}_k/\mathfrak{m}$, and μ the smallest positive integer such that u^μ is totally positive and $\equiv 1 \bmod \mathfrak{m}$. Therefore, if μ is small, we obtain a relatively large class field. For a fixed μ, we have a very restrictive choice for $\mathfrak{m}$, since it must divide $u^\mu - 1$. This is why we have "less" class fields over k. It often happens even that $F_{\mathfrak{m}}$ is the composite of a cyclotomic field with k. A more precise statement can be proved, for example, in the following form:

Proposition 2. *Let q be a rational prime unramified in k, and m a positive integer, and μ the smallest positive integer such that u^μ is totally positive and $u^\mu \equiv 1 \bmod (q^m)$. Suppose that $q^m > 2$, and $u^\mu - 1$ is not divisible by q^{m+1}. Then the maximal abelian extension of k, in which only the factors of $q \cdot P_\infty$ are ramified, can be generated over $F_{\mathfrak{m}}$, with $\mathfrak{m} = q^m \mathfrak{v}_k$, by the q^n-th roots of unity for all n.*

The class field we shall obtain in our later discussion is exactly the extension $F_{\mathfrak{m}}$ of this type. We note that, if $N_{k/\mathbf{Q}}(u) = -1$, then $u^2 - 1 = u \cdot \mathrm{Tr}_{k/\mathbf{Q}}(u)$, so that the relation (I_c) with $w = 2$ implies that $u^2 \equiv 1 \bmod N(\mathfrak{b}_\nu)$. Therefore, if $N(\mathfrak{b}_\nu)$ is an odd prime, and $u^2 - 1$ is not divisible by $N(\mathfrak{b}_\nu)^2$, then the maximal ray class field over k modulo $N(\mathfrak{b}_\nu)P_\infty$ has a special meaning as described in the above proposition. For example, in the case $N = 29$ and $w = 2$, we obtain the maximal ray class field F over $\mathbf{Q}(\sqrt{29})$ of conductor $5 \cdot P_\infty$, which generates, together with the 5^n-th roots of unity for all n, the maximal abelian extension of k in which only the prime factors of $5 \cdot P_\infty$ are ramified.

3. *Detailed discussion in the case* $w = 2$

In this section we assume throughout $w = 2$. Let $\mathfrak{H}$ denote the upper half complex plane $\{z \in \mathbf{C} \mid Im(z) > 0\}$. Then the quotient $\mathfrak{H}/\Gamma_\chi$, suitably compactified, is isomorphic to a projective non-singular curve V, defined over[1] $\mathbf{Q}$. Let J denote the Jacobian variety of V, also defined over $\mathbf{Q}$. The generator of $\Gamma_0(N)/\Gamma_\chi$ defines an automorphism γ of J defined over $\mathbf{Q}$. Put $J_\chi = (1-\gamma)J$. Then J_χ is an abelian subvariety of J, defined over $\mathbf{Q}$, whose tangent space at the origin can be identified with $S_{2,\chi}$. Moreover J is isogenous to the product of J_χ and the Jacobian variety of $\mathfrak{H}/\Gamma_0(N)$.

The operator $T(n)$ on $S_{2,\chi}$ defines an endomorphism of J_χ in a natural way, which we denote by $\tau(n)$. The subspace $e_\nu \cdot S_{2,\chi}$ of $S_{2,\chi}$ corresponds to an abelian subvariety A_ν of J_χ. Further we can define an isomorphism θ_ν of K_ν into $\mathrm{End}(A_\nu)\otimes\mathbf{Q}$ so that $\theta_\nu(a_{\nu n})$ is the restriction of $\tau(n)$ to A_ν for all n. We note that

$$[K_\nu : \mathbf{Q}] = \dim(A_\nu),$$

$$J = (1+\gamma)J + J_\chi ,$$

$$J_\chi = A_1 + \dots + A_r.$$

It can be shown that A_ν and the elements of $\mathrm{End}(A_\nu)\cap\theta_\nu(K_\nu)$ are all defined over $\mathbf{Q}$.

Now the operator H of (1.3) defines an automorphism η of J_χ of order 2, defined over k, satisfying

(3.1) $\eta\cdot\tau(n) = \chi(n)\tau(n)\eta$, if n is prime to N,

(3.2) $\eta^\varepsilon = -\eta$,

where ε is the generator of $\mathrm{Gal}(k/\mathbf{Q})$. Put

(3.3) $A'_\nu = (1+\eta)A_\nu$.

Then A'_ν is an abelian subvariety of A_ν rational over k, and

---

[1] There is a "standard" way to find a model for $\mathfrak{H}/\Gamma_\chi$ defined over $\mathbf{Q}$. For the detailed proofs of this and other facts stated in this section, the reader is referred to [10, Chap. 7].

(3.4) $A_\nu = A'_\nu + A'^{\varepsilon}_\nu, \quad A'^{\varepsilon}_\nu = (1-\eta)A_\nu$.

Let $a \in K_\nu$ and $\theta_\nu(a) \in \mathrm{End}(A_\nu)$. We see from (3.1) that

$$(3.5) \qquad \theta_\nu(a) \text{ sends } A'_\nu \text{ into } \begin{matrix} A'_\nu \\ A'^{\varepsilon}_\nu \end{matrix}, \text{ and } A'^{\varepsilon}_\nu \text{ into } \begin{matrix} A'^{\varepsilon}_\nu \\ A'_\nu \end{matrix} \text{ if } \begin{matrix} a \in K' \\ a^\rho = -a. \end{matrix}$$

(As is defined in §1, ρ is the complex conjugation, and $K'_\nu = \{x \in K_\nu \mid x^\rho = x\}$.) Therefore, we can define an isomorphism θ'_ν of K'_ν into $\mathrm{End}(A'_\nu) \otimes \mathbf{Q}$ so that $\theta'_\nu(a)$ is the restriction of $\theta_\nu(a)$ if $a \in K'_\nu$.

Now the congruence relations for $\tau(p)$ enables us to determine the zeta functions of A_ν and A'_ν :

Proposition 3. *The zeta function of A_ν over $\mathbf{Q}$ and the zeta function of A'_ν over k coincide, up to finitely many Euler factors, with the product $\prod_\sigma L(s, f_{\nu\sigma})$ for all the isomorphisms σ of K into $\mathbf{C}$.*

Although this is implicit in the previous works by Eichler and the author, the present formulation (in a more general case) is due to Miyake [6] (cf. also [10, Chap. 7]). A result of Igusa, combined with a result about good reduction of factors of an abelian variety, implies that the bad Euler factors may exist only for the prime factors of N. On account of this, the above proposition is essentially equivalent with the following statement:

Let p be a rational prime not dividing N, and Φ a Frobenius automorphism of the algebraic closure $\overline{\mathbf{Q}}$ of $\mathbf{Q}$ for any prime divisor of $\overline{\mathbf{Q}}$ extending p. Then

$$(3.6) \qquad t^\Phi + \chi(p) p t^{\Phi^{-1}} = \theta_\nu(a_{\nu p}) t$$

for every point t of A_ν whose order is finite and prime to p.

From now on, we fix our attention to one ν, and write K_ν, K'_ν, A_ν, A'_ν, θ_ν, θ'_ν, $f_\nu = \sum a_{\nu n} e^{2\pi i n z}$ simply as K, K', A, A', θ, θ', $f = \sum a_n e^{2\pi i n z}$, by dropping the subscript ν. Let $\mathfrak{o}$ (resp. $\mathfrak{o}'$) be the ring of algebraic integers in K (resp. K'). Changing A and A' for some abelian varieties isogenous to them, if necessary, we may assume that $\theta(\mathfrak{o}) \subset \mathrm{End}(A)$ and $\theta'(\mathfrak{o}') \subset \mathrm{End}(A')$. It can be shown that this change is made without interfering with (3.2- 3.6).

Let $\mathfrak{b}$ denote the ideal of $\mathfrak{o}$ generated by the elements a of $\mathfrak{o}$ such that $a^\rho = -a$. We assume

(3.7) $\mathfrak{b} \neq \mathfrak{o}$ *and* $\mathfrak{b}$ *is prime to* 2.

Then $\mathfrak{b}$ is a square-free product of several prime ideals of $\mathfrak{o}$ ramified over K'. Therefore one finds a unique ideal $\mathfrak{t}$ of $\mathfrak{o}'$ such that $\mathfrak{t}^2\mathfrak{o} = \mathfrak{b}$ and $\mathfrak{t} = \mathfrak{b} \cap \mathfrak{o}'$; then $\mathfrak{t}$ is also square-free. We can prove

Proposition 4. $-N(\mathfrak{b}) \in N_{k/\mathbb{Q}}(k)$.

For a prime N, Hecke proved that $a_N a_N^\rho = N$, so that $N \in N_{K/K'}(K)$. The above proposition is "reciprocal" to this result.

Now our first main result can be stated as follows:

Theorem 1. *Let* $A[\mathfrak{b}] = \{t \in A \mid \theta(\mathfrak{b})t = 0\}$, *and let* F *be the extension of* k *generated by the coordinates of the points of* $A[\mathfrak{b}]$. *Then the following assertions hold:*

(1) *F is an abelian extension of k, which is unramified at every prime of k not dividing $N(\mathfrak{b})N$.*

(2) *If m is a positive rational integer prime to $N(\mathfrak{b})N$, and* $\sigma = \left(\frac{F/k}{(m)}\right)$, *then* $x^\sigma = mx$ *for every* $x \in A[\mathfrak{b}]$.

(3) *If $\mathfrak{t} \cap \mathbb{Z} = q\mathbb{Z}$ with a positive integer q, and ζ is a primitive q-th root of unity, then $\zeta \in F$, and $F \neq k(\zeta)$.*

The fact that F is abelian can be shown as follows. Let $Y = A[\mathfrak{b}] \cap A'$ and $Y_\varepsilon = A[\mathfrak{b}] \cap A'^{\varepsilon}$. Then the assumption (3.7) and some elementary observations imply that $A[\mathfrak{b}] = Y + Y_\varepsilon$, and both Y and Y_ε are isomorphic to $\mathfrak{o}'/\mathfrak{c}$ as $\mathfrak{o}'$-modules. (Note that $\mathfrak{o}'/\mathfrak{c}$ is $\mathfrak{o}'$-isomorphic to $\mathfrak{o}/\mathfrak{b}$.) Every element of Gal$(F/k)$ induces an automorphism of each of Y and Y_ε. Let $(\mathfrak{o}'/\mathfrak{c})^\times$ denote the group of invertible elements of $\mathfrak{o}'/\mathfrak{c}$. Since the group of all automorphisms of the $\mathfrak{o}'$-module $\mathfrak{o}'/\mathfrak{c}$ is isomorphic to $(\mathfrak{o}'/\mathfrak{c})^\times$, we thus obtain a (canonically defined) homomorphism

$$\text{(3.8)} \qquad \mathrm{Gal}(F/k) \to (\mathfrak{o}'/\mathfrak{c})^\times \times (\mathfrak{o}'/\mathfrak{c})^\times,$$

which is obviously injective. Therefore F is abelian over k. The proof of the remaining assertions is somewhat more involved, but follows, in essence, from (3.6).

We can obtain a more definite result about the conductor of F over k, under the following set of assumptions:

(3.9) (i) *$N(\mathfrak{c})$ is a prime (so that $N(\mathfrak{c}) = q$);* (ii) *N is a square-free product of odd primes which do not divide $q(q-1)$;* (iii) $\chi(a) = \left(\frac{a}{N}\right)$.

Then $k = \mathbb{Q}(\sqrt{N})$.

Theorem 2. *Under the above assumptions, the conductor of F over k is exactly $q \cdot P_\infty$.*

In view of Proposition 4, we see that the prime q becomes the product of two primes $\mathfrak{q}$ and $\mathfrak{q}^\varepsilon$ in k. If $\mathfrak{o}_k$ denotes the ring of integers in k, both $\mathfrak{o}_k/\mathfrak{q}$ and $\mathfrak{o}_k/\mathfrak{q}^\varepsilon$ are isomorphic to $\mathbb{Z}/q\mathbb{Z}$. Now to each totally positive element α of $\mathfrak{o}_k$, we associate an element $\left(\frac{F/k}{\alpha\mathfrak{o}_k}\right)$ of Gal(F/k). Since every element of $\mathfrak{o}_k/q\,\mathfrak{o}_k = \mathfrak{o}_k/\mathfrak{q} \oplus \mathfrak{o}_k/\mathfrak{q}^\varepsilon$ can be represented by such an α, we thus obtain a sequence of homomorphisms

$$\text{(3.10)} \qquad (\mathbb{Z}/q\mathbb{Z})^{\times 2} \to (\mathfrak{o}_k/q\,\mathfrak{o}_k)^\times \to \mathrm{Gal}(F/k) \to (\mathfrak{o}'/\mathfrak{c})^{\times 2} \to (\mathbb{Z}/q\mathbb{Z})^{\times 2}.$$

The first and last arrows are isomorphisms, unique up to the change of factors. Therefore we obtain a homomorphism

$$(3.11)\qquad (\mathbb{Z}/q\mathbb{Z})^{\times 2} \ni (x,y) \mapsto (g(x,y)\,,\, h(x,y)) \in (\mathbb{Z}/q\mathbb{Z})^{\times 2}$$

which is canonically defined up to the change of factors. Now we can naturally ask the following question:

(Q) *Is the map* (3.11) *the identity map, up to the change of* x *and* y?

It is quite likely that this is so if k is of class number one, and N is a prime. Probably one will have to modify the formulation if the class number of k is greater than one. We can at least prove

$$g(x,x) = h(x,x) = x,$$
$$g(x,y)h(x,y) = xy,$$
$$g(x,y) = h(y,x).$$

Further we obtain

Proposition 5. *Let* u *be the fundamental unit of* k. *If the map* (3.11) *is an isomorphism,* $N_{k/\mathbb{Q}}(u) = -1$, *and the class number of* k *is one, then* F *is the maximal ray class field over* k *of conductor* $q\cdot P_\infty$, *and* $u^2 - 1$ *is divisible by* q.

Let p be a rational prime such that $\chi(p) = 1$, and α a totally positive element of $\mathfrak{o}_k$ such that $N_{k/\mathbb{Q}}(\alpha) = p$. If $\alpha \equiv x$ and $\alpha^\varepsilon \equiv y \bmod \mathfrak{q}$ with rational integers x and y, the relation (3.6) implies

$$(3.12)\qquad g(x,y) + h(x,y) \equiv a_p \mod \mathfrak{r}\,.$$

From this fact we obtain easily

Theorem 3. *The answer to* (Q) *is affirmative if and only if*

$$(3.13)\qquad \mathrm{Tr}_{k/\mathbb{Q}}(\alpha) \equiv a_p \mod \mathfrak{r}$$

for every rational prime p, not dividing qN, and for every totally positive element α of $\mathfrak{o}_k$ *such that* $\chi(p) = 1$ *and* $N_{k/\mathbf{Q}}(\alpha) = p$. *Moreover* (3.13) *is satisfied by all such p and α, if it is satisfied by at least one α such that* $\alpha/\alpha^{\varepsilon}$ *is of order* $q-1$ *modulo* $\mathfrak{q}$.

It should be noted that (3.13) is a special case of (I_b).

We now give a table of *K*, *q*, and *u* for a few small primes *N*.

Table I:[2] $w = 2$; $N \equiv 1 \bmod (4)$.

N	$\dim S_{2,\chi}$	K	q	u
29	2	$\mathbf{Q}(\sqrt{-5})$	5	$(5+\sqrt{29})/2$
37	2	$\mathbf{Q}(\sqrt{-1})$	?	$6+\sqrt{37}$
41	2	$\mathbf{Q}(\sqrt{-2})$	?	$32+5\sqrt{41}$
53	4	$\mathbf{Q}(\sqrt{-3+\sqrt{2}})$	7	$(7+\sqrt{53})/2$
61	4	$\mathbf{Q}(\sqrt{-4+\sqrt{3}})$	13	$(39+5\sqrt{61})/2$
73	4	$\mathbf{Q}(\sqrt{(-19+\sqrt{5})/2})$	89	$1068+125\sqrt{73}$
89	6	$t^6+17t^4+83t^2+125=0$	5 (125?)	$500+53\sqrt{89}$
97	6	$t^6+27t^4+204t^2+467=0$	467	$5604+569\sqrt{97}$
101	8	$t^8+13t^6+51t^4+67t^2+20=0$	5	$10+\sqrt{101}$
109	8	$\mathbf{Q}(\sqrt{-3})$	3	$(261+25\sqrt{109})/2$
		$t^6+12t^4+39t^2+29=0$	29	

Each equation is for a generating element of *K*.

At the beginning of this section we decomposed J_χ into the sum of abelian subvarieties $A_1,\ldots,A_r$. For each prime $N \leqq 101$, we have $J_\chi = A = A' + A'^{\varepsilon}$, so that there is only one *K*. However, if $N = 109$, we have $J_\chi = A_1 + A_2$, $A_1 = A_1' + A_1'^{\varepsilon}$, $A_2 = A_2' + A_2'^{\varepsilon}$, $\dim(A_1') = 1$, $\dim(A_2') = 3$; the table indicates that $\mathrm{End}(A_1)\otimes\mathbf{Q}$

[2] Tables I, II, III are based on the numerical values of the eigenvalues of Hecke operators for some small prime degrees, which have been obtained by K. Doi and H. Naganuma (by hand), and H. Trotter (by computer). I wish to thank them very much.

contains $\mathbb{Q}(\sqrt{-3})$, and $\mathrm{End}(A_2)\otimes\mathbb{Q}$ contains a field of degree 6.

Now applying the last assertion of Theorem 3 to these cases, we obtain

Theorem 4. *The following assertions hold (at least) for* $N = 29$, 53, 61, 73, 89, 97.

(1) *The map* (3.11) *is the identity map, up to the change of* x *and* y.

(2) F *is the maximal class field over* k *of conductor* $q\cdot P_\infty$.

(3) *There is no abelian variety defined over* $\mathbb{Q}$ *which is isogenous to* A' *over* $\mathbb{C}$.

(4) A' *is simple, and* $\mathrm{End}(A')\otimes\mathbb{Q} = \theta'(K')$.

The class number of k is one for these six values of N. At present, the existence of α in the last assertion of Theorem 3 can be verified only by the numerical value of a_p. This is why we have to restrict the above theorem only to those N. One can naturally conjecture that the assertions of Theorem 4 hold for many more, probably infinitely many, N.

Recently W. Casselman [1] has proved that the abelian variety A' for each N of Theorem 4 has good reduction for all prime ideals of k. As a consequence of this result, we can show that if $N = 29$, the zeta-function of A' over k is exactly (i.e., without exceptional Euler factors) $L(s,f)L(s,f_\rho) = (\sum_{n=1}^{\infty} a_n n^{-s})(\sum_{n=1}^{\infty} a_n^{\rho} n^{-s})$.

4. *Generalizations and comments*

It is possible to make the same type of discussion for cusp forms of weight $w > 2$ at least if w is even. In fact we can consider the l-adic representations obtained by P. Deligne [2], in place of the l-adic representations on A. His result, suitably generalized and modified, seems to indicate the following[3]:

For a fixed N, and for each rational prime l, there exists a free $\mathbb{Z}_l$-module W_l, with the action of Hecke operators $T(n)$ and $\mathrm{Gal}(\overline{\mathbb{Q}}/\mathbb{Q})$, *satisfying the following conditions*[4]:

(4.1) *$[W_l : \mathbb{Z}_l]$ is twice the dimension of the vector space $S_w(\Gamma_\chi)$ of all cusp forms of weight w, with respect to Γ_χ.*

(4.2) *The action of $T(n)$ commutes with the action of every element of* $\mathrm{Gal}(\overline{\mathbb{Q}}/\mathbb{Q})$.

(4.3) *If p is a prime not dividing lN, and Φ a Frobenius element of* $\mathrm{Gal}(\overline{\mathbb{Q}}/\mathbb{Q})$ *for any prime divisor of $\overline{\mathbb{Q}}$ dividing p, then there is an endomorphism Φ^* of W_l such that*

$$\Phi + \Phi^* R_p = T(p), \quad \Phi\Phi^* = p^{w-1},$$

where R_p is the action of an element γ of $\Gamma_0(N)$ such that $p\cdot\gamma \equiv \begin{pmatrix} 1 & 0 \\ 0 & p^2 \end{pmatrix} \bmod (N)$.

(4.4) *The action H of $\begin{pmatrix} 0 & -1 \\ N & 0 \end{pmatrix}$ on W_l is defined, and*

$$H^2 = (-1)^w, \quad \Phi^* R_p = H\Phi^* H^{-1}.$$

(4.5) *If δ is an element of* $\mathrm{Gal}(\overline{\mathbb{Q}}/\mathbb{Q})$, *then $\delta H = H\delta$, or $\delta H = -H\delta$, according as δ is trivial on k or not.*

(4.6) $\det(x - \Phi)$ *coincides with the determinant of $x^2 - T(p)x + R_p p^{w-1}$ on $S_w(\Gamma_\chi)$.*

Now assuming the existence of such a $\mathbb{Z}_l$-module W_l, we can develop, for every even w, a theory parallel to the case $w = 2$. In

---

[3] At least one has to verify the relations (4.4, 4.5) which are not given in [2], although these do not seem difficult. There is also a different method of constructing l-adic representations, on which I gave a lecture at the conference on automorphic functions and arithmetic groups, Oberwolfach, August 1968. In this theory, the existence of W_l satisfying (4.1 - 4.5) can be shown, but there is some difficulty in proving (4.6).

[4] $\overline{\mathbb{Q}}$ denotes the algebraic closure of $\mathbb{Q}$, and $\mathbb{Z}_l$ the ring of l-adic integers.

fact we fix an eigenfunction $f(z) = \sum_{n=1}^{\infty} a_n e^{2\pi i n z}$ in $S_{w,\chi}$ and the field K generated by the a_n over $\mathbf{Q}$. Then we define $\mathfrak{b}$ and $\mathfrak{r}$ in the same way. Assuming $N(\mathfrak{r})$ to be an odd prime q, we consider the space W_l with $l = q$. Put

$$X = \{t \in e \cdot W_q \mid R_p t = \chi(p)t, \quad \mathfrak{r}t = 0\},$$

where e is the identity element of the simple component of R corresponding to K. Further put

$$Y = (1 + H)X, \quad Y_\varepsilon = (1 - H)X.$$

Then Y and Y_ε are stable under $\mathrm{Gal}(\overline{\mathbf{Q}}/k)$, on account of (4.5). Let M be the subgroup of $\mathrm{Gal}(\overline{\mathbf{Q}}/k)$ consisting of the elements which induce the identity map on $Y + Y_\varepsilon$, and let F be the subfield of $\overline{\mathbf{Q}}$ corresponding to M. Then F is abelian over k; and there is an injective homomorphism

$$\mathrm{Gal}(F/k) \to (\mathfrak{o}'/\mathfrak{r})^{\times 2},$$

from which we obtain, by a procedure similar to (3.10), a homomorphism $(x,y) \mapsto (g_w(x,y), h_w(x,y))$ of $(\mathbf{Z}/q\mathbf{Z})^{\times 2}$ into itself. Then the question (Q) can be generalized to the following form (when k is of class number one):

(Q_w) *Do we have* $g_w(x,y) = x^{w-1}$, $h_w(x,y) = y^{w-1}$, *up to the change of* x *and* y?

It is then easy to prove a generalization of Theorem 3, i.e., that (I_b) is equivalent to the affirmative answer to (Q_w).

Let us now present some numerical evidence.

Table II: w even, $\geqq 4$; $N \equiv 1 \bmod (4)$.

N	w	$\dim S_{w,\chi}$	K	q	u^{w-1}
5	6	2	$\mathbb{Q}(\sqrt{-11})$	11	$(11+5\sqrt{5})/2$
5	8	2	$\mathbb{Q}(\sqrt{-29})$	29	$(29+13\sqrt{5})/2$
5	10	4	$\mathbb{Q}(\sqrt{-854+30\sqrt{809}})$	19	$38+17\sqrt{5}$
13	4	2	$\mathbb{Q}(\sqrt{-1})$	(3?)	$18+5\sqrt{13}$
17	4	4	$\mathbb{Q}(\sqrt{-37+3\sqrt{33}})$	67	$268+65\sqrt{17}$

We notice that

$$(\mathrm{I}_c) \qquad \mathrm{Tr}_{k/\mathbb{Q}}(u^{w-1}) \equiv 0 \mod (q)$$

holds in all these cases. By means of the same principle as in Theorem 3, we can prove, assuming (4.1 - 4.6), that

$$\mathrm{Tr}_{k/\mathbb{Q}}(\alpha^{w-1}) \equiv a_p \mod (q)$$

for every totally positive algebraic integer α such that $N_{k/\mathbb{Q}}(\alpha) = p$ and $\chi(p) = 1$, at least in the two cases: $N = 5$, $w = 6$; $N = 17$, $w = 4$. Then the field F is the maximal class field over k of conductor $q \cdot P_\infty$.

Table III: w odd, $\geqq 3$; $N \equiv 3 \bmod (4)$.

N	w	$\dim S_{w,\chi}$	K	$u^{(w-1)/2}$ (u: fundamental unit of $\mathbb{Q}(\sqrt{N})$)
3	9	2	$\mathbb{Q}(\sqrt{-14})$	$97+56\sqrt{3}$
3	11	2	$\mathbb{Q}(\sqrt{-5})$	$362+209\sqrt{3}$
3	13	3	$\mathbb{Q}(\sqrt{-26})$	$1351+780\sqrt{3}$
7	7	3	$\mathbb{Q}(\sqrt{-510})$	$2024+765\sqrt{7}$
11	5	3	$\mathbb{Q}(\sqrt{-30})$	$199+60\sqrt{11}$
19	3	3	$\mathbb{Q}(\sqrt{-13})$	$170+39\sqrt{19}$

If w is odd and N is a prime $\equiv 3 \bmod (4)$, $S_{w,\chi}$ may contain cusp forms corresponding to a Grössen-character of the *imaginary* field $\mathbb{Q}(\sqrt{-N})$. Table III excludes the simple components of R corresponding to them. For such an N, $\mathbb{Q}(\sqrt{-N})$ is a reasonable field to take as the basic field. However, as the above table shows, there is still some relation between the field K and the power $u^{(w-1)/2}$ of the fundamental unit u of the *real* field $\mathbb{Q}(\sqrt{N})$. Namely, if $u^{(w-1)/2} = a + b\sqrt{N}$ with rational integers a and b, then b and the discriminant of K over $\mathbb{Q}$ have a non-trivial common factor, except for the case $N = 3$, $w = 11$.

We mention that there is room for improvement on the formulation. In the above discussion we have started with an ideal $\mathfrak{b}$ generated by the integers x in K such that $x^{\rho} = -x$. It might be better to take, instead of $\mathfrak{b}$, the ideal $\mathfrak{b}^*$ generated by the elements a_p for all primes p such that $\chi(p) = -1$. Clearly $\mathfrak{b}^* \subset \mathfrak{b}$. For example, in the case $N = 37$, $w = 2$ (resp. $N = 13$, $w = 4$), it seems that a_p is a multiple of $2i$ (resp. $3i$) for all such p. Further, in the case $N = 89$, $w = 2$, there is a possibility that one should consider 5^3 instead of $q = 5$. Also it is desirable to loosen the assumptions (3.7) and (3.9) which are too restrictive.

It should be noted that the relations (I_a, I_b) are analogous to the known relations for the coefficients of the zeta-function with a Grössen-character ψ of an *imaginary* quadratic field M. Suppose that M is of class number one, and let $g(z) = \sum_{n=1}^{\infty} b_n e^{2\pi i n z}$ be the Mellin transform of the zeta function $L(s) = \sum \psi(\mathfrak{a})N(\mathfrak{a})^{-s+(w-1)/2}$. If $\mathfrak{f}$ is the conductor of ψ, we have $\psi((\mu)) = (\mu/|\mu|)^{w-1}$ for every $\mu \in \mathfrak{o}_M$ such that $\mu \equiv 1 \bmod \mathfrak{f}$, where $\mathfrak{o}_M$ denotes the ring of algebraic integers in M. Therefore we see that

$b_p = 0$ *if p remains prime in M,*

$b_p = \mathrm{Tr}_{M/\mathbb{Q}}(\mu^{w-1})$ *if* $p = \mu\bar{\mu}$ *with* $\mu \in \mathfrak{o}_M$, *and* $\mu \equiv 1 \bmod \mathfrak{f}$.

Thus, in this case, we have the exact equalities instead of the congruences (I_a, I_b).

Also one may notice that Proposition 4 resembles the relation between a *CM*-type and its dual in the theory of complex multiplication of abelian varieties.

Instead of quadratic characters, one can consider characters of higher order, although things are rather obscure in this case. To be more specific, for a character χ of $(\mathbb{Z}/N\mathbb{Z})^\times$ of an arbitrary order > 1, we define Γ_χ, $S_{w,\chi}$, $T(n)$, and $R_{\mathbb{C}}$ as in section 1. Again take a common eigen-function $f(z) = \sum_{n=1}^{\infty} a_n e^{2\pi i n z}$ in $S_{w,\chi}$ for all $T(n)$, and let K denote the field generated by the coefficients a_n over $\mathbb{Q}$. If the condition (1.5) is satisfied, we can prove that K is a totally imaginary quadratic extension of a totally real algebraic number field K'. The field K contains the values $\chi(n)$ for all n, and $a_n = \chi(n) a_n^\rho$ for all n. Therefore, assuming that χ is of order > 2, we see easily that

$$a_n/[1+\chi(n)] \in K' \qquad \text{if} \quad \chi(n) \neq -1,$$

$$a_n/(\omega - \omega^{-1}) \in K' \qquad \text{if} \quad \chi(n) = -1,$$

where ω is a root of unity, other than ± 1, contained in K. Therefore, if we form an ideal $\mathfrak{b}_n$ in K generated by the algebraic integers x in K satisfying $x = \chi(n) x^\rho$, then $\mathfrak{b}_n$ is a divisor of either $1 + \chi(n)$ or $\omega - \omega^{-1}$ according as $\chi(n) \neq -1$ or $\chi(n) = -1$. This phenomenon makes the case of characters of higher order quite different from that of quadratic characters. However, we can at least prove the following: If $N = 13$, $w = 2$, and χ is of order 6, then $\dim(S_{2,\chi}) = 1$, so that there is a unique eigen-function $f(z) = \sum_{n=1}^{\infty} a_n e^{2\pi i n z}$ in $S_{2,\chi}$. Then $K = \mathbb{Q}(\zeta)$ with $\zeta = e^{2\pi i/3}$, and one has

$$a_p \equiv 0 \bmod (1-\zeta) \quad \textit{if } p \textit{ remains prime in } \; k = \mathbb{Q}(\sqrt{13}),$$

$$\psi(c)^{-1} a_p \equiv \mathrm{Tr}_{k/\mathbb{Q}}(\alpha) \bmod (3) \quad \textit{if } \; p = N_{k/\mathbb{Q}}(\alpha) \equiv c^2 \bmod (13)$$

with a totally positive integer α *of* k *and a rational integer* c.

The fundamental unit of $\mathbb{Q}(\sqrt{13})$ is $(3+\sqrt{13})/2$. Further we obtain the maximal ray class field over k of conductor $3 \cdot P_\infty$ from certain points of finite order on the Jacobian variety of $\mathfrak{H}/\Gamma_\chi$. It is an open question whether or not this is an exceptional case.

Finally a word about further generalizations. One can also discuss automorphic forms with respect to unit groups in a quaternion algebra, whose relation with zeta-functions of curves was given by the author [8] and T. Miyake [6]. Actually Miyake has obtained the decomposition of J and a generalization of Proposition 3 in the most general case with such automorphic forms and with characters of arbitrary order. In this case, k is replaced by an algebraic number field of higher degree, so that one has to deal with groups of units with more than one generator.

Note 1. Except for [7], [8], [9]. In these papers, it has been shown that the class fields over a totally real algebraic number field occur as fields of definition, or fields of moduli, of arithmetic quotients of bounded symmetric domains. However, this is completely different from the subject of the present article.

Note 2. For an associative ring R with identity element, we denote by $R^\times$ the group of all invertible elements of R. The symbols $\mathbb{Z}$, $\mathbb{Q}$, $\mathbb{R}$, and $\mathbb{C}$ denote, as usual, the ring of rational integers, the fields of rational, real, and complex numbers, respectively.

References

1. W. Casselman, *Some new abelian varieties with good reduction*, to appear.

2. P. Deligne, *Formes modulaires et représentations l-adiques*, Séminaire Bourbaki, février 1969, exposé 355.

3. E. Hecke, *Zur Theorie der elliptischen Funktionen*, Math. Ann. 97 (1926), 210-242 (= Werke, 428-460).

4. ———, *Analytische Arithmetik der positiven quadratischen Formen*, Danske Vid. Selsk. Mat.-Fys. Medd. XVII, 12 (Kobenhavn 1940) (= Werke, 789-918).

5. H. Maass, *Über eine neue Art von nichtanalytischen automorphen Funktionen und die Bestimmung Dirichletscher Reihen durch Funktionalgleichungen*, Math. Ann. 121 (1949), 141-183.

6. T. Miyake, *Decomposition of Jacobian varieties and Dirichlet series of Hecke type*, to appear.

7. G. Shimura, *Class-fields and automorphic functions*, Ann. of Math. (2) 80 (1964), 444-463.

8. ———, *Construction of class fields and zeta functions of algebraic curves*, Ann. of Math. (2) 85 (1967), 58-159.

9. ———, *On canonical models of arithmetic quotients of bounded symmetric domains*, Ann. of Math. (2) 91 (1970), 140-222.

10. ———, *Introduction to the arithmetic theory of automorphic functions*, to appear.

Princeton University, Princeton, New Jersey 08540

AN OSGOOD TYPE EXTENSION THEOREM FOR COHERENT ANALYTIC SHEAVES

Yum-Tong Siu[1]

0. *Introduction*

For an analytic sheaf F on a complex space define the nth *absolute gap-sheaf* $F^{[n]}$ of F by the following presheaf:

$$U \mapsto \text{ind. lim } \Gamma(U-A\ , F),$$

where A runs through all subvarieties of dimension $\leq n$ in U (see [12]).

For an analytic subsheaf G of F the nth *relative gap-sheaf* $G_{[n]F}$ of G relative to F is defined as the subsheaf of F whose stalks are given as follows: an element s of F_x belongs to $(G_{[n]F})_x$ if and only if for some open neighborhood U of x and some subvariety A of dimension $\leq n$ in U, s is induced by some $t \in \Gamma(U,F)$ satisfying $t|U-A \in \Gamma(U-A,G)$ (see [16]).

For $a \in \mathbb{R}^n$ the components of a are denoted by $a_1,\ldots,a_n$. For $a,b \in \mathbb{R}^n$, $a < b$ means $a_i < b_i$ for $1 \leq i \leq n$ and $a \leq b$ means $a_i \leq b_i$ for $1 \leq i \leq n$. $K^n(a)$ denotes

$$\{(z_1,\ldots,z_n) \in \mathbb{C}^n \mid |z_i| < a_i \text{ for } 1 \leq i \leq n\}.$$

$G^n(a,b)$ denotes

$$\{(z_1,\ldots,z_n) \in K^n(b) \mid |z_i| > a_i \text{ for some } 1 \leq i \leq n\}.$$

In [13] the following extension theorem for coherent analytic sheaves is proved:

---

[1] Partially supported by NSF Grant GP-7265.

Theorem (0.1). *Suppose* D *is an open subset of* $\mathbb{C}^n$, $0 \leqq a < b$ *in* $\mathbb{R}^N$, *and* F *is a coherent analytic sheaf on* $D \times G^N(a,b)$ *such that* $F^{[n+1]} = F$. *Then* F *can be uniquely extended to a coherent analytic sheaf* $\tilde{F}$ *on* $D \times K^N(b)$ *satisfying* $\tilde{F}^{[n+1]} = \tilde{F}$.

The subsheaf version of Theorem (0.1) is the following:

Corollary (0.2). *Suppose* D *is an open subset of* $\mathbb{C}^n$, $0 \leqq a < b$ *in* $\mathbb{R}^N$, G *is a coherent analytic sheaf on* $D \times K^N(b)$, *and* F *is a coherent analytic subsheaf of* $G|D \times G^N(a,b)$ *such that* $F_{[n+1]G} = F$. *Then* F *can be uniquely extended to a coherent analytic subsheaf* $\tilde{F}$ *of* G *on* $D \times K^N(b)$ *satisfying* $\tilde{F}_{[n+1]G} = \tilde{F}$.

In [15] the following stronger subsheaf extension theorem is proved:

Theorem (0.3). *Suppose* $0 \leqq a < b$ *in* $\mathbb{R}^N$, D *is a domain in* $\mathbb{C}^n$, *and* D' *is a non-empty open subset of* D. *Suppose* G *is a coherent analytic sheaf on* $D \times K^N(b)$ *and* F *is a coherent analytic subsheaf of* G *on* $(D \times G^N(a,b)) \cup (D' \times K^N(b))$ *such that* $F_{[n]G} = F$. *Then* F *can be uniquely extended to a coherent analytic subsheaf* $\tilde{F}$ *of* G *on* $D \times K^N(b)$ *satisfying* $\tilde{F}_{[n]G} = \tilde{F}$.

The abstract sheaf version of Theorem (0.3) is conjectured but has not been proved.

In this paper we are concerned with another type of extension theorem for coherent analytic sheaves. This theorem is somewhat related in spirit to the Osgood lemma [6, I.A.2] that a continuous function holomorphic in each variable separately is holomorphic. Therefore, we call this theorem an Osgood type extension theorem. A special subsheaf case of this theorem is considered in [15]. In order to state the theorem, we introduce the following notation: If T and X are complex spaces, $t \in T$, and F is a coherent analytic sheaf on $T \times X$, then $F(t)$ denotes F/IF, where I is the ideal-sheaf on $T \times X$ for $\{t\} \times X$.

Main Theorem. Suppose D is an open subset of $\mathbb{C}^n$, $0 \leqq a < b$ in $\mathbb{R}^N$, and F is a coherent analytic sheaf on $D \times G^N(a,b)$ such that $F^{[n]} = F$. Suppose that, for every $t \in D$, $F(t)$ can be extended to a coherent analytic sheaf on $\{t\} \times K^N(b)$. Then F can be uniquely extended to a coherent analytic sheaf $\tilde{F}$ on $D \times K^N(b)$ satisfying $\tilde{F}^{[n]} = \tilde{F}$.

In this paper complex spaces are not necessarily reduced and they have countable topology. A Stein open covering of a complex space means a countable open covering whose members are all Stein subsets. A coherent analytic sheaf F on a complex space X is canonically identified with $F|\text{Supp } F$, where $\text{Supp } F$ is the support of F. We denote by $S_k(F)$ the set

$$\{x \in X \mid \text{codh } F_x \leqq k\},$$

where $\text{codh } F_x$ is the homological codimension of F_x. $S_k(F)$ is always a subvariety of dimension $\leqq k$ [11, p. 81, Satz 5]. If T is a complex space and $t \in T$, then $\{t\} \times X$ is canonically identified with X ${}_nO$ denotes the structure sheaf of $\mathbb{C}^n$.

$\mathbb{R}_+ = \{c \in \mathbb{R} \mid c > 0\}$. $\mathbb{N}$ = the set of all natural numbers. $\mathbb{N}_*$ = the set of all nonnegative integers. All Banach spaces are over $\mathbb{C}$. For a subset E of a topological space, ∂E denotes the boundary of E.

Results concerning absolute and relative gap-sheaves can be found respectively in [12] and [16]. For a coherent analytic sheaf F the condition $F^{[n]} = F$ is equivalent to $\dim S_{k+2}(F) \leqq k$ for $-1 \leqq k < n$ ([11] and [13, p. 134, Prop. 19]; cf. also [17, (2.5)]).

The author wishes to thank G. Trautmann for many useful conversations.

1. *Bundles with Banach space fibers*

If X is a complex manifold and E is a Banach space, then we denote $X \times E$ by E_X. E_X is a trivial bundle with Banach space fibers. For $x \in X$, $\{x\} \times E$ is denoted by E_x and is canonically identified with E. If F is another Banach space, a map $v\colon E_X \to F_X$ is called a *homomorphism* if

(i) v is holomorphic when E_X and F_X are regarded naturally as Banach-analytic manifolds, and

(ii) for every $x \in X$, v induces a linear map $v_x\colon E_x \to F_x$.

In other words, $v = \{v_x\}_{x \in X}$ is a family of continuous linear maps from E to F, holomorphically parametrized by X. v is called *direct* if in addition it satisfies the condition that Ker v_x and Im v_x are respectively direct closed subspaces of E_x and F_x for $x \in X$. If Y is an open subset of X, then we denote by v_Y the map $E_Y \to F_Y$ induced by v.

Proposition (1.1). *Suppose X is a connected complex manifold and E, F are Banach spaces. Suppose $v\colon E_X \to F_X$ is a direct homomorphism such that* dim Ker $v_x < \infty$ *for every $x \in X$. Then there exists a subvariety A of codimension ≥ 1 in X such that* Ker v_{X-A} *is a holomorphic vector bundle over $X - A$ with finite-dimensional fibers.*

Proof. For $k \in \mathbb{N}_*$ let

$$V_k = \{x \in X \mid \dim \operatorname{Ker} v_x \geq k\}.$$

Consider the following sequence of homomorphisms:

$$0_X \xrightarrow{u} E_X \xrightarrow{v} F_X ,$$

where 0_X represents the product of X with the 0-dimensional Banach space and u is the zero map. By [2, p. 337, Prop. (VI. 2)], V_k is a subvariety of X for every $k \in \mathbb{N}_*$. Let k_o be the largest nonnegative integer such that $V_{k_o} = X$. Let $A = V_{k_o+1}$. We claim that A satisfies the requirement. Clearly A is a subvariety of codimension ≥ 1 in X. For $x \in X - A$, $\dim \operatorname{Ker} v_x = k_o$.

To verify the local triviality of $\operatorname{Ker} v_{X-A}$, take $x_o \in X - A$. Let E' be a closed subspace of E such that $E = \operatorname{Ker} v_{x_o} \oplus E'$. Let F' be a closed subspace of F such that $F = \operatorname{Im} v_{x_o} \oplus F'$. Define $\tilde{v}: E'_X \oplus F'_X \to F_X$ by $\tilde{v}_x(a \oplus b) = v_x(a) + b$. Clearly $\tilde{v}_{x_o}$ is an isomorphism. By [1, p. 21, Prop. 3] there exists an open neighborhood U of x_o in $X - A$ such that $\tilde{v}_U: E'_U \oplus F'_U \to F_U$ is an isomorphism, i.e., $\tilde{v}_U^{-1}$ is a homomorphism of trivial bundles with Banach space fibers. For $x \in U$, we have $\operatorname{Ker} v_x \cap E' = 0$. Let $p: E \to \operatorname{Ker} v_{x_o}$ be the projection parallel to E'. Then $p|\operatorname{Ker} v_x$ is injective for $x \in U$. For $x \in U$,

$$\dim p(\operatorname{Ker} v_x) = \dim \operatorname{Ker} v_x = k_o = \dim \operatorname{Ker} v_{x_o}.$$

Therefore $p(\operatorname{Ker} v_x) = \operatorname{Ker} v_{x_o}$ for $x \in U$. Hence $E = \operatorname{Ker} v_x \oplus E'$ for $x \in U$. $v_x(E) = v_x(E')$ for $x \in U$. Let $q: E'_U \oplus F'_U \to F'_U$ be the projection onto the second summand. Consider the following sequence:

$$(*) \qquad E_U \xrightarrow{v_U} F_U \xrightarrow{q(\tilde{v}_U)^{-1}} F'_U .$$

Since $v_x(E) = v_x(E')$ for $x \in U$, we have

$$\tilde{v}_x^{-1}\, v_x(E) = E' \oplus 0 = \operatorname{Ker} q_x$$

for $x \in U$. Hence $v_x(E) = \operatorname{Ker} q_x \tilde{v}_x^{-1}$. The sequence (*) is exact. Since both homomorphisms v_U and $q(\tilde{v}_U)^{-1}$ in (*) are direct, by [1, p. 21, Prop. 3] there exist an open neighborhood W of x_o in U and a Banach space G such that $\operatorname{Ker} v_W$ is isomorphic to G_W. Q.E.D.

2. *Square integrable cohomology*

A. *Lemma* (2.1). *Suppose* D_i *is a bounded open subset of* $\mathbb{C}$ $(1 \leqq i \leqq n)$, G *is a bounded open subset of* $\mathbb{C}^k$, *and* v *is a non-negative continuous plurisubharmonic function on* $G \times D_1 \times \ldots \times D_n$. *Suppose for every* $x \in \partial D_1 \times \ldots \times \partial D_n$ *there exists an open neighborhood* U *of* x *in* $\mathbb{C}^n$ *such that* v *is integrable on* $G \times (U \cap (D_1 \times \ldots \times D_n))$. *Then* v *is integrable on* $G \times D_1 \times \ldots \times D_n$.

Proof. We prove by induction on n. The case $n = 0$ is trivial. Suppose $n \geqq 1$.

Fix $y \in \partial D_n$. By virtue of the assumption on v, for every $t \in \partial D_1 \times \ldots \times \partial D_{n-1}$ we can find an open neighborhood $H_{y,t}$ of y in $\mathbb{C}$ and an open neighborhood U_t of t in $\mathbb{C}^{n-1}$ such that v is integrable on $G \times ((U_t \times H_{y,t}) \cap (D_1 \times \ldots \times D_n))$. Choose $t_1, \ldots, t_l \in \partial D_1 \times \ldots \times \partial D_{n-1}$ such that

$$\partial D_1 \times \ldots \times \partial D_{n-1} \subset \bigcup_{i=1}^{l} U_{t_i} .$$

Let $H_y = \bigcap_{i=1}^{l} H_{y,t_i}$. Since v is integrable on $G \times (U_{t_i} \cap (D_1 \times \ldots \times D_{n-1})) \times (H_y \cap D_n)$ $(1 \leqq i \leqq l)$ and every point of $\partial D_1 \times \ldots \times \partial D_{n-1}$ is contained in some U_{t_i}, by induction hypothesis v is integrable on $G \times D_1 \times \ldots \times D_{n-1} \times (H_y \cap D_n)$. Choose $y_1, \ldots, y_m \in \partial D_n$ such that $\partial D_n \subset \bigcup_{i=1}^{m} H_{y_i}$. Let $Q = D_n \cap (\bigcup_{i=1}^{m} H_{y_i})$. Let $C = \int v dV$, where dV is the Euclidean volume element in $\mathbb{C}^{k+n}$ and the integration is over $G \times D_1 \times \ldots \times D_{n-1} \times Q$. C is a finite number.

Choose $\delta > 0$ such that the distance of $D_n - Q$ from ∂D_n exceeds 3δ. Let D_n' be the set of points in D_n whose distances from ∂D_n exceed δ. Let D_n'' be the set of points in D_n whose distances from ∂D_n exceed 2δ. For $z \in \mathbb{C}$ let $B(z)$ denote the disc of radius δ centered at z. Choose an increasing sequence $\{M_p\}$ of relatively compact open subsets of $G \times D_1 \times \ldots \times D_{n-1}$ whose union is $G \times D_1 \times \ldots \times D_{n-1}$. Define a function $\tilde{v}_p$ on D_n' as follows:

$$\tilde{v}_p(z) = \int_{M_p \times B(z)} v\,dV$$

It is easily verified that $\tilde{v}_p$ is plurisubharmonic on D_n'. Since for $z \in \partial D_n''$

$$M_p \times B(z) \subset G \times D_1 \times \ldots \times D_{n-1} \times Q,$$

we have $\tilde{v}_p(z) \leqq C$ for $z \in \partial D_n''$. Hence $\tilde{v}_p(z) \leqq C$ for $z \in D_n''$. For $z \in D_n''$,

$$\int_{G \times D_1 \times \ldots \times D_{n-1} \times B(z)} v\,dV = \lim_{p \to \infty} \tilde{v}_p(z) \leqq C.$$

Choose $z_1, \ldots, z_r \in D_n''$ such that $D_n - Q \subset \bigcup_{i=1}^{r} B(z_i)$. Then

$$\int_{G \times D_1 \times \ldots \times D_n} v\,dV \leqq \sum_{i=1}^{r} \int_{G \times D_1 \times \ldots \times D_{n-1} \times B(z_i)} v\,dV$$

$$+ \int_{G \times D_1 \times \ldots \times D_{n-1} \times Q} v\,dV \leqq (r+1)C.$$

v is integrable on $G \times D_1 \times \ldots \times D_n$. Q.E.D.

Proposition (2.2). *Suppose D_i is a bounded open subset of $\mathbb{C}$ $(1 \leqq i \leqq n)$ and f is a holomorphic function on $D_1 \times \ldots \times D_n$. Suppose for every $x \in \partial D_1 \times \ldots \times \partial D_n$ there exists an open neighborhood U of x in $\mathbb{C}^n$ such that f is square integrable on $U \cap (D_1 \times \ldots \times D_n)$. Then f is square integrable on $D_1 \times \ldots \times D_n$.*

Proof. Apply Lemma (2.1) with $v = |f|^2$ and $k = 0$. Q.E.D.

Suppose M is an open subset of a complex manifold $(\tilde{M},\mathcal{O})$ of dimension n and $\mathfrak{U} = \{U_\alpha\}$ is an open covering of M satisfying the following conditions:

(i) $U_\alpha \subset\subset \tilde{M}$, and

(ii) there exists a biholomorphic map ϕ_α from an open neighborhood $\tilde{U}_\alpha$ of U_α^- in $\tilde{M}$ onto an open subset of $\mathbb{C}^n$ such that $\phi_\alpha(U_\alpha)$ has the form $D_1^{(\alpha)} \times \dots \times D_n^{(\alpha)}$, where $D_i^{(\alpha)}$ is an open subset of $\mathbb{C}$.

Suppose $\mathcal{F}$ is a locally free analytic sheaf of rank r on $\tilde{M}$ such that there is a sheaf-isomorphism $\lambda_\alpha\colon \mathcal{F} \to \mathcal{O}^r$ on $\tilde{U}_\alpha$. Let $C_0^p(\mathfrak{U},\mathcal{F})$ be the group of all elements $f = \{f_{\alpha_0\dots\alpha_p}\} \in C^p(\mathfrak{U},\mathcal{F})$ such that each of the r components of

$$\lambda_{\alpha_0}(f_{\alpha_0\dots\alpha_p}) \in \Gamma(U_{\alpha_0} \cap \dots \cap U_{\alpha_p}, \mathcal{O}^r)$$

is square integrable on $U_{\alpha_0} \cap \dots \cap U_{\alpha_p}$. Denote

$$\operatorname{Ker}(C_0^p(\mathfrak{U},\mathcal{F}) \xrightarrow{\delta} C_0^{p+1}(\mathfrak{U},\mathcal{F}))$$

by $Z_0^p(\mathfrak{U},\mathcal{F})$ and denote

$$\operatorname{Im}(C_0^{p-1}(\mathfrak{U},\mathcal{F}) \xrightarrow{\delta} C_0^p(\mathfrak{U},\mathcal{F}))$$

by $B_0^p(\mathfrak{U},\mathcal{F})$, where δ is induced by the coboundary operator $C^i(\mathfrak{U},\mathcal{F}) \to C^{i+1}(\mathfrak{U},\mathcal{F})$ $(i = p-1,p)$. Define the *square integrable cohomology groups* $H_0^p(\mathfrak{U},\mathcal{F})$ as follows:

$$H_0^p(\mathfrak{U},\mathcal{F}) = Z_0^p(\mathfrak{U},\mathcal{F})/B_0^p(\mathfrak{U},\mathcal{F}).$$

$H_0^0(\mathfrak{U},\mathcal{F}) = Z_0^0(\mathfrak{U},\mathcal{F})$ is also denoted by $\Gamma_0(\mathfrak{U},\mathcal{F})$. The inclusion map $C_0^p(\mathfrak{U},\mathcal{F}) \hookrightarrow C^p(\mathfrak{U},\mathcal{F})$ induces a natural homomorphism $H_0^p(\mathfrak{U},\mathcal{F}) \to H^p(\mathfrak{U},\mathcal{F})$.

Proposition (2.3) *If the distinguished boundary of* U_α *is contained in M, then the natural homomorphism* $H_o^p(\mathfrak{U},F) \to H^p(\mathfrak{U},F)$ *is injective.*

Proof. Suppose $\{f_{\alpha\beta}\} \in Z_o^1(\mathfrak{U},F)$ and $f_{\alpha\beta} = f_\beta - f_\alpha$ on $U_\alpha \cap U_\beta$ for some $\{f_\alpha\} \in C^0(\mathfrak{U},F)$. We need only prove that $\{f_\alpha\} \in C_o^0(\mathfrak{U},F)$.

Fix U_α and take a point x in the distinguished boundary of U_α. By assumption of the proposition, $x \in U_\beta$ for some $\beta \neq \alpha$. Since $\lambda_\alpha \lambda_\beta^{-1}: \mathcal{O}^r \to \mathcal{O}^r$ is defined on $\tilde{U}_\alpha \cap \tilde{U}_\beta$ which contains $U_\alpha \cap U_\beta$ as a relatively compact open subset, for some $C \in \mathbb{R}_+$ we have

$$\|\lambda_\alpha(f_\beta)\| = \|\lambda_\alpha \lambda_\beta^{-1}(\lambda_\beta(f_\beta))\| \leqq C\|\lambda_\beta(f_\beta)\| \quad \text{on} \quad U_\alpha \cap U_\beta \ ,$$

where $\|g\|$ means $(\sum_{i=1}^{r} |g_i|^2)^{\frac{1}{2}}$ if $g = (g_1,\ldots,g_r)$. Let ω be a C^∞ (n,n)-form on $\tilde{M}$ which serves as the volume element. Choose a relatively compact open neighborhood W of x in U_β. From $f_{\alpha\beta} = f_\beta - f_\alpha$ on $U_\alpha \cap U_\beta$, we obtain

$$\int_{U_\alpha \cap W} \|\lambda_\alpha(f_\alpha)\|^2 \omega \leqq \int_{U_\alpha \cap U_\beta} \|\lambda_\alpha(f_{\alpha\beta})\|^2 \omega + C \int_W \|\lambda_\beta(f_\beta)\|^2 \omega < \infty \ .$$

Since x is an arbitrary point in the distinguished boundary of U_α, by Proposition (2.2), each of the r components of $\lambda_\alpha(f_\alpha)$ is square integrable on U_α. Hence $\{f_\alpha\} \in C_o^0(\mathfrak{U},F)$. Q.E.D.

B. Suppose $0 < a < b < c$ in $\mathbb{R}^2$. Let

$$X_1 = \bigcup_{i=1}^{2} \{(z_1,z_2) \in \mathbb{P}_1 \times \mathbb{P}_1 \mid |z_i| > a_i\} \ ,$$

where z_i can take on the value ∞. Let $X_2 = K^2(b)$. Let $X = X_1 \cup X_2$ and $X_{12} = X_1 \cap X_2$. Then $X = \mathbb{P}_1 \times \mathbb{P}_2$ and $X_{12} = G^2(a,b)$. Let

$$U_1 = \{(z_1,z_2) \in \mathbb{P}_1 \times \mathbb{P}_1 \mid |z_1| > a_1, \ |z_2| < c_2\};$$

$$U_2 = \{(z_1,z_2) \in \mathbb{P}_1 \times \mathbb{P}_1 \mid |z_1| < c_1, \ |z_2| > a_2\};$$

$$U_3 = \{(z_1,z_2) \in \mathbb{P}_1 \times \mathbb{P}_1 \mid |z_1| > b_1, \ |z_2| > b_2\};$$

Let $\mathfrak{U} = \{U_i\}_{i=1}^{3}$. Then $\mathfrak{U}$ is a Stein open covering of X_1. Suppose F is a locally free analytic sheaf on X.

Proposition (2.4). *The quotient topology of* $H^1(\mathfrak{U},F)$ *induced by the Fréchet space* $Z^1(\mathfrak{U},F)$ *is Hausdorff* (i.e., $B^1(\mathfrak{U},F)$ is closed in $Z^1(\mathfrak{U},F)$).

Proof. (a) Let $W_i = \{(z_1,z_2) \in K^2(b) \mid |z_i| > a_i\}$ $(i = 1,2)$. Let $\mathfrak{W} = \{W_1,W_2\}$. Since $(X_1 - X_2)^- \cap (X_2 - X_1)^- = \emptyset$, we can choose Stein open coverings $\mathfrak{V}_i$ of X_i $(i = 1,2)$ such that

(i) $\mathfrak{V}_1$ refines $\mathfrak{U}$,

(ii) $\mathfrak{V}_1 \cap \mathfrak{V}_2$ covers X_{12},

(iii) $\mathfrak{V}_1 \cap \mathfrak{V}_2$ refines $\mathfrak{W}$, and

(iv) if $V_1 \in \mathfrak{V}_1 - \mathfrak{V}_2$ and $V_2 \in \mathfrak{V}_2$ with $V_1 \cap V_2 \neq \emptyset$, then $V_2 \in \mathfrak{V}_1 \cap \mathfrak{V}_2$.

Let $\mathfrak{V} = \mathfrak{V}_1 \cup \mathfrak{V}_2$ and $\mathfrak{V}_{12} = \mathfrak{V}_1 \cap \mathfrak{V}_2$. By virtue of condition (iv), corresponding to the Mayer-Vietoris sequence of F on $X = X_1 \cup X_2$, we have the following exact sequence

$$H^1(\mathfrak{V},F) \to H^1(\mathfrak{V}_1,F) \oplus H^1(\mathfrak{V}_2,F) \to H^1(\mathfrak{V}_{12},F).$$

Since X_2 is Stein, $H^1(\mathfrak{V}_2,F) = 0$. The following sequence

$$H^1(\mathfrak{V},F) \overset{\alpha}{\to} H^1(\mathfrak{V}_1,F) \overset{\beta}{\to} H^1(\mathfrak{V}_{12},F)$$

is exact.

(b) Let $\rho\colon Z^1(\mathfrak{U},F) \to Z^1(\mathfrak{W}_1,F)$ be the restriction map. Let $\eta\colon Z^1(\mathfrak{U},F) \to H^1(\mathfrak{U},F)$ and $\tau\colon Z^1(\mathfrak{W}_1,F) \to H^1(\mathfrak{W}_1,F)$ be the natural maps. Consider the following commutative diagram:

$$\begin{array}{ccc} C^0(\mathfrak{W}_1,F) \oplus Z^1(\mathfrak{U},F) & \xrightarrow{\ \xi\ } & Z^1(\mathfrak{W}_1,F) \\ \sigma\downarrow & & \downarrow\tau \\ H^1(\mathfrak{U},F) & \xrightarrow[\approx]{\ \theta\ } & H^1(\mathfrak{W}_1,F), \end{array}$$

where $\xi(a\oplus b) = \delta(a) + \rho(b)$ and $\sigma(a\oplus b) = \eta(b)$. Since ξ, σ, and τ are continuous and open, the natural isomorphism θ is also a topological isomorphism. To show that $H^1(\mathfrak{U},F)$ is Hausdorff, it suffices to show that $H^1(\mathfrak{W}_1,F)$ is Hausdorff.

(c) We are going to prove that $H^1(\mathfrak{W}_{12},F)$ is Hausdorff. On X_2, F is isomorphic to $\mathcal{O}^r$ for some nonnegative integer r, where $\mathcal{O}$ is the structure sheaf of $\mathbb{P}_1 \times \mathbb{P}_1$ [3, p. 270, Satz 6]. Hence $H^1(\mathfrak{W}_{12},F)$ is topologically isomorphic to $H^1(\mathfrak{W}_{12},\mathcal{O}^r)$. Using an argument entirely analogous to the one used in (b), we derive that $H^1(\mathfrak{W}_{12},\mathcal{O})$ is topologically isomorphic to $H^1(\mathfrak{M},\mathcal{O})$. $H^1(\mathfrak{M},\mathcal{O})$ is Hausdorff, because $Z^1(\mathfrak{M},\mathcal{O})$ is equal to $\Gamma(W_1\cap W_2\,,\mathcal{O})$ and $B^1(\mathfrak{M},\mathcal{O})$ is the set of elements $f\in\Gamma(W_1\cap W_2\,,\mathcal{O})$ whose Laurent series expansions have the form

$$f = \sum \{f_{\nu_1\nu_2}\, z_1^{\nu_1} z_2^{\nu_2} \mid \nu_1 \geqq 0 \ \text{ or } \ \nu_2 \geqq 0\}.$$

Therefore $H^1(\mathfrak{W}_{12},F)$ is Hausdorff.

(d) Since X is compact, $H^1(\mathfrak{W},F)$ is finite-dimensional over $\mathbb{C}$. Since $\beta^{-1}(0) = \operatorname{Im}\alpha$, $\dim \beta^{-1}(0) < \infty$. We can write $\beta^{-1}(0) = \bigoplus_{i=1}^{l} \mathbb{C} v_i$. Let $\tilde\beta\colon Z^1(\mathfrak{W}_1,F) \to Z^1(\mathfrak{W}_{12},F)$ be the restriction map. v_i is induced by some $\tilde v_i \in \tilde\beta^{-1}(B^1(\mathfrak{W}_{12},F))$. Define

$$\gamma\colon \mathbb{C}^l \oplus C^0(\mathcal{W}_1,F) \to \tilde{\beta}^{-1}(B^1(\mathcal{W}_{12},F))$$

by

$$\gamma((a_1,\ldots,a_l) \oplus b) = \sum_{i=1}^{l} a_i\tilde{v}_i + \delta b.$$

Then γ is a continuous linear surjection.

Since $H^1(\mathcal{W}_{12},F)$ is Hausdorff, $\tilde{\beta}^{-1}(B^1(\mathcal{W}_{12},F))$ is a closed subspace of the Fréchet space $Z^1(\mathcal{W}_1,F)$. By the open mapping theorem for Fréchet spaces, $\tilde{\beta}^{-1}(B^1(\mathcal{W}_{12},F))$ is topologically isomorphic to $(\mathbb{C}^l \oplus C^0(\mathcal{W}_1,F))/\text{Ker}\,\gamma$. Hence $B^1(\mathcal{W}_1,F)$ is topologically isomorphic to $((0 \oplus C^0(\mathcal{W}_1,F)) + \text{Ker}\,\gamma)/\text{Ker}\,\gamma$.

Let $p\colon \mathbb{C}^l \oplus C^0(\mathcal{W}_1,F) \to \mathbb{C}^l$ be the projection onto the first summand. Since every linear subspace of $\mathbb{C}^l$ is closed,

$$(0 \oplus C^0(\mathcal{W}_1,F)) + \text{Ker}\,\gamma = p^{-1}(p((0 \oplus C^0(\mathcal{W}_1,F)) + \text{Ker}\,\gamma))$$

is a closed subspace of $\mathbb{C}^l \oplus C^0(\mathcal{W}_1,F)$. Therefore $B^1(\mathcal{W}_1,F)$ is a Fréchet space. $H^1(\mathcal{W}_1,F)$ is Hausdorff. By (b), $H^1(\mathfrak{U},F)$ is Hausdorff. Q.E.D.

We can find a biholomorphic map ϕ_i from an open neighborhood of U_i^- in X onto an open subset of $\mathbb{C}^2$ such that $\phi_i(U_i)$ is a polydisc. Moreover, we can find a sheaf-isomorphism $F \to \mathcal{O}^r$ on an open neighborhood of U_i^- in X. Since the distinguished boundary of U_i is contained in X_1, by combining Propositions (2.3) and (2.4), we obtain the following:

Proposition (2.5). *The quotient topology of* $H_0^1(\mathfrak{U},F)$ *induced by the Hilbert space* $Z_0^1(\mathfrak{U},F)$ *is Hausdorff.*

3. *Uniqueness*

Suppose X is an open subset of a complex space $\tilde{X}$ and F is a coherent analytic sheaf on X. Suppose $\tilde{F}$ is a coherent analytic sheaf on $\tilde{X}$ extending F and satisfying a certain condition (C). We say that the extension $\tilde{F}$ is *unique* (up to isomorphism) if for any other coherent analytic extension $\hat{F}$ of F on $\tilde{X}$ satisfying (C) there exists uniquely a sheaf-isomorphism $\tilde{F} \to \hat{F}$ on $\tilde{X}$ whose restriction to X is the identity map of F.

Suppose $\rho' < \rho$ in $\mathbb{R}^n_+$ and $0 \leqq a < b$ in $\mathbb{R}^N$. Let $Q = K^n(\rho) \times K^N(b)$ and

$$Q' = (K^n(\rho) \times G^N(a,b)) \cup (K^n(\rho') \times K^N(b)).$$

Suppose D is a domain in $\mathbb{C}^n$ and D' is a non-empty open subset of D. Let $R = D \times K^N(b)$ and $R' = (D \times G^N(a,b)) \cup (D' \times K^N(b))$.

From [12, p. 373, Th. 3] and the exhaustion techniques of [10, §8] we obtain easily the following proposition.

Proposition (3.1). *If* F *is a coherent analytic sheaf on* Q *satisfying* $F^{[n-1]} = F$, *then the restriction map* $\Gamma(Q,F) \to \Gamma(Q',F)$ *is bijective.*

Corollary (3.2). *If* F *is a coherent analytic sheaf on* R *satisfying* $F^{[n-1]} = F$, *then the restriction map* $\Gamma(R,F) \to \Gamma(R',F)$ *is bijective.*

Proposition (3.3). *If* F *is a coherent analytic sheaf on* Q' *satisfying* $F^{[n-1]} = F$, *then any coherent analytic extension* $\tilde{F}$ *on* Q *satisfying* $\tilde{F}^{[n-1]} = \tilde{F}$ *is unique.*

Proof. Suppose $\hat{F}$ is another coherent analytic extension of F on Q satisfying $\hat{F}^{[n-1]} = \hat{F}$. By Proposition (3.1) there exists uniquely an isomorphism $\psi\colon \Gamma(Q,\tilde{F}) \to \Gamma(Q,\hat{F})$ such that $\hat{r}\psi = \tilde{r}$, where

$\tilde{r}: \Gamma(Q,\tilde{F}) \to \Gamma(Q',F)$ and $\hat{r}: \Gamma(Q,\hat{F}) \to \Gamma(Q',F)$ are the restriction maps.

Take $x \in Q$ and $s \in \tilde{F}_x$. Then $s = \sum_i a_i(t_i)_x$ for some $t_i \in \Gamma(Q,\tilde{F})$. Define $\phi(s) = \sum_i a_i\psi(t_i)_x$. It is easy to verify that $\phi: \tilde{F} \to \hat{F}$ is well-defined and is the unique sheaf-isomorphism between $\tilde{F}$ and $\hat{F}$ whose restriction to Q' is the identity map of F (cf. the proof of [13, p. 136, Th. I_n(b)]). Q.E.D.

Corollary (3.4). *If F is a coherent analytic sheaf on R' satisfying $F^{[n-1]} = F$, then any coherent analytic extension $\tilde{F}$ on R satisfying $\tilde{F}^{[n-1]} = \tilde{F}$ is unique.*

Proposition (3.5). *Suppose $a \leqq a' < b$ in $\mathbb{R}^N$.*

(a) *Suppose F is a coherent analytic sheaf on $D \times G^N(a,b)$ such that $F^{[n]} = F$. Then the restriction map $\Gamma(D \times G^N(a,b), F) \to$ $\to \Gamma(D \times G^N(a',b), F)$ is bijective.*

(b) *Suppose F is a coherent analytic sheaf on $D \times G^N(a',b)$ such that $F^{[n]} = F$. Then any coherent analytic extension $\tilde{F}$ on $D \times G^N(a,b)$ satisfying $\tilde{F}^{[n]} = \tilde{F}$ is unique.*

Proof. We can assume that $a'_i = a_i$ for $1 \leqq i \leqq N-1$, because the general case follows from repeating the argument for this special case N times. Let $U = \{(z_1,\ldots,z_N) \in \mathbb{C}^N \mid |z_N| > a_N\}$. The special case follows from the following two statements which are consequences of Corollaries (3.2) and (3.4):

(i) For any coherent analytic sheaf F on $D \times (G^N(a,b) \cap U)$ satisfying $F^{[n]} = F$ the restriction map $\Gamma(D \times (G^N(a,b) \cap U), F) \to$ $\to \Gamma(D \times (G^N(a',b) \cap U), F)$ is bijective.

(ii) Any coherent analytic sheaf $\hat{F}$ on $D \times (G^N(a,b) \cap U)$ extending $F | D \times (G^N(a',b) \cap U)$ and satisfying $\hat{F}^{[n]} = \hat{F}$ is unique. Q.E.D.

Corollary (3.6). *Suppose* F *and* G *are coherent analytic sheaves on* $D \times K^N(b)$ *with* $F^{[n]} = F$. *Then any sheaf-homomorphism* $G \to F$ *on* $D \times G^N(a,b)$ *can uniquely be extended to a sheaf-homomorphism* $G \to F$ *on* $D \times K^N(b)$.

Proof. For any relatively compact open subset U of $D \times K^N(b)$ we have an exact sequence of sheaf-homomorphisms on U of the form

$$_{n+N}O^p \to {}_{n+N}O^q \to G \to 0.$$

We obtain from it the following exact sequence

$$0 \to \operatorname{Hom}_{n+N^O}(G,F) \to F_{(q)} \to F_{(p)},$$

where $F_{(\nu)}$ is the direct sum of ν copies of F. From $F^{[n]}_{(q)} = F_{(q)}$ and $0_{[n]}F_{(p)} = 0$ it follows that $\operatorname{Hom}_{n+N^O}(G,F)^{[n]} = \operatorname{Hom}_{n+N^O}(G,F)$. By Proposition (3.5)(a) the restriction map

$$\Gamma(D \times K^N(b), \operatorname{Hom}_{n+N^O}(G,F)) \to \Gamma(D \times G^N(a,b), \operatorname{Hom}_{n+N^O}(G,F))$$

is bijective. Q.E.D.

Corollary (3.7). *Suppose* F *is a coherent analytic sheaf on* $D \times G^N(a,b)$ *satisfying* $F^{[n]} = F$. *Suppose for every* $t \in D$ *there exists an open neighborhood* U *of* t *and some* $a \leqq a' < b' \leqq b$ *in* $\mathbb{R}^N$ *such that* $F|U \times G^N(a',b')$ *admits a coherent analytic extension on* $U \times K^N(b')$. *Then* F *admits a unique coherent analytic extension* $\tilde{F}$ *on* $D \times K^N(b)$ *satisfying* $\tilde{F}^{[n]} = \tilde{F}$.

Proposition (3.8). *Suppose* Y *is a subvariety of codimension* $\geqq 1$ *in* D *and* F *is a coherent analytic sheaf on* $((D - Y) \times K^N(b)) \cup (D \times G^N(a,b))$ *such that* $F^{[n]} = F$. *Then* F *can be extended to a coherent analytic sheaf on* $D \times K^N(b)$.

Proof. Take arbitrarily $x \in Y$. After a linear transformation of coordinates in $\mathbb{C}^n$ we can assume the following:

(i) $x = 0$,

(ii) $K^n(1,\ldots,1) \subset D$,

(iii) $K^{n-1}(1,\ldots,1) \times G^1(\frac{1}{2},1)$ is disjoint from Y.

By Theorem (0.1) the restriction of F to

$$(K^n(1,\ldots,1) \times G^N(a,b)) \cup (K^{n-1}(1,\ldots,1) \times G^1(\tfrac{1}{2},1) \times K^N(b))$$

can be extended to a coherent analytic sheaf $\tilde{F}$ on $K^n(1,\ldots,1) \times K^N(b)$. Since x is arbitrary, by Corollary (3.7), $F|D \times G^N(a,b)$ can be extended to a coherent analytic sheaf $\tilde{F}$ on $D \times K^N(b)$ satisfying $\tilde{F}^{[n]} = \tilde{F}$. By Corollary (3.7), $\tilde{F}$ agrees with F on $(D - Y) \times K^N(b)$.

Q.E.D.

Proposition (3.9). *Suppose* $a \leqq a' < b$ *in* $\mathbb{R}^N$, G *is a coherent analytic sheaf on* $D \times G^N(a,b)$, *and* F *is a coherent analytic subsheaf of* $G|D \times G^N(a',b)$ *such that* $F_{[n]G} = F$. *Then any coherent analytic subsheaf* $\tilde{F}$ *of* G *on* $D \times G^N(a,b)$ *satisfying* $\tilde{F}_{[n]G} = \tilde{F}$ *is unique.*

Proof. If $\hat{F}$ is a coherent analytic subsheaf of G on $D \times G^N(a,b)$ with $\hat{F}_{[n]G} = \hat{F}$ which extends F and differs from $\tilde{F}$, then the subvariety in $D \times G^N(a,b)$ where $\tilde{F}$ and $\hat{F}$ disagree is disjoint from $D \times G^N(a',b)$ and has dimension $\geqq n+1$, contradicting [13, p. 131, Lemma 5].

Q.E.D.

The following corollary follows from Theorem (0.3):

Corollary (3.10). *Suppose* G *is a coherent analytic sheaf on* $D \times K^N(b)$ *and* F *is a coherent analytic subsheaf of* $G|D \times G^N(a,b)$ *such that* $F_{[n]G} = F$. *If for some* $a \leqq a' < b' \leqq b$ *in* $\mathbb{R}^N$,

$F|D' \times G^N(a',b')$ *can be extended to a coherent analytic subsheaf of* $G|D' \times K^N(b')$, *then* F *can be uniquely extended to a coherent analytic subsheaf* $\tilde{F}$ *of* G *on* $D \times K^N(b)$ *satisfying* $\tilde{F}_{[n]G} = \tilde{F}$.

4. *Subvariety case*

A. Suppose $\pi: X \to Y$ is a holomorphic map of complex spaces. The *rank* of π at $x \in X$, denoted by $\text{rank}_x \pi$, is defined as $\dim_x X - \dim_x \pi^{-1}(\pi(x))$. rank π is defined as $\sup_{x \in X} \text{rank}_x \pi$. If X has pure dimension, then $\{x \in X \mid \text{rank}_x \pi \leq k\}$ is a subvariety of X [7, p. 162]. When X is a manifold and Y is reduced, $\text{rank}_x \pi$ is the same as the rank of the linear map from the tangent space of X at x to the tangent space of Y at $\pi(y)$.

Lemma (4.1). *Suppose U is an open neighborhood of 0 in $\mathbb{C}^n$, X is a complex space, and Y is a relatively compact open subset of X. Suppose $\pi: X \to \mathbb{C}^n$ is a holomorphic map and rank $\pi \leq n-1$. Then after a homogeneous linear transformation of the coordinates system of $\mathbb{C}^n$ there exist $d_1, \ldots, d_{n-1} > 0$, $d_n > d'_n > 0$ such that $K^n(d_1, \ldots, d_n) \subset U$ and $K^{n-1}(d_1, \ldots, d_{n-1}) \times G^1(d'_n, d_n)$ is disjoint from $\pi(Y)$.*

Proof. By shrinking X, we can assume that $\dim X < \infty$. We are going to prove by induction on $\dim X$ that $\pi(X)$ is a countable union of *locally closed* submanifolds of dimension $\leq n-1$ in $\mathbb{C}^n$. Let $X = \bigcup_{i \in I} X_i$ be the decomposition into irreducible branches. Let $r_i = \text{rank}\ \pi | X_i$. X_i can be decomposed as the union of the following three sets:

(i) the singular set S_i of X_i,

(ii) the set T_i of points x in $X_i - S_i$ satisfying $\text{rank}_x\ \pi | X_i < r_i$, and

(iii) $X_i - S_i - T_i$.

By induction hypothesis $\pi(S_i)$ and $\pi(T_i)$ are countable unions of locally closed submanifolds of dimension $\leqq n-1$ in $\mathbb{C}^n$. Since $\pi | X_i - S_i - T_i$ has constant rank r_i on the manifold $X_i - S_i - T_i$, $\pi(X_i - S_i - T_i)$ is a countable union of locally closed submanifolds of dimension r_i in $\mathbb{C}^n$. Hence $\pi(X)$ is a countable union of locally closed submanifolds of dimension $\leqq n-1$ in $\mathbb{C}^n$.

By second category arguments, we can choose the coordinates system in $\mathbb{C}^n$ so that $E := \{(t_1,\ldots,t_n) \in \mathbb{C}^n \mid t_1 = \ldots = t_{n-1} = 0\}$ intersects $\pi(X)$ in a countable set. Consider the map $\Phi: X \cap \pi^{-1}(E) \to \mathbb{C}$ defined by $\Phi(x) = t_n(\pi(x))$. The image of Φ is a countable set. Hence Φ is constant on every branch of $X \cap \pi^{-1}(E)$. Since Y^- intersects only a finite number of branches of $X \cap \pi^{-1}(E)$, $\Phi(Y^- \cap \pi^{-1}(E))$ is a finite set. We can choose $d_n > d_n' > 0$ such that

(i) $G^1(d_n',d_n)^- \cap \Phi(Y^- \cap \pi^{-1}(E)) = \emptyset$, and

(ii) $\{(t_1,\ldots,t_n) \in \mathbb{C}^n \mid t_1 = \ldots = t_{n-1} = 0, \ |t_n| \leqq d_n\} \subset U$.

Hence $\{(t_1,\ldots,t_n) \in \mathbb{C}^n \mid t_1 = \ldots = t_{n-1} = 0, \ d_n' < |t_n| < d_n\}$ is disjoint from $\Phi(Y^-)$. By continuity arguments, we can choose $d_1,\ldots,d_{n-1} > 0$ such that $K^{n-1}(d_1,\ldots,d_{n-1}) \times G^1(d_n',d_n)$ is disjoint from $\pi(Y)$ and $K^n(d_1,\ldots,d_n) \subset U$. Q.E.D.

B. Suppose $\pi: X \to D$ is an analytic cover of λ sheets with critical set A, where D is a domain in $\mathbb{C}^n$. The following two statements are well-known [6, III. B. 12 and 23]:

(i) For every holomorphic function f on X there exist naturally unique holomorphic functions a_i on D $(0 \leqq i \leqq \lambda - 1)$ such that $P_f(\pi(x);f(x)) = 0$ on X, where

$$P_f(t;Z) = Z^\lambda + \sum_{i=0}^{\lambda-1} a_i(t) Z^i.$$

(ii) If f, g are holomorphic functions on X such that g separates all points in $\pi^{-1}(t^{\circ})$ for some $t^{\circ} \in D - A$, then there exist naturally unique holomorphic functions b_i on D $(0 \leq i \leq \lambda - 1)$ such that

$$f(x)\, P_g'(\pi(x); g(x)) = T_{f,g}(\pi(x); g(x)),$$

where $P_g'(t;Z) = \frac{\partial}{\partial Z} P_g(t;Z)$ and

$$T_{f,g}(t;Z) = \sum_{i=0}^{\lambda-1} b_i(t) Z^i.$$

From the above two statements we obtain the following lemma:

Lemma (4.2). *Suppose D is a domain in $\mathbb{C}^n$, X is a subvariety of an open subset Ω of $\mathbb{C}^N$, and $\Pi: \Omega \to D$ is a holomorphic map such that $\Pi | X$ makes X an analytic cover over D. Suppose g is a holomorphic function on Ω which separates all points in $X \cap \Pi^{-1}(t^{\circ})$ for some $t^{\circ} \in D$ not contained in the critical set. Then X is the n-dimensional component of the subvariety of Ω defined by the following equations (where $z = (z_1, \ldots, z_N) \in \Omega$):*

$$P_{g|X}(\Pi(z); g(z)) = 0,$$

$$P_{z_i|X}(\Pi(z); z_i) = 0,$$

$$z_i\, P_{g|X}'(\Pi(z); g(z)) = T_{z_i|X, g|X}(\Pi(z); g(z)), \quad 1 \leq i \leq N.$$

Lemma (4.3). *Suppose $\Omega = \{z \in \mathbb{C}^N \mid |f_i(z)| < 1,\ 1 \leq i \leq k\}$, where f_i is a holomorphic function on $\mathbb{C}^N$. Suppose D is a domain in $\mathbb{C}^m$ and $H \subset \tilde{H}$ are domains in $\mathbb{C}^n$ such that $\tilde{H} - H$ is compact. Suppose $\Pi: \Omega \to \mathbb{C}^{m+n}$ is a holomorphic map and X is a subvariety in $\Pi^{-1}(D \times H)$ such that $\Pi | X$ makes X an analytic cover over $D \times H$. Suppose g is a holomorphic function on $\mathbb{C}^N$ which separates all points in $X \cap \Pi^{-1}(t^{\circ})$ for some $t^{\circ} \in D \times H$ not contained in the critical set.*

Then X can be extended to a subvariety $\tilde{X}$ *in* $\Pi^{-1}(D\times\tilde{H})$ *such that* $\Pi|\tilde{X}$ *makes* $\tilde{X}$ *an analytic cover over* $D\times\tilde{H}$ *if and only if all the coefficients of powers of Z in* $P_{g|X}(t;Z)$, $P_{z_i|X}(t;Z)$, *and* $T_{z_i|X,g|X}(t;Z)$ *can be extended to holomorphic functions on* $D\times\tilde{H}$ *(where* $z_1,\ldots,z_N$ *are the coordinate functions of* $\mathbb{C}^N$ *and* $t\in D\times H$*). If so,* $\tilde{X}$ *is unique.*

Proof. Consider first the special case where $\Omega=\mathbb{C}^{m+n}\times\mathbb{C}^N$ and $\Pi\colon \mathbb{C}^{m+n}\times\mathbb{C}^N\to\mathbb{C}^{m+n}$ is the projection onto the first factor. This special case follows from Lemma (4.2).

Next, consider the case when $\Omega=\mathbb{C}^{m+n}\times Q$ and $\Pi\colon\mathbb{C}^{m+n}\times Q\to\mathbb{C}^{m+n}$ is the projection onto the first factor with $Q=\{z\in\mathbb{C}^N \mid |h_i(z)|<1,\ 1\leqq i\leqq l\}$, where h_i is a holomorphic function on $\mathbb{C}^N$. This case follows from the first special case and from applying the maximum modulus principle and making use of the compactness of $\tilde{H}-H$.

The general case follows from applying the second case to $\Phi(X)$, where $\Phi\colon\Omega\to\mathbb{C}^{m+n}\times\Omega$ is defined by $\Phi(z)=(\Pi(z),z)$. Q.E.D.

Lemma (4.4). *Suppose D is a domain in* $\mathbb{C}^n$, $b\in\mathbb{R}^N_+$, *and* $\Phi\colon D\times K^N(b)\to D\times\mathbb{C}$ *is a holomorphic map defined by* $\Phi(t,z)=(t,f(z))$, *where f is a holomorphic function on* $D\times K^N(b)$. *Suppose X is a subvariety in* $\Phi^{-1}(D\times G^1(\alpha,\beta))$ *(where* $0\leqq\alpha<\beta$ *in* $\mathbb{R}$*) such that* $\Phi|X$ *makes X an analytic cover over* $D\times G^1(\alpha,\beta)$ *with critical set A. Suppose for every* $t\in D$, $(\{t\}\times G^1(\alpha,\beta))\cap A$ *is at most discrete and* $(\{t\}\times K^N(b))\cap X$ *can be extended to a subvariety of* $\Phi^{-1}(\{t\}\times K^1(\beta))$ *which under* Φ *is an analytic cover over* $\{t\}\times K^1(\beta)$. *Then X can be extended uniquely to a subvariety* $\tilde{X}$ *of* $\Phi^{-1}(D\times K^1(\beta))$ *which under* Φ *is an analytic cover over* $D\times K^1(\beta)$. *Moreover, for every compact subset C in D,* $(\tilde{X}-X)\cap(C\times K^N(b))$ *is compact.*

Proof. The existence and uniqueness of $\tilde{X}$ follow from Lemma (4.4) and the fact that a holomorphic function h on $D \times G^1(\alpha,\beta)$ can be extended to a holomorphic function on $D \times K^1(\beta)$ if $h|\{t\} \times G^1(\alpha,\beta)$ can be extended to a holomorphic function on $\{t\} \times K^1(\beta)$ for every $t \in D$. The last statement follows from $(\tilde{X} - X) \cap (C \times K^N(b)) = \Phi^{-1}(C \times (K^1(\beta) - G^1(\alpha,\beta)))$. Q.E.D.

Proposition (4.5). *Suppose D is a domain in $\mathbb{C}^n$ and $0 \leqq a < b$ in $\mathbb{R}^N$. Suppose V is a subvariety of pure dimension $n+1$ in $D \times G^N(a,b)$ such that, for every $t \in D$, $V(t) := V \cap (\{t\} \times G^N(a,b))$ can be extended to a subvariety $\tilde{V}(t)$ in $\{t\} \times K^N(b)$. Then V can be extended uniquely to a subvariety of pure dimension $n+1$.*

Proof. By Corollary (3.10) or [10] we need only show that, for some nonempty open subset U of D and some $a \leqq a' < b' \leqq b$ in $\mathbb{R}^N$, $V \cap (U \times G^N(a',b'))$ can be extended to a subvariety V' in $U \times K^N(b')$.

Let $\pi\colon D \times G^N(a,b) \to D$ be the projection onto the first factor. Let W be the subvariety of V where the rank of $\pi|V$ is $\leqq n-1$. Since the rank of $\pi|W$ is $\leqq n-1$, according to Lemma (4.1), by replacing D by a suitable nonempty open subset and by replacing a and b by some a' and b' satisfying $a < a' < b' < b$, we can assume that $W = \emptyset$. $V(t)$ has pure dimension 1 for $t \in D$. We can assume that $\tilde{V}(t)$ has pure dimension 1 for $t \in D$.

Introduce the following notations: For $t \in D$ and $a \leqq c < d \leqq b$ in $\mathbb{R}^N$

$$V_c^d = V \cap (D \times G^N(c,d)), \qquad \tilde{V}^d(t) = \tilde{V}(t) \cap (D \times K^N(d)),$$

$$\overline{V}_c^d = V \cap (D \times G^N(c,d)^-), \qquad \overline{V}_c^d(t) = \overline{V}_c^d \cap V(t),$$

$$\dot{V}_c = V \cap (D \times \partial K^N(c)), \qquad \dot{V}_c(t) = \dot{V}_c \cap V(t).$$

Fix $t^\circ \in D$ and $a < a' < b' < b$ in $\mathbb{R}^N$. By applying [6, VII. B. 3] to the closed subset $\dot{V}_{a'}(t^\circ)$ of the analytic polyhedron $\tilde{V}^{b'}(t)$ in the complex space $\tilde{V}(t^{\circ})$, we can find a holomorphic function $f_{\circ}$ on $\tilde{V}(t^{\circ})$ and $\beta \in \mathbb{R}_{+}$ such that $f_{\circ}(\dot{V}_{a'}(t^{\circ})) \subset K^{1}(\beta)$ and $f_{\circ}(\dot{V}_{b'}(t^{\circ})) \cap K^{1}(\beta) = \emptyset$. For some $0 < \alpha < \beta$, $f_{\circ}(\dot{V}_{a'}(t^{\circ})) \subset K^{1}(\alpha)$. By replacing D by a Stein open neighborhood of t°, we can extend $f_{\circ}$ to a holomorphic function f on $D \times K^{N}(b)$. By continuity arguments, after replacing D by a suitable open neighborhood of t°, we have $f(\dot{V}_{a'}) \subset K^{1}(\alpha)$ and $f(\dot{V}_{b'}) \cap K^{1}(\beta) = \emptyset$. Since $f(\dot{V}_{a'}) \subset K^{1}(\alpha)$, by the maximum modulus principle applied to $f|\tilde{V}(t)$, we obtain $f(\bar{V}_{a}^{a'}(t)) \subset K^{1}(\alpha)$ for $t \in D$. Hence $f(\bar{V}_{a}^{a'}) \subset K^{1}(\alpha)$.

Let $\Phi\colon D \times K^{N}(\beta) \to D \times \mathbb{C}$ be defined by $\Phi(t,z) = (t, f(z))$. Φ makes $\Phi^{-1}(D \times G^{1}(\alpha,\beta)) \cap V_{a}^{b'}$ an analytic cover over $D \times G^{1}(\alpha,\beta)$. Let A be the critical set. $\Phi^{-1}(A)$ has pure dimension n or is empty. Let Z be the subvariety of $\Phi^{-1}(A)$ where the rank of $\pi|\Phi^{-1}(A)$ is $\leqq n-1$. Since rank $\pi|Z \leqq n-1$, according to Lemma (4.1), by replacing D by a suitable nonempty open subset and by replacing α and β by α' and β' satisfying $\alpha < \alpha' < \beta' < \beta$, we can assume that $Z = \emptyset$. $(\{t\} \times G^{1}(\alpha,\beta)) \cap A$ is at most discrete for $t \in D$. Since $f(\dot{V}_{b'}) \cap K^{1}(\beta) = \emptyset$, Φ makes $\Phi^{-1}(D \times K^{1}(\beta)) \cap \tilde{V}^{b'}(t)$ an analytic cover over $\{t\} \times K^{1}(\beta)$. By Lemma (4.4), $\Phi^{-1}(D \times G^{1}(\alpha,\beta)) \cap (D \times K^{N}(b')) \cap V$ can be extended to a subvariety V^{*} of pure dimension $n+1$ in $\Phi^{-1}(D \times K^{1}(\beta)) \cap (D \times K^{N}(b'))$ which is an analytic cover over $D \times K^{1}(\beta)$ under Φ. Because of the uniqueness statement in Lemma (4.3), for $t \in D$ we have

$$V^{*} \cap (\{t\} \times K^{N}(b')) = \Phi^{-1}(D \times K^{1}(\beta)) \cap \tilde{V}^{b'}(t).$$

Hence $\Phi^{-1}(D \times K^{1}(\beta)) \cap (D \times K^{N}(b')) \cap V \subset V^{*}$. We claim that $\tilde{V} := V \cup V^{*}$ is a subvariety of $D \times K^{N}(b)$ extending $V_{b'}^{b}$.

Since V is a subvariety of $D \times G^{N}(a,b)$ and, for every compact subset C of D, $(C \times K^{N}(b)) \cap (V^{*} - V)$ is a compact subset of

$D \times K^N(b')$, we need only verify that $\hat{V} := V \cap (D \times K^N(b'))$ is a subvariety of $D \times K^N(b')$. Let $L = \{w \in \mathbb{C} \mid |w| > \alpha\}$. Let

$$H_1 = \Phi^{-1}(D \times K^1(\beta)) \cap (D \times K^N(b')),$$

$$H_2 = \Phi^{-1}(D \times L) \cap (D \times K^N(b')).$$

$H_1 \cap \hat{V} = V^*$ is a subvariety of H_1. Since $f(\overline{V}^{a'}_a) \subset K^1(\alpha)$, we have $H_2 \cap \hat{V} = H_2 \cap V = H_2 \cap V^{b'}_{a'}$. $H_2 \cap V^{b'}_{a'}$ is a subvariety of H_2, because $V^{b'}_{a'}$ is a subvariety of $D \times K^N(b') - \dot{V}_{a'}$ and $H_2 \subset D \times K^N(b') - \dot{V}_{a'}$. Since $D \times K^N(b') = H_1 \cup H_2$, $\hat{V}$ is a subvariety of $D \times K^N(b')$. The claim is proved. It is clear that $\tilde{V}$ extends $V^b_{b'}$. Q.E.D.

5. *Subsheaf case*

Suppose $\pi: X \to (Y, {}_Y O)$ is a holomorphic map and F is a coherent analytic sheaf on X. F is said to be *π-flat* at $x \in X$ if F_x, when naturally regarded through π as a ${}_Y O_{\pi(x)}$-module, is a flat ${}_Y O_{\pi(x)}$-module.

Lemma (5.1). *Suppose D and G are respectively open subsets of $\mathbb{C}^n$ and $\mathbb{C}^N$ and $\pi: D \times G \to D$ is the natural projection. Suppose F is a coherent analytic sheaf on $D \times G$ and Z is the set of points in $D \times G$ where F is not π-flat. Then Z is a subvariety in $D \times G$ and* rank $\pi | Z \leqq n - 1$.

Proof. From [14, Lemmas 3 and 4] and [1, p. 66, Prop. 1] we conclude that F is π-flat at $x \in D \times G$ if and only if

$$\dim_x S_k(F) \cap \pi^{-1}\pi(x) \leqq k - n$$

for every k.

For a pure-dimensional subvariety V of $D \times G$ denote by $\sigma_k(V)$

the subvariety of V where the rank of $\pi|V$ is $< k$. It follows that

$$Z = \bigcup\{\sigma_{l-k+n}(V) \mid V \text{ is an } l\text{-dimensional branch of } S_k(F) \text{ for some } k \geqq 0\}.$$

Q.E.D.

Suppose $G \subset \tilde{G}$ are domains in $\mathbb{C}^n$. Suppose S is a coherent analytic subsheaf of ${}_n\mathcal{O}^p|G$ satisfying $S_{[n-1]_n\mathcal{O}^p} = S$. Let $F = {}_n\mathcal{O}^p / S$. F is a torsion-free coherent analytic sheaf on G. Let $s_i \in \Gamma(G,F)$ be the image of $(0,\ldots,0,1,0,\ldots,0)$ under the natural sheaf-epimorphism ${}_n\mathcal{O}^p \to F$, where 1 is in the ith position. It is clear that S can be extended coherently to $\tilde{G}$ as a subsheaf of ${}_n\mathcal{O}^p$ if and only if

(i) F can be extended to a coherent analytic sheaf $\tilde{F}$ on $\tilde{G}$, and

(ii) s_i can be extended to some element of $\Gamma(\tilde{G},\tilde{F})$ $(1 \leqq i \leqq p)$.

Let $r = \operatorname{rank} F$. Select $s_{i_1},\ldots,s_{i_r}$ such that the sheaf-homomorphism $\phi\colon {}_n\mathcal{O}^r \to F$ on G defined by these r sections is injective. Let I be the maximum ideal-sheaf on G such that $IF \subset \operatorname{Im}\phi$. I is coherent and the zero-set A of I has dimension $\leqq n-1$. We call A an *associated pole set* of S. A depends on the choice of $i_1,\ldots,i_r$. Since $IF \subset \operatorname{Im}\phi$, every local section of F can be lifted back through ϕ to a meromorphic section of the trivial vector bundle associated to ${}_n\mathcal{O}^r$. Therefore we have a sheaf-monomorphism $\psi\colon F \to {}_n\mathcal{M}^r$ on G such that

$$\begin{array}{ccc} {}_n\mathcal{O}^r & \xrightarrow{\ \phi\ } & F \\ \downarrow & \swarrow_{\psi} & \\ {}_n\mathcal{M}^r & & \end{array}$$

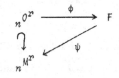

is commutative, where ${}_n\mathcal{M}$ is the sheaf of germs of meromorphic functions on $\mathbb{C}^n$. Let $v_i = \psi(s_i)$. We call $v_1,\ldots,v_p$ a set of *associated meromorphic vector-functions* for S. $v_1,\ldots,v_p$ depend on the choice of the indices $i_1,\dots,i_r$.

If S can be extended to a coherent analytic subsheaf $\tilde{S}$ of ${}_n\mathcal{O}^p|\tilde{G}$, then we can assume that $\tilde{S}_{[n-1]_n\mathcal{O}^p} = \tilde{S}$. We repeat the preceding argument with $\tilde{G} = {}_n\mathcal{O}^p/\tilde{S}$ instead of F (using the same indices $i_1,\dots,i_r$) and obtain associated meromorphic vector-functions $\tilde{v}_1,\dots,\tilde{v}_p$ for $\tilde{S}$. $\tilde{v}_i$ is an r-tuple of meromorphic functions on $\tilde{G}$ extending v_i $(1 \leqq i \leqq p)$.

Conversely, if v_i can be extended to an r-tuple v_i^* of meromorphic functions on G $(1 \leqq i \leqq p)$, then the subsheaf F^* of ${}_nM^r|\tilde{G}$ generated by $v_1^*,\dots,v_p^*$ is coherent. For, if f is a non-identically-zero holomorphic function on some connected open subset U of $\tilde{G}$ such that fv_i^* is an r-tuple of holomorphic functions on U, then $F^* \approx fF^*$ on U and fF^* is the subsheaf of ${}_n\mathcal{O}^r|U$ generated by $fv_1^*,\dots,fv_p^*$. If we identify F with $\psi(F)$, then F^* extends F and v_i^* extends s_i $(1 \leqq i \leqq p)$.

Hence, S can be extended coherently to $\tilde{G}$ as a subsheaf of ${}_n\mathcal{O}^p$ if and only if $v_1,\dots,v_p$ can be extended to r-tuples of meromorphic functions on $\tilde{G}$.

Lemma (5.2). *Suppose D is a domain in $\mathbb{C}^n$ and $0 \leqq \alpha < \beta$ in $\mathbb{R}$. Suppose S is a coherent analytic subsheaf of ${}_{n+1}\mathcal{O}^p$ on $D \times G^1(\alpha,\beta)$ such that $S_{[n]_{n+1}\mathcal{O}^p} = S$ and ${}_{n+1}\mathcal{O}^p/S$ is flat with respect to the natural projection $D \times G^1(\alpha,\beta) \to D$. Suppose A is an associated pole set of S and, for every $t \in D$, $(\{t\} \times G^1(\alpha,\beta)) \cap A$ is at most discrete. If for every $t \in D$, $S(t)$ can be extended coherently to $\{t\} \times K^1(\beta)$ as a subsheaf of ${}_{n+1}\mathcal{O}^p(t)$, then S can be extended coherently to $D \times K^1(\beta)$ as a subsheaf of ${}_{n+1}\mathcal{O}^p$.*

Proof. The lemma follows from the fact that, if a meromorphic function v on $D \times G^1(\alpha,\beta)$ whose restriction to $\{t\} \times G^1(\alpha,\beta)$ can be extended to a meromorphic function on $\{t\} \times K^1(\beta)$ for every

$t \in D$, then v can be extended to a meromorphic function $D \times K^1(\beta)$ [9, p. 84, Hauptsatz]. Q.E.D.

Theorem (5.3). *Suppose D is a domain in $\mathbb{C}^n$ and $0 \leqq a < b$ in $\mathbb{R}^N$. Suppose G is a coherent analytic sheaf on $D \times K^N(b)$ and F is a coherent analytic subsheaf of G on $D \times G^N(a,b)$ such that $F_{[n]G} = F$. If for every $t \in D$, $\operatorname{Im}(F(t) \to G(t))$ can be extended to $\{t\} \times K^N(b)$ as a subsheaf of $G(t)$, then F can be uniquely extended to a coherent analytic subsheaf $\tilde{F}$ of G on $D \times K^N(b)$ satisfying $\tilde{F}_{[n]G} = \tilde{F}$.*

Proof. By Corollary (3.10), we need only prove that for some nonempty open subset U of D and some $a \leqq a' < b' \leqq b$ in $\mathbb{R}^N$, $F|U \times G^N(a',b')$ can be extended coherently to $U \times K^N(b')$ as a subsheaf of G.

By Theorem (0.3), $F_{[n+1]G}$ can be extended to a coherent analytic subsheaf F' of G on $D \times K^N(b)$ satisfying $F'_{[n+1]G} = F'$. By replacing G by F', we can assume that $F_{[n+1]G} = G$. Let $X = \operatorname{Supp} G/F$. X is either empty or a subvariety of pure dimension $n+1$ in $D \times G^N(a,b)$. We assume that $X \neq \emptyset$, because the case where $X = \emptyset$ is trivial. Since $X \cap (\{t\} \times G^N(a,b)) = \operatorname{Supp} G(t)/\operatorname{Im}(F(t) \to G(t))$, $X \cap (\{t\} \times G^N(a,b))$ can be extended uniquely to a subvariety of $\{t\} \times K^N(b)$ for every $t \in D$. By Proposition (4.5), X can be extended to a subvariety $\tilde{X}$ of pure dimension $n+1$ in $D \times K^N(b)$. Let I be the ideal-sheaf of $\tilde{X}$. By replacing D by a nonempty relatively compact open subset and by replacing a and b by some a' and b' satisfying $a < a' < b' < b$, we can assume that $I^k G \subset F$ on $D \times G^N(a,b)$ for some $k \geqq 1$. Let $G' = G/I^k G$ and $F' = F/I^k G$. Then G' (when restricted to $\tilde{X}$) can be regarded as a coherent analytic sheaf on $(\tilde{X}, {}_{n+1}\mathcal{O}/I^k)$ and F' (when restricted to X) can be regarded as a coherent analytic subsheaf of $G'|X$. By replacing F and G by F' and G' respectively, we can assume that $\operatorname{Supp} G = \tilde{X}$.

Let $\pi: D \times K^N(b) \to D$ be the natural projection. Let Y be the subvariety where the rank of $\pi|\tilde{X}$ is $\leqq n-1$. Since rank $\pi|Y \leqq n-1$, by Lemma (4.1), by replacing D by a suitable nonempty open subset and by shrinking b, we can assume that $Y = \emptyset$.

Fix $t^o \in D$ and $a < b^* < b$ in $\mathbb{R}^N$. $\tilde{X} \cap (\{t^o\} \times K^N(b))$ has pure dimension 1. As in the proof of Proposition (4.5), by applying [6, VII. B. 3] to $\tilde{X} \cap (\{t^o\} \times K^N(b))$ and by replacing D by a suitable open neighborhood of t^o, we can find a holomorphic map $F: \tilde{X} \to \mathbb{C}$ and $\alpha < \beta$ in $\mathbb{R}_+$ such that

(i) $F(\tilde{X} \cap (D \times K^N(a)^-)) \subset K^1(\alpha)$ and

(ii) $F(\tilde{X} \cap (D \times \partial K^N(b^*))) \cap K^1(\beta) = \emptyset$.

Let $\Phi: \tilde{X} \to D \times \mathbb{C}$ be defined by $\Phi(x) = (\pi(x), F(x))$. Let

$$Q = \Phi^{-1}(D \times G^1(\alpha,\beta)) \cap (D \times K^N(b^*));$$

$$\tilde{Q} = \Phi^{-1}(D \times K^1(\beta)) \cap (D \times K^N(b^*)).$$

Let $\phi = \Phi|\tilde{Q}$. Then $\phi: \tilde{Q} \to D \times K^1(\beta)$ is proper and nowhere degenerate. Consider the zeroth direct images F^* and G^* of $F|Q$ and $G|\tilde{Q}$ respectively under ϕ. G^* is a coherent analytic sheaf on $D \times K^1(\beta)$ and F^* is a coherent analytic subsheaf of $G^*|D \times G^1(\alpha,\beta)$.

By replacing D by a suitable nonempty open subset and by shrinking β, we can assume that we have a sheaf-epimorphism $\lambda: {}_{n+1}\mathcal{O}^p \to G^*$ on $D \times K^1(\beta)$. Let $S = \lambda^{-1}(F^*)$. $S_{[n]_{n+1}\mathcal{O}^p} = S$. Let A be the associated pole set of S and let Z be the subvariety of $D \times G^1(\alpha,\beta)$ where ${}_{n+1}\mathcal{O}^p/S$ is not p-flat, where $p: D \times G^1(\alpha,\beta) \to D$ is the natural projection. A is either empty or of pure dimension n. Let B be the subvariety of $D \times G'(\alpha,\beta)$ where the rank of $p|A$ is $\leqq n-1$. Since rank $p|Z \cup B \leqq n-1$, according to Lemma (4.1), after replacing D by a suitable nonempty open subset and replacing α and β by some α' and β' satisfying $\alpha < \alpha' < \beta' < \beta$, we can assume that

$Z \cup B = \emptyset$. ${}_{n+1}\mathcal{O}^p/S$ is p-flat and $A \cap (\{t\} \times G^1(\alpha,\beta))$ is at most discrete for every $t \in D$. For $t \in D$, $S(t)$ can be extended coherently to $\{t\} \times K^1(\beta)$ as a subsheaf of ${}_{n+1}\mathcal{O}^p(t)$. By Lemma (5.2), S can be extended coherently to $D \times K^1(\beta)$ as a subsheaf of ${}_{n+1}\mathcal{O}^p$. F^* can be extended coherently to $D \times K^1(\beta)$ as a subsheaf of G^*. F^* is generated on $D \times G^1(\alpha,\beta)$ by elements ξ of $\Gamma(D \times K^1(\beta), G^*)$ which satisfy

$$\xi | D \times G^1(\alpha,\beta) \in \Gamma(D \times G^1(\alpha,\beta), F^*).$$

Hence, $F|Q$ is generated on Q by elements η of $\Gamma(\tilde{Q},G)$ which satisfy $\eta|Q \in \Gamma(Q,F)$. $F|Q$ can be extended to a coherent analytic subsheaf $\hat{F}$ of $G|\tilde{Q}$.

Let $L = \{w \in \mathbb{C} \mid |w| > \alpha\}$. Let $\tilde{F}$ be the coherent analytic subsheaf of G on $\tilde{X}$ which agrees with F on $\Phi^{-1}(D \times L) \cap (D \times G^N(a,b))$ and agrees with $\hat{F}$ on $\Phi^{-1}(D \times K^1(\beta)) \cap (D \times K^N(b^*))$. $\tilde{F}$ is well-defined and extends $F|D \times G^N(b^*,b)$. Q.E.D.

Corollary (5.4). *Suppose D is a domain in $\mathbb{C}^n$ and $0 \leqq a < b$ in $\mathbb{R}^N$. Suppose G is a coherent analytic sheaf on $D \times K^N(b)$ and F is a coherent analytic subsheaf of G on $D \times G^N(a,b)$ such that $G^{[n]} = G$ and $F^{[n]} = F$. If for every $t \in D$, $F(t)$ can be extended to a coherent analytic sheaf $\tilde{F}(t)$ on $\{t\} \times K^N(b)$, then F can be extended to a coherent analytic sheaf on $D \times K^N(b)$.*

Proof. Let Z be the subvariety in $D \times K^N(b)$ where G is not flat with respect to the projection $D \times K^N(b) \to D$. Fix $a < b^* < b$. Fix arbitrarily $t^* \in D$. According to Lemma (4.1), after a linear coordinates transformation in $\mathbb{C}^n$, we can find $d_1,\ldots,d_{n-1} > 0$, $d_n > d_n' > 0$ such that

(i) $t^* = 0$,

(ii) $K^n(d_1,\ldots,d_n) \subset D$,

(iii) $(H \times K^N(b^*)) \cap Z = \emptyset$, where $H = K^{n-1}(d_1,\ldots,d_{n-1}) \times G^1(d_n',d_n)$.

By [13, p. 134, Prop. 19], $G(t)^{[0]} = G(t)$ on $\{t\} \times K^N(b^*)$ for $t \in H$. By Corollary (3.6), for $t \in H$, the sheaf-homomorphism $F(t) \to G(t)$ on $\{t\} \times G^N(a,b^*)$ induced by $F \hookrightarrow G$ can be extended to a sheaf-homomorphism $\tilde{F}(t) \to G(t)$ on $\{t\} \times K^N(b^*)$. Hence, $\operatorname{Im}(F(t) \to G(t))$ can be extended coherently to $\{t\} \times K^N(b)$ as a subsheaf of $G(t)$ for $t \in H$. By Theorem (5.3), $F | H \times G^N(a,b)$ can be extended to a coherent analytic subsheaf S of G on $D \times K^N(b)$ satisfying $S_{[n]G} = S$. Let F^* be the coherent analytic sheaf on $(H \times K^N(b)) \cup (K^n(d_1,\ldots,d_n) \times G^N(a,b))$ which agrees with F on $K^n(d_1,\ldots,d_n) \times G^N(a,b)$ and agrees with S on $H \times K^N(b)$. $(F^*)^{[n]} = F^*$. By Theorem (0.1), F^* can be extended to a coherent analytic sheaf on $K^n(d_1,\ldots,d_n) \times K^N(b)$. Since t^* is an arbitrary point of D, by Proposition (3.7), F can be extended to a coherent analytic sheaf on $D \times K^N(b)$. Q.E.D.

Proposition (5.5). *Suppose D is a domain in $\mathbb{C}^n$ and $0 \leq a < b$ in $\mathbb{R}^N$. Suppose $t^\circ \in D$ and V is a subvariety of dimension $\leq n+1$ in $D \times G^N(a,b)$ such that $V \cap (\{t^\circ\} \times G^N(a,b)) = \emptyset$. Suppose F is a coherent analytic sheaf on $D \times G^N(a,b)$ such that $F^{[n]} = F$. Suppose for every $t \in D$, $F(t)$ can be extended to a coherent analytic sheaf on $\{t\} \times K^N(b)$ and F is generated on $(D \times G^N(a,b)) - V$ by $\Gamma(D \times G^N(a,b), F)$. Then F can be extended to a coherent analytic sheaf on $D \times K^N(b)$.*

Proof. By Proposition (3.7) we can assume without loss of generality the following:

(i) D is Stein,

(ii) there is a sheaf-homomorphism $\lambda\colon {}_{n+N}\mathcal{O}^p \to F$ on $D \times G^N(a,b)$ such that λ is a sheaf-epimorphism on $D \times G^N(a,b) - V$, and

(iii) there is an open neighborhood U of t° such that $V \cap (U \times G^N(a,b)) = \emptyset$.

Let $S = \lambda^{-1}(F)$. Then $S_{[n+1]_{n+N}O^p} = S$. By Corollary (0.2), S can be extended to a coherent analytic subsheaf $\tilde{S}$ of ${}_{n+N}O^p$ on $D \times K^N(b)$ satisfying $\tilde{S}_{[n+1]_{n+N}O^p} = \tilde{S}$. $G := ({}_{n+N}O^p/\tilde{S})^{[n]}$ is a coherent analytic sheaf on $D \times K^N(b)$. $G|D \times G^N(a,b)$ can be canonically embedded as a subsheaf of F and it agrees with F on $D \times G^N(a,b) - V'$, where V' is the $(n+1)$-dimensional component of V. By Theorem (0.3) or [10], condition (iii) implies that V' can be extended to a subvariety $\tilde{V}$ of pure dimension $n+1$ in $D \times K^N(b)$.

Choose a holomorphic function f on $D \times K^N(b)$ such that f vanishes identically on $\tilde{V}$ and does not vanish identically on any branch of $\operatorname{Supp} 0_{[k]F}$ for $k \geqq 0$. This is possible, because all branches of $\operatorname{Supp} 0_{[k]F}$ are either empty or of dimension $\geqq n+2$ and hence can be extended to $D \times K^N(b)$. By replacing D by an *arbitrary* relatively compact open subset and replacing a and b by some a' and b' satisfying $a < a' < b' < b$ and after replacing f by its sufficiently high power, we can assume that $fF \subset G$ on $D \times G^N(a,b)$. Since F is isomorphic to fF under the sheaf-homomorphism defined by multiplication by f [14, Lemma 3], $(fF)^{[n]} = fF$ and $(fF)(t)$ can be extended to a coherent analytic sheaf on $\{t\} \times K^N(b)$ for $t \in D$. By Corollary (5.4), fF can be extended to a coherent analytic sheaf on $D \times K^N(b)$. The proposition follows from $F \approx fF$. Q.E.D.

6. *Locally free case*

A. For $\alpha \in \mathbb{R}^2_+$ let $X_\alpha = \bigcup_{i=1}^{2} \{(z_1,z_2) \in \mathbb{P}_1 \times \mathbb{P}_1 \mid |z_i| > \alpha_i\}$. Fix $a \in \mathbb{R}^2_+$. Let $X = X_a$.

Proposition (6.1). *Suppose D is a Stein domain in $\mathbb{C}^n$ and F is a locally free analytic sheaf on $D \times X$ such that, for every $t \in D$, $F(t)$ can be extended to a coherent analytic sheaf $\tilde{F}(t)$ on $\{t\} \times \mathbb{P}_1 \times \mathbb{P}_1$. Then there exists a subvariety Y of pure dimension 1 in D such that, for $t \in D - Y$, the natural map*

$$\Gamma((D-Y) \times X , F) \to \Gamma(\{t\} \times X , F(t))$$

is surjective.

Proof. We can assume that $\tilde{F}(t)^{[0]} = \tilde{F}(t)$ for $t \in D$. By [13, p. 134, Prop. 19], $\tilde{F}(t)$ is locally free on $\{t\} \times \mathbb{P}_1 \times \mathbb{P}_1$ for $t \in D$.

By [13, p. 133, Lemma 3], for $a < a^*$ in $\mathbb{R}^2_+$ and any open subset D^* of D the following two restriction maps are isomorphisms:

$$\Gamma(D^* \times X , F) \to \Gamma(D^* \times X_{a^*} , F),$$

$$\Gamma(\{t\} \times X , F(t)) \to \Gamma(\{t\} \times X_{a^*} , F(t)), \quad t \in D.$$

Hence, by replacing a by some $a < a^*$ in $\mathbb{R}^2_+$, we can assume that F is the restriction to $D \times X$ of a locally free analytic sheaf $\hat{F}$ on $D \times X_{\hat{a}}$ for some $\hat{a} < a$ in $\mathbb{R}^2_+$.

Let $\mathcal{O}$ and $\tilde{\mathcal{O}}$ be respectively the structure sheaves of $\mathbb{P}_1 \times \mathbb{P}_1$ and $\mathbb{C}^n \times \mathbb{P}_1 \times \mathbb{P}_1$.

Take $a < b < c$ in $\mathbb{R}^2_+$. Let

$$U_1 = \{(z_1,z_2) \in \mathbb{P}_1 \times \mathbb{P}_1 \mid |z_1| > a_1, \ |z_2| < c_2\},$$

$$U_2 = \{(z_1,z_2) \in \mathbb{P}_1 \times \mathbb{P}_1 \mid |z_1| < c_1, \ |z_2| > a_2\},$$

$$U_3 = \{(z_1,z_2) \in \mathbb{P}_1 \times \mathbb{P}_1 \mid |z_1| > b_1, \ |z_2| > b_2\}.$$

Let $\mathfrak{U} = \{U_i\}_{i=1}^3$. Take a contractible Stein open neighborhood $\hat{U}_i$ of U_i^- in $X(\hat{a})$. By [3, p. 270, Satz 6], there exists a sheaf-isomorphism $\phi_i\colon \tilde{O}^r \to \hat{F}$ on $D \times \hat{U}_i$, where $r = \operatorname{rank} F$. For $t \in D$, let $\phi_i(t)\colon \tilde{O}^r(t) \to \hat{F}(t)$ be induced by ϕ_i.

Let $E^0 = \bigoplus_{i=1}^{3} \Gamma_o(U_i, O)$ and $E^1 = \bigoplus_{i<j} \Gamma_o(U_i \cap U_j, O)$. For $t \in D$ consider the following commutative diagram:

$$\begin{array}{ccc} C_o^0(\mathfrak{U}, F(t)) & \xrightarrow{\ \delta_t\ } & C_o^1(\mathfrak{U}, F(t)) \\ \Big\downarrow{\scriptstyle \eta_t^0} & & \Big\downarrow{\scriptstyle \eta_t^1} \\ E^0 & \xrightarrow{\ v_t\ } & E^1 \end{array}$$

where

(i) δ_t is the coboundary operator,

(ii) η_t^0 maps $\{f_i\} \in C_o^0(\mathfrak{U}, F(t))$ to $\bigoplus_{i=1}^{3} \phi_i(t)^{-1} f_i$,

(iii) η_0^1 maps $\{f_{ij}\} \in C_o^1(\mathfrak{U}, F(t))$ to $\bigoplus_{i<j} \phi_i(t)^{-1} f_{ij}$,

(iv) v_t maps $\bigoplus_{i=1}^{3} g_i \in E^0$ to $\bigoplus_{i<j} h_{ij} \in E^1$ with $h_{ij} =$
$= (\phi_i(t)^{-1}\phi_j(t) g_j - g_i) \mid U_i \cap U_j$.

It is easily seen that η_t^i is a topological isomorphism of Hilbert spaces. Identify the *disjoint* union $\bigcup_{t \in D} C_o^i(\mathfrak{U}, F(t))$ with $D \times E^i$ through the bijection which maps $f \in C_o^i(\mathfrak{U}, F(t))$ to $(t, \eta_t^i f) \in D \times E^i$. After this identification, $\bigcup_{t \in D} C_o^i(\mathfrak{U}, F(t))$ is a trivial bundle over D with Hilbert space fibers.

Since $v_t\colon E^0 \to E^1$ is clearly a family of continuous linear maps holomorphically parametrized by $t \in D$, the map

$$\delta\colon \bigcup_{t \in D} C_o^0(\mathfrak{U}, F(t)) \to \bigcup_{t \in D} C_o^1(\mathfrak{U}, F(t))$$

which maps $f \in C_o^0(\mathfrak{U}, F(t))$ to $\delta_t f \in C_o^1(\mathfrak{U}, F(t))$ is a homomorphism of trivial bundles with Hilbert space fibers.

Since $F(t)$ can be extended to a locally free analytic sheaf on $\{t\} \times \mathbb{P}_1 \times \mathbb{P}_1$, by Proposition (2.5), $H_0^1(\mathfrak{U}, F(t))$ is Hausdorff. Hence $\operatorname{Im} \delta_t$ is a closed subspace of the Hilbert space $C_0^1(\mathfrak{U}, F(t))$. Since clearly $\operatorname{Ker} \delta_t$ is a closed subspace of the Hilbert space $C_0^0(\mathfrak{U}, F(t))$, $\operatorname{Im} \delta_t$ and $\operatorname{Ker} \delta_t$ are respectively direct closed subspaces of $C_0^1(\mathfrak{U}, F(t))$ and $C_0^0(\mathfrak{U}, F(t))$. Hence δ is a direct homomorphism of trivial bundles with Hilbert space fibers.

$\operatorname{Ker} \delta_t$ is isomorphic to $\Gamma_0(X, F(t))$. By [13, p. 133, Lemma 6], $\Gamma(X, F(t))$ is isomorphic to $\Gamma(\mathbb{P}_1 \times \mathbb{P}_1, F(t))$. Hence $\operatorname{Ker} \delta_t$ is isomorphic to $\Gamma(X, F(t))$ and $\dim \operatorname{Ker} \delta_t < \infty$.

By Proposition (1.1) there exists a subvariety Y of codimension ≥ 1 in D such that $\operatorname{Ker} \delta_{D-Y}$ is a holomorphic vector bundle over $D - Y$ with finite-dimensional fibers. By replacing Y by a subvariety of pure codimension 1 in D which contains Y, we can assume that Y has pure codimension 1 in D. Since $D - Y$ is Stein, by Cartan's Theorem A, the restriction map

$$\sigma_t : \Gamma(D-Y, \operatorname{Ker} \delta_{D-Y}) \to \operatorname{Ker} \delta_t$$

is surjective for $t \in D - Y$.

Since the restriction map $\Gamma((D-Y) \times X_{\hat{a}}, \hat{F}) \to \Gamma((D-Y) \times X, F)$ is surjective (Proposition (3.5)(a)), $\Gamma(D-Y, \operatorname{Ker} \delta_{D-Y})$ is isomorphic to $\Gamma((D-Y) \times X, F)$. The proposition follows from the following commutative diagram for $t \in D - Y$:

$$\begin{array}{ccc} \Gamma(D-Y, \operatorname{Ker} \delta_{D-Y}) & \xrightarrow{\sigma_t} & \operatorname{Ker} \delta_t \\ \wr\wr & & \wr\wr \\ \Gamma((D-Y) \times X, F) & \xrightarrow{\rho_t} & \Gamma(\{t\} \times X, F(t)), \end{array}$$

where ρ_t is the natural map. Q.E.D.

Suppose L is the sheaf associated to the line bundle of a hyperplane section of $\mathbb{P}_1$. Let L_i be the inverse image of L under $\sigma_i: \mathbb{P}_1 \times \mathbb{P}_1 \to \mathbb{P}_1$, where $\sigma_1: \mathbb{P}_1 \times \mathbb{P}_1 \to \mathbb{P}_1$ and $\sigma_2: \mathbb{P}_1 \times \mathbb{P}_1 \to \mathbb{P}_1$ are respectively projections onto the first and second factor.

Lemma (6.2). *If F is a coherent analytic sheaf on $\mathbb{P}_1 \times \mathbb{P}_1$, then there exist nonnegative integers m_1 and m_2 such that $F \otimes L_1^{m_1} \otimes L_2^{m_2}$ is generated on $\mathbb{P}_1 \times \mathbb{P}_1$ by $\Gamma(\mathbb{P}_1 \times \mathbb{P}_1, F \otimes L_1^{m_1} \otimes L_2^{m_2})$ where $L_i^{m_i}$ denotes the tensor product of L_i tensoring with itself m_i times.*

Proof. By [4, p. 425, Hilfsatz 1] there exists a nonnegative integer m_1 such that local sections of the zeroth direct image G of $F \otimes L_1^{m_1}$ under σ_2 generate $F \otimes L_1^{m_1}$ on $\mathbb{P}_1 \times \mathbb{P}_1$.

Since G is coherent [4, p. 424, Satz II_n^m], by [4, p. 394, Theorem von Serre] there exists a nonnegative integer m_2 such that $\Gamma(\mathbb{P}_1, G \otimes L^{m_2})$ generates $G \otimes L^{m_2}$ on $\mathbb{P}_1$. It is easily verified that m_1 and m_2 satisfy the requirement. Q.E.D.

Proposition (6.3). *Suppose D is an open subset of $\mathbb{C}^n$ and F is a locally free analytic sheaf on $D \times X$ such that, for every $t \in D$, $F(t)$ can be extended to a coherent analytic sheaf on $\{t\} \times \mathbb{P}_1 \times \mathbb{P}_1$. Then F can be extended to a coherent analytic sheaf on $D \times \mathbb{P}_1 \times \mathbb{P}_1$.*

Proof. By virtue of Corollary (3.7) we can assume that D is Stein.

Let $\tilde{L}_i$ be the inverse image of L_i under the natural projection $D \times \mathbb{P}_1 \times \mathbb{P}_1 \to \mathbb{P}_1 \times \mathbb{P}_1$. For $m = (m_1, m_2) \in \mathbb{N}_*^2$, let $F^{(m)} = F \otimes \tilde{L}_1^{m_1} \otimes \tilde{L}_2^{m_2}$. Clearly, for $t \in D$, $F^{(m)}(t)$ can be extended to a coherent analytic sheaf on $\{t\} \times \mathbb{P}_1 \times \mathbb{P}_1$. By Proposition (6.1) there exists a subvariety Y_m of pure codimension 1 in D such that, for $t \in D - Y_m$, the natural map

$$\Gamma((D - Y_m) \times X, F^{(m)}) \to \Gamma(\{t\} \times X, F^{(m)}(t))$$

is surjective.

For $m \in \mathbb{N}_*^2$ let

$$E_m = \{t \in D \mid \Gamma(\{t\} \times X, F^{(m)}(t)) \text{ generates } F^{(m)}(t) \text{ on } \{t\} \times X\}.$$

Since $F(t)$ can be extended to a coherent analytic sheaf on $\{t\} \times \mathbb{P}_1 \times \mathbb{P}_1$, by Lemma (6.2), we have $\bigcup_{m \in \mathbb{N}_*^2} E_m = D$. Since D is a topological space of second category, there exists $m^* \in \mathbb{N}_*^2$ such that $E_{m_*}^-$ contains a nonempty open subset of D. Hence $E_{m_*} \not\subset Y_{m^*}$. Choose $t^* \in E_{m^*} - Y_{m^*}$. Since

$$\Gamma((D - Y_{m_*}) \times X, F^{(m^*)}) \to \Gamma(\{t^*\} \times X, F^{(m^*)}(t^*))$$

is surjective and $\Gamma(\{t^*\} \times X, F^{(m^*)}(t^*))$ generates $F^{(m^*)}(t^*)$ on $\{t^*\} \times X$, it follows from Nakayama's lemma that $\Gamma((D - Y_{m^*}) \times X, F^{(m^*)})$ generates $F^{(m^*)}$ on $\{t^*\} \times X$.

Let V be the subvariety of $(D - Y_{m^*}) \times X$ where $\Gamma((D - Y_{m^*}) \times X, F^{(m^*)})$ fails to generate $F^{(m^*)}$. Then $\dim V \leqq n + 1$ and V is disjoint from $\{t^*\} \times X$. By Proposition (5.5), $F^{(m^*)}$ can be extended to a coherent analytic sheaf $\hat{F}$ on $(D - Y_{m_*}) \times \mathbb{P}_1 \times \mathbb{P}_1$. Without loss of generality we can assume that $\hat{F}^{[n]} = \hat{F}$.

Let $\tilde{F}$ be the coherent analytic sheaf on $(D \times X) \cup ((D - Y_{m^*}) \times \mathbb{P}_1 \times \mathbb{P}_1)$ which agrees with F on $D \times X$ and agrees with $\hat{F}$ on $(D - Y_{m^*}) \times \mathbb{P}_1 \times \mathbb{P}_1$. Since $\tilde{F}^{[n]} = \tilde{F}$, by Proposition (3.8), $\tilde{F}$ can be extended to a coherent analytic sheaf on $D \times \mathbb{P}_1 \times \mathbb{P}_1$. Q.E.D.

8. Suppose G_1 and G_2 are open subsets of $\mathbb{C}^m$ and $S = (s_{ij})$ is a nonsingular $r \times r$ matrix of holomorphic functions on $G_1 \cap G_2$. Denote by $L_S(G_1, G_2)$ the locally free analytic sheaf on $G_1 \cup G_2$ associated to the vector bundle with $\{G_1, G_2\}$ as the covering for a bundle atlas and with S as the transition function. In other words, $L_S(G_1, G_2)$ is characterized by the existence of sheaf-isomorphisms $\phi_i : {}_m\mathcal{O}^r \to L_S(G_1, G_2)$ on G_i $(i = 1,2)$ such that the sheaf-isomorphism $\phi_1^{-1}\phi_2 : {}_m\mathcal{O}^r \to {}_m\mathcal{O}^r$ on $G_1 \cap G_2$ is represented by S (i.e., $\phi_1^{-1}\phi_2$ maps $(u_1, \ldots, u_r) \in {}_m\mathcal{O}^r_x$ to

$$\left(\sum_{j=1}^{r} (s_{1j})_x u_j, \ldots, \sum_{j=1}^{r} (s_{rj})_x u_j\right) \in {}_m\mathcal{O}^r_x$$

for $x \in G_1 \cap G_2$).

Suppose $a < b$ in $\mathbb{R}^2_+$. Let

$$X = \bigcup_{i=1}^{2} \{(z_1, z_2) \in \mathbb{P}_1 \times \mathbb{P}_1 \mid |z_i| > a_i\},$$

$$U_1 = \{(z_1, z_2) \in \mathbb{P}_1 \times \mathbb{P}_1 \mid |z_1| > a_1, \ |z_2| < b_2\},$$

$$U_2 = \{(z_1, z_2) \in \mathbb{P}_1 \times \mathbb{P}_1 \mid |z_1| < b_1, \ |z_2| > a_2\},$$

$$U_3 = \{(z_1, z_2) \in \mathbb{P}_1 \times \mathbb{P}_1 \mid |z_1| > a_1, \ |z_2| > a_2\},$$

$$W_1 = G^1(a_1, b_1) \times K^1(b_2), \qquad W_2 = K^1(b_1) \times G^1(a_2, b_2).$$

Suppose D is a complex manifold. Let $\tilde{U}_i = D \times U_i$ and $\tilde{W}_i = D \times W_i$.

Lemma (6.4). *Suppose D is a contractible Stein domain in $\mathbb{C}^n$. Suppose M is a nonsingular $r \times r$ matrix of holomorphic functions on $\tilde{W}_1 \cap \tilde{W}_2$. Then $L_M(\tilde{W}_1, \tilde{W}_2)$ can be extended to a locally free analytic sheaf on $D \times X$ if and only if there exist nonsingular $r \times r$ matrices of holomorphic functions A, B, P, Q on $\tilde{W}_1$, $\tilde{W}_2$, $\tilde{U}_1 \cap \tilde{U}_3$, $\tilde{U}_2 \cap \tilde{U}_3$ respectively such that $AMB = PQ$ on $\tilde{W}_1 \cap \tilde{W}_2$.*

Proof. Every locally free analytic sheaf F on $D \times X$ corresponds uniquely to a holomorphic vector bundle F over $D \times X$. Since $D \times X = \bigcup_{i=1}^{3} \tilde{U}_i$ and $\tilde{U}_i$ is contractible and Stein, to define a holomorphic vector bundle F with r-dimensional fibers over $D \times X$, $\{\tilde{U}_i\}_{i=1}^{3}$ can be used as the covering for a bundle atlas [3, p. 270, Satz 6]. The transition functions N_{12}, N_{23}, N_{31} are respectively nonsingular $r \times r$ matrices of holomorphic functions on $\tilde{U}_1 \cap \tilde{U}_2$, $\tilde{U}_2 \cap \tilde{U}_3$, $\tilde{U}_3 \cap \tilde{U}_1$ satisfying $N_{12} = N_{13}N_{32}$ on $\tilde{U}_1 \cap \tilde{U}_2 \cap \tilde{U}_3$.

Since $\tilde{W}_i \subset \tilde{U}_i$ $(i = 1,2)$ and $\tilde{W}_1 \cap \tilde{W}_2 = \tilde{U}_1 \cap \tilde{U}_2$, it is clear that $L_{N_{12}}(\tilde{W}_1,\tilde{W}_2) = F|\tilde{W}_1 \cup \tilde{W}_2$. On the other hand, $L_{N_{12}}(\tilde{W}_1,\tilde{W}_2) \approx L_M(\tilde{W}_1,\tilde{W}_2)$ if and only if there exist nonsingular $r \times r$ matrices of holomorphic functions A, B on $\tilde{W}_1$, $\tilde{W}_2$ respectively such that $AMB = N_{12}$ on $\tilde{W}_1 \cap \tilde{W}_2$. The lemma follows if we set $P = N_{13}$ and $Q = N_{32}$. Q.E.D.

Suppose $\gamma < \gamma'$ in $\mathbb{R}_+$. Let $L_1 = K^1(\gamma')$ and $L_2 = \{z \in \mathbb{P} \mid |z| > \gamma\}$.

For an r-tuple $(n_1,\ldots,n_r)$ of integers let $\Lambda_{n_1\ldots n_r}$ denote the following $r \times r$ diagonal matrix of meromorphic functions:

$$\begin{pmatrix} z^{n_1} & & \\ & \ddots & \\ & & z^{n_r} \end{pmatrix}$$

on $\mathbb{P}_1$, where z is the coordinate of $\mathbb{P}_1$.

Define a group $\mathbb{G} = \mathbb{G}(n_1,\ldots,n_r)$ as follows:

(i) $\mathbb{G}$ consists of all couples (p,q), where p and q are nonsingular $r \times r$ matrices of holomorphic functions on L_1 and L_2 respectively, such that $p\Lambda_{n_1\ldots n_r} = \Lambda_{n_1\ldots n_r}q$ on $L_1 \cap L_2$.

(ii) the product of (p,q), $(p',q') \in \mathbb{G}$ is $(pp',qq') \in \mathbb{G}$.

Clearly $\mathbb{G}$ does not depend on the choices of γ and γ'.

For $p = (p_{ij})$ and $q = (q_{ij})$, the equation $p\Lambda_{n_1 \ldots n_r} = \Lambda_{n_1 \ldots n_r} q$ is equivalent to $p_{ij} = z^{n_i - n_j} q_{ij}$ for all i, j. Moreover, $p\Lambda_{n_1 \ldots n_r} = \Lambda_{n_1 \ldots n_r} q$ implies that $\det p = \det q$. Hence $p = (p_{ij})$ is the first member of an element of $\mathbb{G}$ if and only if

(i) p_{ij} is a meromorphic function on $\mathbb{P}_1$ which is holomorphic on $\mathbb{P}_1 - \{\infty\}$ and whose order at ∞ is $\geqq n_j - n_i$, and

(ii) $\det p$ is a nonzero complex constant.

Note that, whenever p satisfies the first of the above two conditions, $\det p$ is always a complex constant, because, if we define an $r \times r$ matrix $q = (q_{ij})$ of holomorphic functions on L_2 by $q_{ij} = z^{n_j - n_i} p_{ij}$ then $\det p = \det q$ is a holomorphic function on $\mathbb{P}_1$.

Lemma (6.5). *$\mathbb{G}$ is a connected complex Lie group.*

Proof. Because of the above considerations, $\mathbb{G}$ is clearly a complex Lie group.

Take arbitrarily $(p,q) \in \mathbb{G}$. Let I be the $r \times r$ identity matrix. For $\lambda \in \mathbb{C}$ define $(p_\lambda, q_\lambda) = (I + \lambda p, I + \lambda q)$. As our earlier observation shows, $\det p_\lambda$ is a complex constant for every fixed $\lambda \in \mathbb{C}$. Hence $\det p_\lambda$ is a holomorphic function on $\mathbb{C}$ when λ is regarded as the variable. Let Z be the zero-set of the holomorphic function $\det p_\lambda$ on $\mathbb{C}$. Since Z does not contain 0 and 1, Z is a closed discrete set in $\mathbb{C}$. We can find a continuous map τ: $[0,1] \to \mathbb{C} - Z$ such that $\tau(0) = 0$ and $\tau(1) = 1$. The continuous map $s \mapsto (p_{\tau(s)}, q_{\tau(s)})$ from $[0,1]$ to $\mathbb{G}$ is a path in $\mathbb{G}$ joining the identity element (I,I) to (p,q). $\mathbb{G}$ is therefore connected.

Q.E.D.

Suppose Ω is a complex manifold. A *holomorphic family* $\mathcal{B}$ *of vector bundles* over $\mathbb{P}_1$ with parameter space Ω means a holomorphic vector bundle over $\Omega \times \mathbb{P}_1$. $\mathcal{B}$ is said to be a *globally trivial family* if $\mathcal{B}$ is obtained by lifting a holomorphic vector bundle over $\mathbb{P}_1$ through the projection $\Omega \times \mathbb{P}_1 \to \mathbb{P}_1$. $\mathcal{B}$ is said to be a *locally trivial family* if every point of Ω admits an open neighborhood U such that the family $\mathcal{B}|U \times \mathbb{P}_1$ with parameter space U is a globally trivial family.

Lemma (6.6). *Suppose* D *is a Stein complex manifold which has the same homotopy type as the 1-sphere. Suppose* $\mathcal{B}$ *is a holomorphic family of vector bundles of rank* r *over* $\mathbb{P}_1$ *with parameter space* D. *If* $\mathcal{B}$ *is a locally trivial family, then* $\mathcal{B}$ *is a globally trivial family.*

Proof. Since every topological *complex* vector bundle over $\tilde{L}_i$ is trivial, by [3, p. 268, Satz I] there exists a nonsingular $r \times r$ matrix M of holomorphic functions on $\tilde{L}_1 \cap \tilde{L}_2$ such that the associated sheaf of $\mathcal{B}$ is $L_M(\tilde{L}_1, \tilde{L}_2)$.

Since $\mathcal{B}$ is a locally trivial family, by [5, p. 126, Cor. to Th. 2.1] we can find

(i) contractible Stein open subsets U_i of D ($i \in \mathbb{N}$),

(ii) integers $n_1 \geqq n_2 \geqq \dots \geqq n_r$,

(iii) nonsingular $r \times r$ matrices of holomorphic functions P_i and Q_i on $U_i \times L_1$ and $U_i \times L_2$ respectively ($i \in \mathbb{N}$) such that $D = \bigcup_{i \in \mathbb{N}} U_i$ and $M = P_i \Lambda_{n_1 \dots n_r} Q_i^{-1}$ on $U_i \times (L_1 \cap L_2)$, where $\Lambda_{n_1 \dots n_r}$ is naturally regarded as a matrix of meromorphic functions on $D \times \mathbb{P}_1$ through the projection $D \times \mathbb{P}_1 \to \mathbb{P}_1$.

On $(U_i \cap U_j) \times (L_1 \cap L_2)$ we have $P_i^{-1} P_j \Lambda_{n_1 \dots n_r} = \Lambda_{n_1 \dots n_r} Q_i^{-1} Q_j$. Hence we can regard $(P_i^{-1} P_j, Q_i^{-1} Q_j)$ as a $\mathbb{G}$-valued holomorphic function on $U_i \cap U_j$. Clearly on $U_i \cap U_j \cap U_k$ the product of the two $\mathbb{G}$-valued holomorphic functions $(P_i^{-1}P_j, Q_i^{-1}Q_j)$ and $(P_j^{-1}P_k, Q_j^{-1}Q_k)$ is the $\mathbb{G}$-valued holomorphic function $(P_i^{-1}P_k, Q_i^{-1}Q_k)$. Therefore, using $\{U_i\}_{i \in \mathbb{N}}$ as the covering for a bundle atlas and $\{(P_i^{-1}P_j, Q_i^{-1}Q_j)\}_{i,j \in \mathbb{N}}$ as ($\mathbb{G}$-valued) transition functions, we can define a holomorphic principal bundle E over D with structure group $\mathbb{G}$. Since $\mathbb{G}$ is connected and D has the same homotopy type as the 1-sphere, all topological principal bundles over D with structure group $\mathbb{G}$ are trivial. By [3, p. 268, Satz I], E is holomorphically trivial. Hence, there exist $\mathbb{G}$-valued holomorphic functions (R_i, T_i) on U_i ($i \in \mathbb{N}$) such that $R_i^{-1}R_j = P_i^{-1}P_j$ on $(U_i \cap U_j) \times L_1$ and $T_i^{-1}T_j = Q_i^{-1}Q_j$ on $(U_i \cap U_j) \times L_2$.

Let P be the nonsingular $r \times r$ matrix of holomorphic functions on $\tilde{L}_1$ which agrees with $P_i R_i^{-1}$ on $U_i \times L_1$. Let Q be the nonsingular $r \times r$ matrix of holomorphic functions on $\tilde{L}_2$ which agrees with $Q_i T_i^{-1}$ on $U_i \times L_2$. Then $M = P\Lambda_{n_1 \ldots n_r} Q^{-1}$ on $\tilde{L}_1 \cap \tilde{L}_2$. B is therefore a globally trivial family. Q.E.D.

Proposition (6.7). *Suppose D is an open subset of $\mathbb{C}^n$ and F is a locally free analytic sheaf of rank r on $D \times G^2(a,b)$. Then, for every $t^0 \in D$, after a linear transformation of coordinates in $\mathbb{C}^n$, there exist $d_1, \ldots, d_{n-1} > 0$, $d_n > d_n' > 0$ satisfying the following conditions.*

(i) $t^0 = 0$,

(ii) $K^n(d_1, \ldots, d_n) \subset D$,

(iii) *for every $t \in K^{n-1}(d_1, \ldots, d_{n-1}) \times G^1(d_n', d_n)$ there exist $a_1 < a_1'(t) < b_1'(t) < b_1$ and an open neighborhood $U(t)$ of t in D such that $F|U(t) \times G^2((a_1'(t), a_2), (b_1'(t), b_2))$ can be extended to a locally free analytic sheaf of rank r on $U(t) \times X_t$, where $X_t =$ $= \{(z_1, z_2) \in \mathbb{P}_1 \times \mathbb{P}_1 \mid |z_1| > a_1'(t)$ or $|z_2| > b_2\}$.*

Proof. Fix $t^{\circ} \in D$. By replacing D by a polydisc neighborhood of t° in D, we can assume that D is Stein and contractible. There exists a nonsingular $r \times r$ matrix M of holomorphic functions on $\tilde{W}_1 \cap \tilde{W}_2$ such that $F = L_M(\tilde{W}_1, \tilde{W}_2)$.

Let $G = L_M(\tilde{W}_1, \tilde{U}_2 \cap \tilde{U}_3)$. G is a locally free analytic sheaf of rank r on $D \times G^1(a_1,b_1) \times \mathbb{P}_1$. Consider the holomorphic vector bundle B over $D \times G^1(a_1,b_1) \times \mathbb{P}_1$ associated to G as a holomorphic family of vector bundles over $\mathbb{P}_1$ with parameter space $D \times G^1(a_1,b_1)$. By [8, p. 451, Cor. 1] there exists a subvariety A of pure codimension 1 in $D \times G^1(a_1,b_1)$ such that the family $B|(D \times G^1(a_1,b_1) - A) \times \mathbb{P}_1$ with parameter space $D \times G^1(a_1,b_1) - A$ is a locally trivial family.

Let $\pi: D \times G^1(a_1,b_1) \to D$ be the natural projection. Take $a_1 < a_1^* < b_1^* < b_1$. Let A' be the subvariety of A where the rank of $\pi|A$ is $\leqq n-1$, by Lemma (4.1), after a linear transformation of coordinates in $\mathbb{C}^n$, we can find $d_1,\ldots,d_{n-1} > 0$, $d_n > d_n' > 0$ such that

(i) $t^{\circ} = 0$,

(ii) $K^n(d_1,\ldots,d_n) \subset D$,

(iii) $A' \cap (H \times G^1(a_1^*,b_1^*)) = \emptyset$, where $H = K^{n-1}(d_1,\ldots,d_{n-1}) \times G^1(d_n',d_n)$.

Take $t \in H$. $A \cap (\{t\} \times G^1(a_1^*,b_1^*))$ is discrete. Hence we can find $a_1^* < a_1'(t) < b_1'(t) < b_1^*$ such that $A \cap (\{t\} \times G^1(a_1'(t),b_1'(t))^-) = \emptyset$. There exists a polydisc neighborhood $U(t)$ of t in D such that $A \cap (U(t) \times G^1(a_1'(t),b_1'(t))^-) = \emptyset$.

Let $L = \{z \in \mathbb{P}_1 \mid |z| > a_2\}$. Since the family $B|U(t) \times G^1(a_1'(t),b_1'(t)) \times \mathbb{P}_1$ with parameter space $U(t) \times G^1(a_1'(t),b_1'(t))$ is a locally trivial family and $U(t) \times G^1(a_1'(t),b_1'(t))$ has the same homotopy type as the 1-sphere, by Lemma (6.6) and its proof, there exist

(i) integers $n_1 \geqq n_2 \geqq \dots \geqq n_r$, and

(ii) nonsingular $r \times r$ matrices of holomorphic functions P and Q on $U(t) \times G^1(a_1'(t), b_1'(t)) \times K^1(b_2)$ and $U(t) \times G^1(a_1'(t), b_1'(t)) \times L$ respectively, such that $M = P(\Lambda_{n_1 \dots n_r} \circ \tau)Q^{-1}$ on $U(t) \times G^1(a_1'(t), b_1'(t)) \times G^1(a_2, b_2)$, where $\tau\colon D \times \mathbb{P}_1 \times \mathbb{P}_1 \to \mathbb{P}_1$ is the projection onto the last factor.

$M = P(\Lambda_{n_1 \dots n_r} \circ \tau)Q^{-1}$ can be written as $P^{-1}MI = (\Lambda_{n_1 \dots n_r} \circ \tau)Q^{-1}$, where I is the $r \times r$ identity matrix. Hence, by Lemma (6.4), $F|U(t) \times G^2((a_1'(t), a_2), (b_1'(t), b_2))$ can be extended to a locally free analytic sheaf on $U(t) \times X_t$. Q.E.D.

Proposition (6.8). *Suppose D is an open subset of $\mathbb{C}^n$ and F is a locally free analytic sheaf on $D \times G^2(a,b)$ such that for every $t \in D$, $F(t)$ can be extended to a coherent analytic sheaf on $\{t\} \times K^2(b)$. Then F can be extended to a coherent analytic sheaf on $D \times K^2(b)$.*

Proof. Fix arbitrarily $t^0 \in D$. By Proposition (6.7), after a linear transformation of coordinates in $\mathbb{C}^n$, we can find $d_1, \dots, d_{n-1} > 0$, $d_n > d_n' > 0$ satisfying the requirements of Proposition (6.7). We use the notations of Proposition (6.7).

Fix arbitrarily $t^* \in H$. $F|U(t^*) \times G^2((a_1'(t^*), a_2), (b_1'(t^*), b_2))$ is extendible to a locally free analytic sheaf G on $U(t^*) \times X_{t^*}$. By Proposition (6.3), G can be extended to a coherent analytic sheaf $\tilde{G}$ on $U(t^*) \times \mathbb{P}_1 \times \mathbb{P}_1$. Since t^* is an arbitrary point of H, by Corollary (3.7), $F|H \times G^2(a,b)$ can be extended to a coherent analytic sheaf $\hat{F}$ on $H \times K^2(b)$ satisfying $\hat{F}^{[n]} = \hat{F}$.

Let $\tilde{F}$ be the coherent analytic sheaf on

$$(K^n(d_1, \dots, d_n) \times G^2(a,b)) \cup (H \times K^2(b))$$

which agrees with F on $K^n(d_1,\ldots,d_n) \times G^2(a,b)$ and agrees with $\hat{F}$ on $H \times K^2(b)$. Since $\tilde{F}^{[n]} = \tilde{F}$, by Theorem (0.1), $\tilde{F}$ can be extended to a coherent analytic sheaf on $K^n(d_1,\ldots,d_n) \times K^2(b)$. Since t° is an arbitrary point of D, by Corollary (3.7), F can be extended to a coherent analytic sheaf on $D \times K^2(b)$. Q.E.D.

Proposition (6.9). *Suppose D is an open subset of $\mathbb{C}^n$ and F is a coherent analytic sheaf on $D \times G^2(a,b)$ with $F^{[n]} = F$ such that, for every $t \in D$, $F(t)$ can be extended to a coherent analytic sheaf on $\{t\} \times K^2(b)$. Then F can be extended to a coherent analytic sheaf on $D \times K^2(b)$.*

Proof. Let V be the subvariety of $D \times G^2(a,b)$ where F is not locally free. Since $F^{[n]} = F$, $\dim V \leq n-1$ [13, p. 134, Prop. 19]. Fix $a < a^* < b^* < b$ in $\mathbb{R}^2$.

Take arbitrarily $t^{\circ} \in D$. By Lemma (4.1), after a linear change of coordinates in $\mathbb{C}^n$, we can find $d_1,\ldots,d_{n-1} > 0$, $d_n > d_n' > 0$ such that

(i) $t^{\circ} = 0$,

(ii) $K^n(d_1,\ldots,d_n) \subset D$, and

(iii) $(H \times G^2(a^*,b^*)) \cap V = \emptyset$, where $H = K^{n-1}(d_1,\ldots,d_{n-1}) \times G^1(d_n',d_n)$.

By Proposition (6.8), $F|H \times G^2(a^*,b^*)$ can be extended to a coherent analytic sheaf $\hat{F}$ on $H \times K^2(b^*)$ satisfying $\hat{F}^{[n]} = \hat{F}$. Let $\tilde{F}$ be the coherent analytic sheaf on $(K^n(d_1,\ldots,d_n) \times G^2(a^*,b^*)) \cup (H \times K^2(b^*))$ which agrees with F on $K^n(d_1,\ldots,d_n) \times G^2(a^*,b^*)$ and agrees with $\hat{F}$ on $H \times K^2(b^*)$. By Theorem (0.1), $\tilde{F}$ can be extended to a coherent analytic sheaf on $K^n(d_1,\ldots,d_n) \times K^2(b^*)$. Since t° is an arbitrary point of D, the proposition follows from Corollary (3.7)

Q.E.D.

7. *General case*

Proposition (7.1). *Suppose D is an open subset of $\mathbb{C}^n$ and $0 \leq a < b$ in $\mathbb{R}^N$. Suppose F is a coherent analytic sheaf on $D \times G^N(a,b)$ such that $\mathsf{F}^{[n]} = \mathsf{F}$ and $0_{[n+2]\mathsf{F}} = \mathsf{F}$. Suppose for every $t \in D$, $\mathsf{F}(t)$ can be extended to a coherent analytic sheaf on $\{t\} \times K^N(b)$. Then F can be extended to a coherent analytic sheaf on $D \times K^N(b)$.*

Proof. We can assume that $\mathsf{F} \neq 0$. Let $X = \operatorname{Supp} \mathsf{F}$. Since $0_{[n+1]\mathsf{F}} = 0$ and $0_{[n+2]\mathsf{F}} = \mathsf{F}$, X has pure dimension $n+2$. X can be extended to a subvariety $\tilde{X}$ of pure dimension $n+2$ in $D \times K^N(b)$. Fix $a < b^* < b$ in $\mathbb{R}^N$.

(a) First consider the case where $\tilde{X} \cap (\{t\} \times K^N(b))$ has pure dimension 2 for $t \in D$. Take arbitrarily $t^\circ \in D$. As in the proof of Proposition (4.5), by applying [6, VII. B. 3] to $\tilde{X} \cap (\{t^\circ\} \times K^N(b))$, we can find

(i) a Stein open neighborhood U of t° in D,

(ii) $\alpha < \beta$ in $\mathbb{R}^2_+$, and

(iii) a holomorphic map $F: U \times K^N(b) \to \mathbb{C}^2$

such that

(i) $F(\tilde{X} \cap (U \times K^N(a)^-)) \subset K^2(\alpha)$, and

(ii) $F(\tilde{X} \cap (U \times \partial K^N(b^*))) \cap K^2(\beta) = \emptyset$.

Let $\Phi: \tilde{X} \cap (U \times K^N(b)) \to U \times \mathbb{C}^2$ be defined by $\Phi(x) = (p(x), F(x))$, where $p: U \times K^N(b) \to U$ is the natural projection. Let

$$Q = \Phi^{-1}(U \times G^2(\alpha,\beta)) \cap (U \times K^N(b^*)),$$

$$\tilde{Q} = \Phi^{-1}(U \times K^2(\beta)) \cap (U \times K^N(b^*)).$$

Let $\phi = \Phi|\tilde{Q}$. Then $\phi\colon \tilde{Q} \to U \times K^2(\beta)$ is proper and nowhere degenerate. Let F^* be the zeroth direct image of $F|Q$ under ϕ. F^* is a coherent analytic sheaf on $U \times G^2(\alpha,\beta)$ and $(F^*)^{[n]} = F^*$. For every $t \in U$, $F^*(t)$ can be extended to a coherent analytic sheaf on $\{t\} \times K^2(\beta)$. By Proposition (6.9), F^* can be extended to a coherent analytic sheaf on $U \times K^2(\beta)$. Hence, $\Gamma(U \times G^2(\alpha,\beta), F^*)$ generates F^* on $U \times G^2(\alpha,\beta)$. $\Gamma(Q,F)$ generates F on Q.

Let I be the ideal-sheaf on $D \times K^N(b)$ for $\tilde{X}$. By shrinking U and replacing α and β by some α' and β' satisfying $\alpha < \alpha' < \beta' < \beta$, we can assume that a finite subset of $\Gamma(Q,F)$ generates F on Q and $I^kF = 0$ on Q for some $k \geqq 1$. Let $O = {}_{n+N}O/I^k$. We have a sheaf-epimorphism $\sigma\colon O^l \to F$ on Q. Consider the zeroth direct images S and T of respectively $\operatorname{Ker}\sigma$ and $O^l|\tilde{Q}$ under ϕ. T is a coherent analytic sheaf on $U \times K^2(\beta)$ and S is a coherent analytic subsheaf of $T|U \times G^2(\alpha,\beta)$. Since $0_{[n+1]T} = 0$, $S_{[n+1]T} = S$. By Corollary (0.2), S can be extended to a coherent analytic subsheaf of T on $U \times K^2(\beta)$. S is generated on $U \times G^2(\alpha,\beta)$ by elements ξ of $\Gamma(U \times K^2(\beta), T)$ satisfying $\xi|U \times G^2(\alpha,\beta) \in \Gamma(U \times G^2(\alpha,\beta), S)$. Hence, $\operatorname{Ker}\sigma$ is generated on Q by elements η of $\Gamma(\tilde{Q},O^l)$ satisfying $\eta|Q \in \Gamma(Q,\operatorname{Ker}\sigma)$. Consequently, $\operatorname{Ker}\sigma$ can be extended to a coherent subsheaf K of $O^l|\tilde{Q}$.

Let $\tilde{F}$ be the coherent analytic sheaf on $\tilde{X} \cap (U \times K^N(b))$ which agrees with F on $\Phi^{-1}(U \times (\mathbb{C}^2 - K^2(\alpha)^-)) \cap (U \times G^N(a,b))$ and agrees with O^l/K on $\tilde{Q}$. $\tilde{F}$ is well-defined and extends $F|U \times G^N(b^*,b)$. Since t° is an arbitrary point of D, it follows from Corollary (3.7) that F can be extended to a coherent analytic sheaf on $D \times K^N(b)$.

(b) For the general case, let $\pi\colon D \times K^N(b) \to D$ be the natural projection. Let A be the subvariety of X where the rank of $\pi|X$ is $\leqq n-1$. Take arbitrarily $t^\circ \in D$. Since $\operatorname{rank} \pi|A \leqq n-1$, by Lemma (4.1), after a linear transformation of coordinates in $\mathbb{C}^n$, we can find $d_1,\ldots,d_{n-1} > 0$, $d_n > d_n' > 0$ such that

(i) $t^\circ = 0$,

(ii) $K^n(d_1,\ldots,d_n) \subset D$,

(iii) $(H \times K^N(b^*)) \cap A = \emptyset$, where $H = K^{n-1}(d_1,\ldots,d_{n-1}) \times G^1(d_n',d_n)$.

For every $t \in H$, $(\{t\} \times K^N(b^*)) \cap \tilde{X}$ has pure dimension 2. By (a), $F|H \times G(a,b^*)$ can be extended to a coherent analytic sheaf on $H \times K^N(b^*)$. As in the proof of Proposition (5.9), it follows from Theorem (0.1) and Corollary (3.7) that F can be extended to a coherent analytic sheaf on $D \times K^N(b)$. Q.E.D.

Suppose $0 \leqq a < b$ in $\mathbb{R}^N$, D is a contractible Stein open subset of $\mathbb{C}^n$, and $\omega \in H^{N-1}(D \times G^N(a,b), {}_{n+N}\mathcal{O})$. For $t \in D$, let $\omega_t \in H^{N-1}(\{t\} \times G, {}_N\mathcal{O})$ be the image of ω under the map

$$H^{N-1}(D \times G, {}_{n+N}\mathcal{O}) \to H^{N-1}(\{t\} \times G, {}_N\mathcal{O})$$

induced by the natural map ${}_{n+N}\mathcal{O} \to {}_{n+N}\mathcal{O}(t) = {}_N\mathcal{O}$. Let $\pi\colon \mathbb{C}^N \to \mathbb{C}$ be defined by $\pi(z_1,\ldots,z_n) = z_1$. Let $\Pi\colon D \times \mathbb{C}^N \to D \times \mathbb{C}$ be defined by $\Pi(t,z) = (t,\pi(z))$ for $t \in D$ and $z \in \mathbb{C}^N$.

Proposition (7.2). *Suppose for every $t \in D$ there exists a holomorphic function f_t on $K^1(b_1)$ such that f_t is nowhere zero on $G^1(a_1,b_1)$ and $(f_t \circ \pi)\omega_t = 0$. Then there exists a holomorphic function F on $D \times K^1(b_1)$ such that F is nowhere zero on $D \times G^1(a_1,b_1)$ and $(F \circ \Pi)\omega = 0$.*

Proof. Let $H = G^1(a_1,b_1) \times \ldots \times G^1(a_N,b_N)$. ω can be represented by a holomorphic function θ on $D \times H$ whose Laurent series expansion in $z_2,\ldots,z_N$ has the form

$$\theta = \sum_{\nu_2,\ldots,\nu_N \leqq -1} \theta_{\nu_2\ldots\nu_N} z_2^{\nu_2} \ldots z_N^{\nu_N},$$

where $z_1,\dots,z_N$ are coordinates of $\mathbb{C}^N$. For $t\in D$, let Θ_t be the restriction of Θ to $\{t\}\times H$. Since $(f_t\circ\pi)\omega_t = 0$, there exist holomorphic functions $g_t^{(i)}$ on

$$\{(z_1,\dots,z_N)\in K^N(b) \mid a_j < |z_j| < b_j \quad \text{for} \quad j\neq i\}$$

such that

$$(*) \qquad (f_t\circ\pi)\Theta_t = \sum_{i=1}^{N} g_t^{(i)} \quad \text{on } H.$$

We can assume that the Laurent series expansion of $g_t^{(1)}$ in $z_2,\dots,z_N$ has the form

$$g_t^{(1)} = \sum_{\nu_2,\dots,\nu_N \leqq -1} g_{t,\nu_2\dots\nu_N}^{(1)} z_2^{\nu_2} \dots z_N^{\nu_N}.$$

Expanding both sides of (*) into Laurent series in $z_2,\dots,z_N$ and equating coefficients, we obtain $(f_t\circ\pi)\Theta_t = g_t^{(1)}$ on H. Hence $\Theta_t = (f_t\circ\pi)^{-1}g_t^{(1)}$ can be extended to a meromorphic function $\tilde{\Theta}_t$ on $K^1(b_1)\times H'$, where

$$H' = G^1(a_2,b_2)\times G^1(a_3,b_3)\times\dots\times G^1(a_N,b_N).$$

The pole set of $\tilde{\Theta}_t$ has the form $E\times H'$, where $E\subset K^1(b_1) - G^1(a_1,b_1)$. Consequently, by [9, p. 84, Hauptsatz], Θ can be extended to a meromorphic function $\tilde{\Theta}$ on $D\times K^1(b_1)\times H'$ and the pole set of $\tilde{\Theta}$ has the form $\tilde{E}\times H'$, where $E'\subset D\times(K^1(b_1) - G^1(a_1,b_1))$.

Choose a holomorphic function F_1 on $D\times K^1(b_1)$ whose zero set is precisely $\tilde{E}$. Take relatively compact nonempty open subsets D^* and H^* of D and H' respectively. There exists a natural number k such that $(F_1\circ\Pi)^k\tilde{\Theta}$ is holomorphic on $D^*\times K^1(b_1)\times H^*$, because $(\tilde{E}\times H')\cap(D^*\times K^1(b_1)\times H^*)$ is relatively compact in $D\times K^1(b_1)\times H'$. Hence, $(F_1\circ\Pi)^k\tilde{\Theta}$ is holomorphic on $D\times K^1(b_1)\times H'$. Let $F = F_1^k$. Then $(F\circ\Pi)\omega = 0$. Q.E.D.

Suppose F is a coherent analytic sheaf on $D \times K^N(b)$ which is flat with respect to the projection $p\colon D \times K^N(b) \to D$ and

$$0 \to {}_{n+N}\mathcal{O}^{q_{N-2}} \to \cdots \to {}_{n+N}\mathcal{O}^{q_1} \to {}_{n+N}\mathcal{O}^{q_0} \to F \to 0$$

is an exact sequence of sheaf-homomorphisms on $D \times K^N(b)$.

Suppose $\xi \in H^1(D \times G^N(a,b), F)$. For $t \in D$, let $\xi_t \in H^1(\{t\} \times G^N(a,b), F(t))$ be the image of ξ under the map

$$\alpha_t\colon H^1(D \times G^N(a,b), F) \to H^1(\{t\} \times G^N(a,b), F(t))$$

induced by the natural map $\beta_t\colon F \to F(t)$.

Proposition (7.3). *Suppose for every* $t \in D$ *there exists a holomorphic function* f_t *on* $K^1(b_1)$ *such that* f_t *is nowhere zero on* $G^1(a_1,b_1)$ *and* $(f_t \circ \pi)\xi_t = 0$. *Then there exists a holomorphic function* F *on* $D \times K^1(b_1)$ *such that* F *is nowhere zero on* $D \times G^1(a_1,b_1)$ *and* $(F \circ \Pi)\xi = 0$.

Proof. Consider the following commutative diagram:

$$\begin{array}{ccccccccccc} 0 & \to & {}_{n+N}\mathcal{O}^{q_{N-2}} & \to & \cdots & \to & {}_{n+N}\mathcal{O}^{q_1} & \to & {}_{n+N}\mathcal{O}^{q_0} & \to & F & \to & 0 \\ & & \Big\downarrow \beta_t^{(N-2)} & & & & \Big\downarrow \beta_t^{(1)} & & \Big\downarrow \beta_t^{(0)} & & \Big\downarrow \beta_t & & \\ 0 & \to & {}_{N}\mathcal{O}^{q_{N-2}} & \to & \cdots & \to & {}_{N}\mathcal{O}^{q_1} & \to & {}_{N}\mathcal{O}^{q_0} & \to & F(t) & \to & 0, \end{array}$$

where $\beta_t^{(i)}\colon {}_{n+N}\mathcal{O}^{q_i} \to {}_{n+N}\mathcal{O}^{q_i}(t) = {}_{N}\mathcal{O}^{q_i}$ is the natural map. The second row is exact, because F is p-flat. Since $H^r(D \times G^N(a,b), {}_{n+N}\mathcal{O}) = H^r(G^N(a,b), {}_{N}\mathcal{O}) = 0$ for $1 \leqq r \leqq n-2$, we obtain the following commutative diagram:

$$\begin{array}{ccc} H^1(D \times G^N(a,b)\,,\,F) & \overset{\approx}{\to} & H^{N-1}(D \times G^N(a,b)\,,\,{}_{n+N}0^{p_{N-2}}) \\ \alpha_t \big\downarrow & & \big\downarrow \gamma_t \\ H^1(\{t\} \times G^N(a,b)\,,\,F(t)) & \overset{\approx}{\to} & H^{N-1}(\{t\} \times G^N(a,b),{}_N 0^{p_{N-2}}), \end{array}$$

where γ_t is induced by $\beta_t^{(N-2)}$. The proposition follows from Proposition (7.2). Q.E.D.

Proof of the Main Theorem. Let $G = 0_{[n+2]F}$ and let $R = F/G$. Let $\pi\colon D \times K^N(b) \to D$ be the natural projection. Fix $a < b^* < b$ in $\mathbb{R}^N$.

(a) Consider first the case where F is π-flat and $R^{[n]} = R$. Since F is π-flat, by [14, Lemmas 2 and 3] and [13, p. 134, Prop. 19] we have $F(t)^{[0]} = F(t)$, $G(t) = 0_{[2]F(t)}$, and $R(t) = F(t)/G(t)$. Hence, for every $t \in D$, $G(t)$ and $R(t)$ can be extended to coherent analytic sheaves on $\{t\} \times K^N(b)$.

$R^{[n+1]}$ is coherent. By Theorem (0.1), $R^{[n+1]}$ can be extended to a coherent analytic sheaf R^* on $D \times K^N(b)$. R can be regarded canonically as a subsheaf of $R^*|D \times G^N(a,b)$. By Corollary (5.4), R can be extended to a coherent analytic sheaf $\tilde{R}$ on $D \times K^N(b)$ satisfying $\tilde{R}^{[n]} = \tilde{R}$.

Since $G^{[n]} = G$ and $0_{[n+2]G} = G$, by Proposition (7.1), G can be extended to a coherent analytic sheaf $\tilde{G}$ on $D \times K^N(b)$ satisfying $\tilde{G}^{[n]} = \tilde{G}$.

Let A' be the subvariety in $D \times K^N(b)$ where $\tilde{R}$ or $\tilde{G}$ is not π-flat. rank $\pi|A' \leqq n-1$ (Lemma (5.1)). Let $A = A' \cup S_{n+1}(\tilde{G})$. Since $\tilde{G}^{[n]} = \tilde{G}$, $\dim S_{n+1}(\tilde{G}) \leqq n-1$. Hence rank $\pi|A \leqq n-1$.

Fix arbitrarily $t^* \in D$. After a linear transformation of coordinates in $\mathbb{C}^n$, we can find $d_1,\dots,d_{n-1} > 0$, $d_n > d_n' > 0$ such that

(i) $t^* = 0$,

(ii) $K^n(d_1,\ldots,d_n) \subset D$, and

(iii) $(H \times G^N(a,b^*)) \cap A = \emptyset$, where $H = K^{n-1}(d_1,\ldots,d_{n-1}) \times G^1(d_n',d_n)$.

Take arbitrarily $t^\circ \in H$. Choose a relatively compact polydisc neighborhood U of t° in H and $a < b' < b^*$ in $\mathbb{R}^N$. Since $\operatorname{codh} \tilde{G} \geqq n+2$ on $H \times K^N(b^*)$, we have an exact sequence of sheaf-homomorphisms

$$0 \to {}_{n+N}\mathcal{O}^{q_{N-2}} \to \cdots \to {}_{n+N}\mathcal{O}^{q_1} \to {}_{n+N}\mathcal{O}^{q_0} \to \tilde{G} \to 0$$

on $U \times K^N(b')$.

Since $\tilde{R}(t)^{[0]} = \tilde{R}(t)$ on $\{t\} \times K^N(b')$ for $t \in H$, by Corollary (3.6) the sheaf-homomorphism $\lambda(t)\colon F(t) \to R(t)$ on $\{t\} \times G^N(a,b')$ induced by the natural map $F \to R$ can be uniquely extended to a sheaf-homomorphism $\tilde{\lambda}(t)\colon \tilde{F}(t) \to \tilde{R}(t)$ on $\{t\} \times K^N(b')$ for $t \in H$, where $\tilde{F}(t)$ is the unique coherent analytic sheaf on $\{t\} \times K^N(b)$ extending $F(t)$ and satisfying $\tilde{F}(t)^{[0]} = \tilde{F}(t)$. For $t \in H$, $\tilde{\lambda}(t)$ yields the following exact sequence:

$$\Gamma(\{t\} \times K^N(b'), \tilde{F}(t)) \xrightarrow{\lambda^*(t)} \Gamma(\{t\} \times K^N(b'), \tilde{R}(t)) \to$$

$$\to \Gamma(\{t\} \times K^N(b'), \operatorname{Coker} \tilde{\lambda}(t)).$$

Since $\lambda(t)$ is surjective, $\operatorname{Supp} \operatorname{Coker} \tilde{\lambda}(t)$ is at most a finite set for $t \in H$. For every $1 \leqq i \leqq N$ and $t \in H$, we can find a holomorphic function $f_t^{(i)}$ on $\mathbb{C}^N$ depending only on z_i (where $z_1,\ldots,z_N$ are coordinates of $\mathbb{C}^N$), vanishing nowhere on $T := G^1(a_1,b_1') \times \cdots \times G^1(a_N,b_N')$, and satisfying $f_t^{(i)} \operatorname{Coker} \tilde{\lambda}(t) = 0$ on $\{t\} \times K^N(b')$. Hence, $f_t^{(i)} \operatorname{Coker} \lambda^*(t) = 0$ for $t \in H$.

For $t \in U$, the following commutative diagram with exact rows on $U \times G^N(a,b')$:

$$\begin{array}{ccccccccc} 0 & \to & G & \to & F & \to & R & \to & 0 \\ & & \downarrow & & \downarrow & & \downarrow & & \\ 0 & \to & G(t) & \to & F(t) & \to & R(t) & \to & 0 \end{array}$$

yields the following commutative diagram with exact rows:

$$\begin{array}{ccccc} \Gamma(U\times G^N(a,b'),F) & \xrightarrow{\beta} & \Gamma(U\times G^N(a,b'),R) & \xrightarrow{\alpha} & H^1(U\times G^N(a,b'),G) \\ & & \Big| & & \Big| \\ \Gamma(\{t\}\times G^N(a,b'),F(t)) & \longrightarrow & \Gamma(\{t\}\times G^N(a,b'),R(t)) & \xrightarrow{\alpha(t)} & H^1(\{t\}\times G^N(a,b'),G(t)) \\ \wr\wr & & \wr\wr & & \\ \Gamma(\{t\}\times K^N(b'),F(t)) & \xrightarrow{\lambda^*(t)} & \Gamma(\{t\}\times K^N(b'),\tilde{R}(t)), & & \end{array}$$

where the two vertical isomorphisms follow from $\tilde{F}(t)^{[0]} = \tilde{F}(t)$ and $\tilde{R}(t)^{[0]} = \tilde{R}(t)$ (Corollary (3.6)). Consequently $f_t^{(i)}$ Im $\alpha(t) = 0$ for $t \in U$ and $1 \leqq i \leqq N$. By Proposition (7.2), for $1 \leqq i \leqq N$ there exists a holomorphic function $F^{(i)}$ on $U \times K^N(b')$ depending only on $t_1,\ldots,t_n,z_i$ (where $t_1,\ldots,t_n$ are coordinates of $\mathbb{C}^n$), vanishing nowhere on $U \times T$, with $F^{(i)}$ Im $\alpha = 0$. Hence $F^{(i)}$ Coker $\beta = 0$ for $1 \leqq i \leqq{<} N$. Since $\tilde{G}$ and $\tilde{R}$ are generated on $U \times K^N(b')$ by global sections and $U \times G^N(a,b')$ contains no common zero of $F^{(i)},\ldots,F^{(N)}$, it follows that F is generated on $U \times G^N(a,b')$ by $\Gamma(U \times G^N(a,b'),F)$. By Proposition (5.5), $F|U \times G^N(a,b')$ can be extended to a coherent analytic sheaf on $U \times K^N(b')$. By Proposition (3.7), $F|H \times G^N(a,b')$ can be extended to a coherent analytic sheaf on $H \times K^N(b')$. As in the proof of Proposition (5.9), it follows from Theorem (0.1) and (3.7) that F can be extended to a coherent analytic sheaf on $D \times K^N(b)$.

(b) For the general case, let Z_1 be the subvariety in $D \times G^N(a,b)$ where F is not π-flat. Let Z_2 be the subvariety in $D \times G^N(a,b)$ where $R^{[n]}$ disagrees with R. Let

$$Z_3 = \bigcup \{x \in Z_2 \mid \text{rank}_x \; \pi|V \leqq n-1 \text{ for some branch } V$$
$$\text{of } Z_2 \text{ containing } x\}.$$

rank $\pi|Z_1 \cup Z_3 \leqq n-1$.

Take arbitrarily $t^* \in D$. By Lemma (4.1), after a linear transformation of coordinates in $\mathbb{C}^n$, we can find $d_1,\ldots,d_{n-1} > 0$, $d_n > d_n' > 0$ such that

(i) $t^* = 0$,

(ii) $K^n(d_1,\ldots,d_n) \subset D$,

(iii) $(H \times G^N(a,b^*)) \cap (Z_1 \quad Z_3) = \emptyset$, where $H =$ $= K^{n-1}(d_1,\ldots,d_{n-1}) \times G^1(d_n',d_n)$.

Take arbitrarily $t^\circ \in D$. Since $(\{t^\circ\} \times G^N(a,b^*)) \cap Z_2$ is at most discrete, we can find an open neighborhood U of t° in H and $a < a' < b' < b^*$ in $\mathbb{R}^N$ such that $(U \times G^N(a',b')) \cap Z_2 = \emptyset$. By (a), $F|U \times G^N(a',b')$ can be extended to a coherent analytic sheaf on $U \times K^N(b')$. By Corollary (3.7), $F|H \times G^N(a,b)$ can be extended to a coherent analytic sheaf on $H \times K^N(b)$. Again, it follows from Theorem (0.1) and Corollary (3.7) that F can be extended to a coherent analytic sheaf on $D \times K^N(b)$. Q.E.D.

References

1. A. Douady, *Le problème des modules pour les sous-espaces analytiques compacts d'un espace analytique donné*, Ann. Inst. Fourier (Grenoble) 16 (1966), 1-95.

2. J. Frisch and J. Guenot, *Prolongement de faisceaux analytiques cohérents*, Invent. Math. 7 (1969), 321-343.

3. H. Grauert, *Analytische Faserungen über holomorph-vollständigen Räumen*, Math. Ann. 135 (1958), 263-273.

4. H. Grauert and R. Remmert, *Bilder und Urbilder analytischer Garben*, Ann. of Math. (2) 68 (1958), 393-443.

5. A. Grothendieck, *Sur la classification des fibrés holomorphes sur la sphère de Riemann*, Amer. J. Math. 79 (1957), 121-138.

6. R. C. Gunning and H. Rossi, *Analytic functions of several complex variables*, Prentice-Hall, Englewood Cliffs, N. J., 1965.

7. N. Kuhlmann, *Algebraic function fields on complex analytic spaces*, Proceedings of the Conference on Complex Analysis (Minneapolis, 1964), Springer-Verlag, Berlin-Heidelberg-New York, 1965, pp. 155-172.

8. H. Röhrl, *On holomorphic families of fiber bundles over the Riemann sphere*, Mem. Coll. Sci., Univ. Kyoto, Ser. A, 33 (1961), 435-477.

9. W. Rothstein, *Ein neuer Beweis des Hartogsschen Hauptsatzes und seine Ausdehnung auf meromorphe Funktionen*, Math. Z. 53 (1950), 84-95.

10. ———, *Zur Theorie der analytischen Mannigfaltigkeiten im Raume von n komplexen Veränderlichen*, Math. Ann. 129 (1955), 96-138.

11. G. Scheja, *Fortsetzungssätze der komplex-analytischen Cohomologie und ihre algebraische Charakterisierung*, Math. Ann. 157 (1964), 75-94.

12. Y.-T. Siu, *Absolute gap-sheaves and extensions of coherent analytic sheaves*, Trans. Amer. Math. Soc. 141 (1969), 361-376.

13. ———, *Extending coherent analytic sheaves*, Ann. of Math. (2) 90 (1969), 108-143.

14. ———, *Analytic sheaves of local cohomology*, Trans. Amer. Math. Soc. 148 (1970), 347-366.

15. Y.-T. Siu and G. Trautmann, *Extension of coherent analytic subsheaves*, Math. Ann. 188 (1970), 128-142.

16. W. Thimm, *Lückengarben von kohärenten analytischen Modulgarben*, Math. Ann. 148 (1962), 372-394.

17. G. Trautmann, *Ein Endlichkeitssatz in der analytischen Geometrie*, Invent. Math. 8 (1969), 143-174.

University of Notre Dame, Notre Dame, Indiana 46556

COMPLEX MANIFOLDS AND UNITARY REPRESENTATIONS

Joseph A. Wolf

1. *Introduction*

Let G_0 be a locally compact group, $\hat{G}_0$ the set of its equivalence classes of irreducible unitary representations. Under certain circumstances one can work out some connections between $\hat{G}_0$ and "partially complex" homogeneous spaces X_0 of G_0, as follows. Let $\mathbb{E} \to X_0$ be a hermitian, G_0-homogeneous, "partially holomorphic" complex vector bundle. Then G_0 acts on the spaces $H_2^{0,q}(\mathbb{E})$ which consist of all measurable q-forms ω on X_0, values in $\mathbb{E}$, such that

(i) ω is harmonic and of type $(0,q)$, with finite L_2-norm, on almost all complex analytic pieces of X_0, and

(ii) the above L_2-norms give a square integrable function on the space of all complex analytic pieces of X_0.

Then, if matters are correctly arranged, the action of G_0 on $H_2^{0,q}(E)$ is an irreducible unitary representation $\pi_{\mathbb{E}}^q$. The immediate problems are the following.

1. Find all partially complex G_0-homogeneous spaces X_0 that carry partially holomorphic hermitian G_0-homogeneous complex vector bundles $\mathbb{E} \to X_0$ such that some $\pi_{\mathbb{E}}^q$ is irreducible and unitary.

2. Given X_0 as above, find all $\mathbb{E} \to X_0$ and q as above.

3. Given X_0, $\mathbb{E}$ and q as above, identify the class $[\pi_{\mathbb{E}}^q] \in \hat{G}_0$.

4. Describe the subset of $\hat{G}_0$ consisting of all classes $[\pi]$ containing an element of the form $\pi_{\mathbb{E}}^q$ above.

5. Given a class $[\pi]$ in that subset of $\hat{G}_0$, find all $(X_0, \mathbb{E}, q)$ such that $\pi^q_{\mathbb{E}} \in [\pi]$.

If G_0 is a compact connected Lie group, these questions are settled by the Bott-Borel-Weil theorem ([2], [13]). If G_0 is a connected semisimple Lie group with finite center, then the case of discrete series representations has almost been settled by Harish-Chandra [7], Narasimhan-Okamoto [16] and (principally) W. Schmid [18].

I will explore these questions for reductive real Lie groups with only finitely many components in [22]. There it turns out that the subset of $\hat{G}_0$ consisting of the $[\pi^q_{\mathbb{E}}]$ is "almost" large enough to support Plancherel measure. Here I want to give a rough but self-contained description of the results of [22] for the case where G_0 is a connected linear semisimple Lie group. That avoids technical problems with the center and the component group of G_0, with the precise character theory developed by Harish-Chandra ([8], [9]), and with the still tentative matter of the various "complementary" series of representations of G_0.

In section 2 we recall some results from [21] concerning G_0-orbits on complex flag manifolds $X = G/P$, where G is the complexification of G_0. The spaces X_0 will be orbits $G_0(x_0) \subset X = G/P$, for various choices of the parabolic subgroup $P \subset G$.

In section 3 we give a brief description of the realization of discrete series representations of G_0 on spaces $H_2^{0,q}(\mathbb{E})$ where $\mathbb{E}$ is a holomorphic vector bundle over an open G_0-orbit.

In section 4 we work out the geometric realization of principal series representations of G_0 on spaces $H_2^{0,0}(\mathbb{E})$ where $\mathbb{E}$ is a hermitian, partially holomorphic, G_0-homogeneous, vector bundle over the closed G_0-orbit on X.

In section 5 we recall Harish-Chandra's construction of certain unitary representations $\pi_{\mu,\eta}$ of G_0, whose classes $[\pi_{\mu,\eta}] \in \hat{G}_0$ support the Plancherel measure there. The discrete and the principal series representations are the two extreme cases.

In section 6 we work out the geometric realization, for *most* of the classes $[\pi_{\mu,\eta}] \in \hat{G}_0$ mentioned just above, on spaces $H_2^{0,q}(\mathbb{E})$, where the bundle $\mathbb{E}$ sits over a measurable integrable G_0-orbit on X.

The incompleteness of our results on geometric realization of the $\pi_{\mu,\eta}$ comes down to some serious analytic problems in the geometric realization of discrete series representations whose lowest highest weight (on a maximal compact subgroup) fails to be "sufficiently" nonsingular. See [16] and [18].

2. *Real group orbits on complex flag manifolds*

The "partially complex" homogeneous spaces of semisimple groups G_0, mentioned in the Introduction, will be G_0-orbits on complex homogeneous spaces of the complexification of G_0. In this section 2 the pertinent results on those G_0-orbits are extracted from [21], omitting most proofs and structural results designed for applications ([22], [23]) more delicate than those that concern us here.

Let G be a complex connected semisimple Lie group. If $P \subset G$ is a complex Lie subgroup then the following conditions are equivalent.

(2.1a) the coset space $X = G/P$ is compact.

(2.1b) $X = G/P$ is a compact simply connected kaehler manifold.

(2.1c) $X = G/P$ is a projective algebraic variety.

(2.1d) $X = G/P$ is a closed G-orbit in a projective representation.

(2.1e) P contains a Borel subgroup of G.

In that case we define

(2.2a) P is a *parabolic subgroup* of G, and

(2.2b) $X = G/P$ is a *complex flag manifold* of G.

We use script for Lie algebras. Thus $\mathcal{G}$ denotes the Lie algebra of G. If $P \subset G$ Lie subgroup then $\mathcal{P} \subset \mathcal{G}$ is the corresponding subalgebra. We say that

(2.3) $\mathcal{P}$ is a *parabolic subalgebra* of $\mathcal{G}$ if P is a parabolic subgroup of G.

In that case P is the analytic subgroup of G for $\mathcal{P}$.

We recall the structure of parabolic subgroups of G and parabolic subalgebras of $\mathcal{G}$. Choose

(2.4a) a Cartan subalgebra $\mathcal{H}$ of $\mathcal{G}$ and

(2.4b) a system Π of simple $\mathcal{H}$-roots of $\mathcal{G}$.

Given a subset $\Phi \subset \Pi$ we denote

(2.5a) Φ^r: all roots that are linear combinations of elements of Φ;

(2.5b) Φ^u: all positive roots not contained in Φ^r; and

(2.5c) $\mathcal{P}^r_\Phi = \mathcal{H} + \sum_{\Phi^r} \mathcal{G}^\phi, \quad \mathcal{P}^u_\Phi = \sum_{\Phi^u} \mathcal{G}^\phi \quad$ and $\quad \mathcal{P}_\Phi = \mathcal{P}^r_\Phi + \mathcal{P}^u_\Phi.$

Then G has complex analytic subgroups

(2.5d) P^r_Φ for $\mathcal{P}^r_\Phi$, $\quad P^u_\Phi$ for $\mathcal{P}^u_\Phi$, $\quad P_\Phi = P^r_\Phi \cdot P^u_\Phi$ for $\mathcal{P}_\Phi$.

Now

(2.6a) P_Φ is a parabolic subgroup of G;

(2.6b) P_Φ has unipotent radical P^u_Φ, reductive part P^r_Φ ;

(2.6c) Φ is a system of simple $\mathcal{H}$-roots for $\mathcal{P}^r_\Phi$; and

(2.6d) every parabolic subgroup of G is conjugate to just one P_Φ.

Let $X = G/P$ complex flag manifold. As P is its own normalizer in G, we have a bijective correspondence $x \leftrightarrow P_x$ between X and the set of G-conjugates of P, given by

(2.7) $$P_x = \{g \in G : g(x) = x\}.$$

We will constantly use (2.7) without further reference.

Let G_0 be a real form of G. Thus G_0 is the real analytic subgroup of G for a real form $\mathcal{G}_0$ of the Lie algebra $\mathcal{G}$. Denote

(2.8) τ: complex conjugation of G over G_0, $\mathcal{G}$ over $\mathcal{G}_0$.

Now consider an orbit $G_0(x) \subset X = G/P$. The isotropy subgroup of G_0 at x is $G_0 \cap P_x$, which has Lie algebra

(2.9a) $\mathcal{G}_0 \cap \mathcal{P}_x = \mathcal{G}_0 \cap (\mathcal{P}_x \cap \tau\mathcal{P}_x)$ real form of $\mathcal{P}_x \cap \tau\mathcal{P}_x$.

The intersection of two Borel subgroups of G contains a Cartan subgroup; it follows that there exists

(2.9b) $\mathcal{H}_0 \subset \mathcal{G}_0 \cap \mathcal{P}_x$ such that $\mathcal{H} = \mathcal{H}_0^{\mathbb{C}}$ is a Cartan subalgebra of $\mathcal{G}$.

Now choose

(2.9c) Π simple $\mathcal{H}$-root system, $\Phi \subset \Pi$, such that $\mathcal{P}_x = \mathcal{P}_\Phi$.

That done, we have

(2.9d)
$$\mathcal{P}_x \cap \tau\mathcal{P}_x = \{\mathcal{H} + \sum_{\Phi^r \cap \tau\Phi^r} \mathcal{G}^\phi\} + \{(\sum_{\Phi^r \cap \tau\Phi^u} + \sum_{\Phi^u \cap \tau\Phi^r} + \sum_{\Phi^u \cap \tau\Phi^u}) \mathcal{G}^\phi\}$$

where the second term in {braces} is the unipotent radical and the first term is its reductive complement.

The first consequence of (2.9) is the fact ([21], Theorem 2.6) that

(2.10) there are only finitely many G_0-orbits on X.

For there are only finitely many G_0-conjugacy classes of real Cartan subalgebras $\mathcal{H}_0$; to each $\mathcal{H}_0$ there are only finitely many systems Π of simple $\mathcal{H}$-roots; to each $(\mathcal{H}_0, \Pi)$ there is just one set $\Phi \subset \Pi$ with P_Φ conjugate (in G) to P. In particular, by dimension,

(2.11) there are open G_0-orbits and closed G_0-orbits on X.

The second consequence of (2.9) is the fact ([21], Theorem 2.12) that

(2.12) $G_0(x)$ has real codimension $|\Phi^u \cap \tau\Phi^u|$ in X.

In particular,

(2.13) $G_0(x)$ is open in X if, and only if, $\Phi^u \cap \tau\Phi^u$ is empty.

Let K be a maximal compact subgroup of G_0 and consider the Cartan decomposition

(2.14) $$G_0 = K + M_0.$$

Every Cartan subalgebra of G_0 is conjugate to one of the form

(2.15) $$H_0 = H_T + H_V\,; \quad H_T = H_0 \cap K, \quad H_V = H_0 \cap M_0.$$

Given H_0 in that form, the following conditions are equivalent ([21], Lemma 4.1).

(2.16a) H_T is a Cartan subalgebra of K.

(2.16b) H_T contains a regular element of G.

(2.16c) Some system Π of simple $H_0^{\mathbb{C}}$-roots has $\tau\Pi = -\Pi$.

Under conditions (2.16abc) we say that

(2.16d) H_0 is a *maximally compact* Cartan subalgebra of G_0.

By careful application of (2.13) one can prove ([21], Theorem 4.5) that an orbit $G_0(x) \subset X$ is open if, and only if, there exist

(2.17a) a maximally compact Cartan subalgebra $H_0 \subset G_0$ and

(2.17b) a simple $H_0^{\mathbb{C}}$-root system Π with $\tau\Pi = -\Pi$

such that

(2.17c) $P_x = P_\Phi$ for some subset $\Phi \subset \Pi$.

That criterion allows us to enumerate the open G_0-orbits as follows ([21], Theorem 4.9). Fix one open orbit $G_0(x)$ and H_0, Π and Φ as in (2.17). Denote

(2.18a) W_G: Weyl group of G relative to $H_0^{\mathbb{C}}$.

(2.18b) W_K: Weyl group of K for H_T, subgroup of W_G.

(2.18c) $W_G^{H_0} = \{w \in W_G : w(H_0) = H_0\}$.

$$(2.18d) \qquad W^{H_0}_{P^r_x \cap \tau P^r_x} = \{w = \mathrm{ad}(g) \in W^{H_0}_G : g \in P^r_x \cap \tau P^r_x\}.$$

Then the open G_0-orbits on X are in one to one correspondence with the elements of the double coset space

$$(2.18e) \qquad W_K \setminus W^{H_0}_G \,/\, W^{H_0}_{P^r_x \cap \tau P^r_x} .$$

Here we will be interested, for purposes of discrete series representations of G_0, in the case rank K = rank G_0. In that case the maximally compact Cartan subalgebras $H_0 = H_T + H_V$ of G_0 are *compact* in the sense that $H_0 = H_T \subset K$, so $\tau\phi = -\phi$ for every root, whence $\tau\Pi = -\Pi$ for every simple root system. Thus $G_0(x) \subset X$ is open if, and only if, $G_0 \cap P_x$ contains a compact Cartan subalgebra H_0 of G_0; in that case $P_x \cap \tau P_x = P^r_x$ reductive, and the open orbits are enumerated by $W_K \setminus W_G / W_{P^r_x}$.

It is for the nondiscrete series representations of G_0 that we need *partially* complex manifolds and *partially* holomorphic vector bundles. We go into that now.

Let D be a subset of a complex analytic space V. We see the extent to which D inherits complex analytic structure from V. For that, define *holomorphic arc in D*:

(2.19a) holomorphic map $f: \{z \in \mathbb{C} : |z| < 1\} \to V$ with image in D;

chain of holomorphic arcs in D:

(2.19b) sequence $\{f_1, \ldots, f_k\}$ of holomorphic arcs in D such that the image of f_j meets image of f_{j+1} for $1 \leq j < k$; and

holomorphic arc component of D:

(2.19c) equivalence class of elements of D under the relation $v_1 \sim v_2$ if there is a chain $\{f_1, \ldots, f_k\}$ of holomorphic arcs in D with $v_1 \in \mathrm{image}\ f_1$ and $v_2 \in \mathrm{image}\ f_k$.

The main point is that the notion of holomorphic arc component is intrinsic. Thus, for example, suppose that A is a group of holomorphic diffeomorphisms of V, that D is an A-stable subset, and that S is a holomorphic arc component of D. Define

(2.20a) $N_A(S) = \{a \in A : a(S) = S\}$, *normalizer of S in A.*

Then ([21], Lemma 8.2)

(2.20b) $N_A(S) = \{a \in A : a(S) \text{ meets } S\}$,

(2.20c) if D is an A-orbit then $N_A(S)$ is transitive on S,

(2.20d) if A is a Lie transformation group on V and D is an A-orbit, then $N_A(S)$ is a Lie subgroup of A, and $S \subset D \subset V$ are embedded real analytic submanifolds.

It does *not* follow that S is a complex analytic submanifold of V (cf. [21], Example 8.12).

We look at the holomorphic arc components of G_0-orbits on the complex flag $X = G/P$. It will be convenient to have the following notation.

(2.21a) $S_{[x]}$: holomorphic arc component of $G_0(x) \subset X$ through x;

(2.21b) $N_{[x],0}$: identity component of $\{g \in G_0 : gS_{[x]} = S_{[x]}\}$;

(2.21c) $\mathfrak{N}_{[x],0}$ is the Lie algebra of $N_{[x],0}$ and $\mathfrak{N}_{[x]} = \mathfrak{N}_{[x],0}^{\mathbb{C}}$;

(2.21d) $N_{[x]}$ is the complex analytic subgroup of G for $\mathfrak{N}_{[x]}$.

The first step in studying the holomorphic arc components of $G_0(x)$ is the construction of a certain parabolic subalgebra $\mathfrak{Q}_{[x]} \subset \mathfrak{N}_{[x]}$ of $\mathfrak{G}$. Let $\mathfrak{P}_x = \mathfrak{P}_\Phi$ as in (2.9) and define

(2.22a) $$\delta_x = \sum_{\Phi^u \cap \tau\Phi^u} \phi = \tau\delta_x ,$$

(2.22b) $$\mathfrak{Q}_{[x]} = \mathfrak{H} + \sum_{\langle\phi,\delta_x\rangle \geqq 0} \mathfrak{G}^\phi = \tau\mathfrak{Q}_{[x]}.$$

Then ([21], Theorem 8.5)

(2.23a) $P_x^u \cap \tau P_x^u \subset Q_{[x]} \subset \{N_{[x]} \cap (P_x + \tau P_x)\}$, so

(2.23b) $N_{[x]}$ is a τ-stable parabolic subalgebra of G and

(2.23c) $N_{[x],0}$ is the identity component of the parabolic $N_{[x]} \cap G_0$ of G_0.

For the second step, define

(2.24a) $\Gamma^0 = \{\text{roots } \phi : -\phi \notin \Phi^u \cap \tau\Phi^u, \ \langle \phi, \delta_x \rangle < 0, \ \phi + \tau\phi \text{ not root}\}$;

(2.24b) $M_{[x]} = Q_{[x]} + \sum_{\Gamma^0} G^\phi$.

Then ([21], Proposition 8.7)

(2.24c) $Q_{[x]} \subset M_{[x]} \subset N_{[x]} \cap (P_x + \tau P_x)$.

The point of (2.24) is ([21], Theorem 8.9) that the following conditions are equivalent.

(2.25a) The holomorphic arc components of $G_0(x)$ are complex manifolds.

(2.25b) $N_{[x]} \subset P_x + \tau P_x$.

(2.25c) $N_{[x]} = M_{[x]}$ subspace of G defined in (2.24b).

(2.25d) $M_{[x]}$ is a subalgebra of G.

Let us agree to say that the orbit $G_0(x) \subset X$ is

(2.26) *partially complex* if its holomorphic arc components are complex submanifolds of X, i.e., if $S_{[x]}$ is a complex manifold;

(2.27) *flag type* if the $N_{[x']}(x') \subset G_0(x)$ are complex flag submanifolds of X, i.e., if $N_{[x]}(x)$ is a flag;

(2.28) *measurable* if its holomorphic arc components carry Radon measures invariant under their normalizers, i.e., if $S_{[x]}$ has an $N_{[x],0}$-invariant Radon measure;

(2.29) *integrable* if the distribution $x' \mapsto G_0 \cap (P_{x'} + \tau P_{x'})$ on $G_0(x)$ is integrable, i.e., if $P_x + \tau P_x$ is a subalgebra of G.

The orbits with which we will be concerned are those that are partially complex, measurable and integrable. The criterion for partial complexity is (2.25), which is rather difficult to check. However, it is subsumed by the measurability criterion below, which is rather more easily verified.

Let $P_x = P_\Phi$ as in (2.9) and denote

(2.30a) $V_x^+ = \sum_{\Phi^u \cap -\tau\Phi^u} G^\phi, \quad V_x^- = \tau V_x^+ = \sum_{-\Phi^u \cap \tau\Phi^u} G^\phi$ and

$V_x = V_x^+ + V_x^-$.

Then ([21], Theorem 9.2) the orbit $G_0(x)$ is

(2.30b) measurable $\Leftrightarrow N_{[x]} = (P_x \cap \tau P_x) + V_x$.

Further, if $G_0(x)$ is measurable then

(2.30c) $G_0(x)$ is partially complex and of flag type,

(2.30d) $G_0(x)$ is integrable if and only if $\tau P_x^r = P_x^r$, and

(2.30e) the invariant Radon measure on $S_{[x]}$ is the volume element of an $N_{[x],0}$-invariant indefinite-kaehler metric.

From a slightly different starting point, suppose that $\tau P_x^r = P_x^r$. Then ([21], Theorem 9.9) the following are equivalent.

(2.31a) $G_0(x)$ is measurable.

(2.31b) $G_0(x)$ is integrable.

(2.31c) $G_0(x)$ is partially complex and of flag type.

Under those circumstances,

(2.31d) $N_{[x]} = (P_x \cap \tau P_x) + V_x = Q_{[x]} = P_x + \tau P_x$.

In fact ([21], Corollary 9.11) integrability of $G_0(x)$ implies $N_{[x]} = P_x + \tau P_x = Q_{[x]}$, implies that $G_0(x)$ is partially complex and of flag type, and implies that $G_0(x)$ is measurable if and only if $\tau P_x^r = P_x^r$.

To understand the interplay (2.30) and (2.31) of these various global conditions on $G_0(x)$, we will examine them in the cases

(2.32a) $G_0(x)$ open in X,

(2.32b) $G_0(x)$ closed in X, and

(2.32c) $G_0(x)$ integrable in X.

Open orbits obviously are integrable, partially complex and of flag type. More precisely, if $G_0(x) \subset X$ is open then $P_x + \tau P_x = G = N_{[x]}$, so $S_{[x]} = G_0(x)$ and $N_{[x]}(x) = G(x) = X$.

Measurability of open orbits can be detected without the machinery of holomorphic arc components. The result ([21], Theorem 6.3) is that the following conditions on an orbit $G_0(x) \subset X$ are equivalent.

(2.33a) $G_0(x)$ is a measurable open orbit.

(2.33b) $G_0(x)$ has an invariant volume element.

(2.33c) $G_0(x)$ has an invariant indefinite-kaehler metric.

(2.33d) $G_0 \cap P_x$ is the centralizer of a (compact) torus.

(2.33e) $P_x \cap \tau P_x$ is reductive, i.e., $P_x \cap \tau P_x = P_x^r \cap \tau P_x^r$.

(2.33f)[1] $P_x \cap \tau P_x = P_x^r$, i.e., $\tau P_x^u = P_x^{-u}$, i.e., $G = P_x + \tau P_x^u$.

(2.33g) $P_x = P_\Phi$ with $\tau\Phi^r = \Phi^r$ and $\tau\Phi^u = -\Phi^u$.

---

[1] See footnote next page.

In particular (see discussion after (2.18e)),

(2.34) if rank K = rank G_0 then every open orbit it measurable.

The phenomenon (2.34) is not isolated. Denote

(2.35a)[1] if $P = P_\Phi$ then $P^- = P^r_\Phi + P^{-u}_\Phi$ *opposite* of P.

Then ([21], Theorem 6.7) the following conditions are equivalent.

(2.35b) Some open G_0-orbit on X is measurable.

(2.35c) Every open G_0-orbit on X is measurable.

(2.35d) P^- and τP are ad(G)-conjugate.

Closed orbits are somewhat more tractable. The first main fact ([21], Theorem 3.3) is that

(2.36a) there is just one closed G_0-orbit on X,

(2.36b) every maximal compact subgroup of G_0 is transitive on it,

(2.36c) it is in the closure of every G_0-orbit, and

(2.36d) it is the lowest-dimensional G_0-orbit.

For example, if $X = G/P$ is a hermitian symmetric space of compact type under the compact real form of G, i.e. ([21], Lemma 9.22), if P^u is abelian, and if there exists $x_0 \in X$ such that $G_0(x_0) \subset X$ is the Borel embedding of the dual bounded symmetric domain, then it follows ([21], Corollary 3.9) that the closed G_0-orbit is the Bergman-Šilov boundary of $G_0(x_0)$ in X.

Minimal dimensionality (2.36d) of the closed orbit has an interesting interpretation ([21], Theorem 3.6). Let $x \in X$ and $P_x = P_\Phi$ as in (2.9). Then $\dim_R G_0(x) \geq \dim_{\mathbb{C}} X = \frac{1}{2}\dim_R X$, and the following conditions are equivalent.

---

[1] If $P_x = P_\Phi$ then $P^{-u}_x = P^{-u}_\Phi$ denotes $\sum_{\Phi^u} G^{-\phi}$, vector space complement to $P_x = P_\Phi$ in G composed of root spaces.

(2.37a) $\dim_R G_0(x) = \dim_{\mathbb{C}} X$.

(2.37b) $\tau\Phi^u = \Phi^u$, i.e., $\delta_x = \sum_{\Phi^u} \phi$, i.e., $Q_{[x]} = P_x$.

(2.37c) View G as a linear algebraic group def/$\mathbb{R}$, with G_0 the topological identity component of the group G_R of real points; then P_x is def/$\mathbb{R}$, i.e., $\tau P_x = P_x$, i.e., $\tau P_x = P_x$.

(2.37d) $G_0(x)$ is the closed orbit, and some conjugate of P is def/$\mathbb{R}$.

(2.37e) X is a complex projective variety def/$\mathbb{R}$ in such a way that $G_0(x) = X_{\mathbb{R}}$.

In the hermitian symmetric case mentioned above, the various conditions of (2.37) all are equivalent to: $G_0(x_0)$ is a tube domain over a self-dual cone and $G_0(x)$ is its Bergman-Šilov boundary.

The second main fact ([21], Theorem 9.12) on closed orbits is:

(2.38) the closed G_0-orbit is measurable, partially complex, and of flag type.

That is the key to geometric realization of principal series representations of G_0. It allows characterization ([21], Corollary 9.16) of the closed orbit by the fact that its holomorphic arc components are compact and shows that those components are "tiny" complex flag manifolds; it also shows that the measure induced from the K-normalizer is invariant under the G_0-normalizer.

Finally we examine our global conditions for integrable orbits on the complex flag $X = G/P$. Let[2]

(2.39a) Q: τ-stable parabolic subalgebra of G

that is a candidate for $P_x + \tau P_x$ in the sense that

(2.39b) Q contains an ad(G)-conjugate of P.

---

[2] Of course one can start with Q and select $X = G/P$, i.e., select P, by means of (2.39b).

Then ([21], Theorem 7.10) there exist, and one can enumerate, orbits $G_0(x) \subset X$ such that

(2.39c) $$\mathcal{Q} = P_x + \tau P_x.$$

From our earlier discussion, these orbits satisfy

(2.40a) $N_{[x]} = \mathcal{Q}_{[x]} = P_x + \tau P_x = \mathcal{Q}$,

(2.40b) $G_0(x)$ is partially complex and of flag type,

(2.40c) $G_0(x)$ is measurable if, and only if, $\tau P_x^r = P_x^r$.

This completes our description of the "partially complex" homogeneous spaces of G_0 mentioned in the Introduction. There we also mentioned a notion of integration over the "space of complex analytic pieces" of a partially complex G_0-homogeneous space. Thus we must look at the space of holomorphic arc components of an orbit $G_0(x) \subset X$.

Consider an orbit $G_0(x) \subset X$. Then ([21], Theorem 8.15) the normalizers of its holomorphic arc components are specified by

(2.41a) $$G_0 \cap N_{[x]} = \{g \in G_0 : gS_{[x]} = S_{[x]}\},$$

so they are parabolic subgroups of G_0; that defines a G_0-equivariant fibration

(2.41b) $$\sigma: G_0(x) \to G_0/(G_0 \cap N_{[x]})$$

of the orbit over its "space of holomorphic arc components", whose

(2.41c) fibres: the holomorphic arc components of $G_0(x)$,

(2.41d) base: compact space $G_0/G_0 \cap N_{[x]} = K/K \cap N_{[x]}$.

Thus Haar measure of K induces a K-invariant Radon measure on the space $K/K \cap N_{[x]}$ of holomorphic arc components of $G_0(x)$. In passing we note that this space of holomorphic arc components is a real flag manifold.

3. *Open orbits and discrete series*

In this section we assume that G_0 has a compact Cartan subgroup, i.e., has rank equal to that of a maximal compact subgroup. For that is the condition (Harish-Chandra [8]) for the existence of discrete series representations of G_0.

$X = G/P$ is a complex flag manifold with $G = G_0^{\mathbb{C}}$. Suppose that we have an orbit

(3.1) $\quad X_0 = G_0(x_0) \subset X$ open with $G_0 \cap P_{x_0}$ compact.

Then we consider finite dimensional representations

(3.2) $\quad \lambda: G_0 \cap P_{x_0} \to GL(r, \mathbb{C})$ irreducible unitary.

Each such representation λ has a unique extension

(3.3) $\quad \lambda: P_{x_0} \to GL(r, \mathbb{C})$ holomorphic.

Thus we have associated G-homogeneous holomorphic vector bundles $\tilde{\pi}: \tilde{\mathbb{E}}_\lambda \to X$, with restrictions to X_0 that we denote

(3.4) $\quad \pi: \mathbb{E}_\lambda \to X_0 \quad G_0$-homogeneous, hermitian, holomorphic.

Consider the spaces

(3.5) $\quad A^{p,q}(\mathbb{E}_\lambda) = \{C^\infty \ (p,q)\text{-forms on } X_0 \text{ with values in } \mathbb{E}_\lambda\}$

and the maps

(3.6) $$\bar{\partial}: A^{p,q}(\mathbb{E}_\lambda) \to A^{p,q+1}(\mathbb{E}_\lambda).$$

Now the Dolbeault cohomology groups

(3.7) $\quad H^{0,q}(X_0, \mathbb{E}_\lambda) = \{\omega \in A^{0,q}(\mathbb{E}_\lambda) : \bar{\partial}\omega = 0\}/\bar{\partial}A^{0,q-1}(\mathbb{E}_\lambda)$

are G_0-modules isomorphic to the cohomology modules of the sheaf of germs of holomorphic sections of $\mathbb{E}_\lambda$. However we are looking for unitary representations so we must find "square integrable cohomology groups".

As the isotropy subgroup $G_0 \cap P_{x_0}$ of G_0 at x_0 was assumed compact (3.1) we have

(3.8) ds^2: G_0-invariant hermitian metric on X_0.

Also λ is unitary on $G_0 \cap P_{x_0}$ so, as mentioned in (3.4),

(3.9) h: G_0-invariant hermitian metric on $\mathbb{E}_\lambda$.

Now we have the Hodge-Kodaira operators

(3.10) $$A^{p,q}(\mathbb{E}_\lambda) \xrightarrow{\#} A^{n-p,n-q}(\mathbb{E}^*_\lambda) \xrightarrow{\tilde{\#}} A^{p,q}(\mathbb{E}_\lambda), \quad n = \dim X.$$

They give the linear space

(3.11) $$A_2^{p,q}(\mathbb{E}_\lambda) = \{\alpha \in A^{p,q}(\mathbb{E}_\lambda) : \int \alpha \wedge \#\alpha < \infty\}$$

which is a pre Hilbert space with inner product

(3.12) $$\langle\alpha,\beta\rangle = \int \alpha \wedge \#\beta.$$

Now denote

(3.13) $L_2^{p,q}(\mathbb{E}_\lambda)$: Hilbert space completion of $A_2^{p,q}(\mathbb{E}_\lambda)$.

$\bar{\partial}$ is densely defined on $L_2^{p,q}(\mathbb{E}_\lambda)$, and

(3.14) $\bar{\delta} = -\tilde{\#}\bar{\partial}\#$ is the formal adjoint of $\bar{\partial}$.

Now, as expected, we define

(3.15) $$\square = (\bar{\partial} + \bar{\delta})^2 = \bar{\partial}\bar{\delta} + \bar{\delta}\bar{\partial} \quad \text{Kodaira-Hodge-Laplacian.}$$

As ds^2 is positive definite and complete, the work of Andreotti-Vesentini [1] shows that

(3.16a) $H_2^{p,q}(\mathbb{E}_\lambda) = \{\omega \in L_2^{p,q}(\mathbb{E}_\lambda) : \square\omega = 0\}$ square integrable harmonic forms

is a closed subspace contained in the space $A_2^{p,q}(\mathbb{E}_\lambda)$ of smooth square integrable forms, and that

$$(3.16b)\qquad L_2^{p,q}(\mathbb{E}_\lambda) = \bar\partial L_2^{p,q-1}(\mathbb{E}_\lambda) \oplus \bar\delta L_2^{p,q+1}(\mathbb{E}_\lambda) \oplus H_2^{p,q}(\mathbb{E}_\lambda)$$

orthogonal direct sum such that

(3.16c) $\bar\partial L_2^{p,q-1} + H_2^{p,q}(\mathbb{E}_\lambda)$ is the kernel of $\bar\partial$,

(3.16d) $\bar\delta L_2^{p,q+1} + H_2^{p,q}(\mathbb{E}_\lambda)$ is the kernel of $\bar\delta$.

Thus the "square integrable cohomology" is given by the square integrable harmonic forms.

The Hilbert space structure of $L_2^{p,q}(\mathbb{E}_\lambda)$ depends on the G_0-homogeneous bundle $\mathbb{E}_\lambda$ and the G_0-invariant hermitian metrics h and ds^2. Thus G_0 has a natural unitary action on $L_2^{p,q}(\mathbb{E}_\lambda)$. Consider the matrix coefficients

$$(3.17)\qquad f_{\alpha\beta}(g) = \langle g\alpha,\beta\rangle \ ; \quad g\in G_0, \quad \alpha,\beta\in L_2^{p,q}(\mathbb{E}_\lambda).$$

If T_0 denotes the holomorphic tangent space of X_0 at x_0, and if E_λ denotes the representation space of λ, then we may view

$$(3.18a)\qquad \alpha,\beta\colon G_0 \to E_\lambda^{p,q} = \Lambda^p T_0^* \otimes \Lambda^q \bar T_0^* \otimes E_\lambda$$

with the appropriate transformation condition for right translation by elements of $G_0 \cap P_{x_0}$. The square integrability condition on α and β becomes

$$(3.18b)\qquad \alpha,\beta\in L_2(G_0) \otimes E_\lambda^{p,q}.$$

G_0 acts on the first tensor factor. Thus, from the corresponding fact about the left regular representation of G_0, we have

$$(3.19)\qquad f_{\alpha\beta}\in L_1(G_0)\cap L_2(G_0) \quad \text{for} \quad \alpha,\beta\in L_1^{p,q}(\mathbb{E}_\lambda)\cap L_2^{p,q}(\mathbb{E}_\lambda).$$

Now define

$$(3.20)\qquad \pi_\lambda^{p,q}\colon \text{unitary representation of } G_0 \text{ on } H_2^{p,q}(\mathbb{E}_\lambda).$$

Then (3.17) and (3.19) show, by L_1-approximation, that

(3.21) all matrix coefficients of $\pi_\lambda^{p,q}$ are in $L_2(G_0)$.

Recall that the following conditions are equivalent for an irreducible unitary representation π of G_0.

(3.22a) The class of π, element of $\hat{G}_0$, has positive Plancherel measure.

(3.22b) π is a subrepresentation of the left regular representation of G_0.

(3.22c) π has a square integrable matrix coefficient.

(3.22d) All matrix coefficients of π are square integrable.

In that case π is said to be a *discrete series* representation of G_0. Thus the discrete series of G_0 is the subset of $\hat{G}_0$ consisting of all classes $[\pi]$ of (necessarily finite) positive Plancherel mass. Now (3.21) says

(3.23) if $\pi_\lambda^{p,q}$ is irreducible it is a discrete series representation of G_0.

The goal in geometric realization of discrete series is to exhibit every discrete series representation in the form $\pi_\lambda^{0,q}$. Schmid [18] has made considerable progress in that direction; one may reasonably expect that he will carry it through, at least for L_1 representations.

4. *Closed orbit and principal series*

We recall the construction of the "principal series" of unitary representations of G_0. Denote

(4.1) K: maximal compact subgroup of G_0,

(4.2) $G_0 = KAN$: Iwasawa decomposition of G_0,

(4.3) MAN: minimal parabolic subgroup of G_0.

In more detail, consider the Cartan involution σ of G_0 with fixed point set K. Choose

(4.4) A: maximal abelian subspace of (-1)-eigenspace of σ on G_0.

Then we have a finite set Δ_A of nonzero real linear functionals on A and a direct sum decomposition

(4.5a) $G_0 = Z_A + \sum_{\Delta_A} G_0^{\phi}$ where

(4.5b) Z_A is the centralizer of A in G_0 and

(4.5c) $G_0^{\phi} = \{u \in G_0 : [a,u] = \phi(a)u \text{ for } a \in A\} \neq 0.$

The elements of Δ_A are the "A-roots" or "restricted roots" of G_0. Every $\phi \in \Delta_A$ determines a hyperplane of A by

(4.6a) $\phi^{\perp} = \{a \in A : \phi(a) = 0\}.$

Now $A - \bigcup \phi^{\perp}$ is a finite union of convex cones on which every restricted root avoids the value 0. By *Weyl chamber* of A we mean a topological component of $A - \bigcup \phi^{\perp}$. Choose a Weyl chamber D of A. We define

(4.6b) $\Delta_A^+ = \{\phi \in \Delta_A : \phi > 0 \text{ on } D\}$ positive restricted root system.

Now observe

(4.6c) $\Delta_A = \Delta_A^+ \cup \Delta_A^-$ disjoint where $\Delta_A^- = -\Delta_A^+$.

Finally, define

(4.7a) $N = \sum_{\Delta_A^+} G_0^{\phi}$ and $N^- = \sum_{\Delta_A^-} G_0^{\phi}$,

(4.7b) N: analytic subgroup of G_0 for N,

(4.7c) A: analytic subgroup of G_0 for A,

(4.7d) M: centralizer of A in K.

That explains (4.2) and (4.3). Note that $MA = M \times A$ is the centralizer of A in G_0, that N is a unipotent group, and that MA normalizes N.

Let λ be an irreducible complex representation of MAN. As $\lambda(N)$ is a nilpotent group normal in the irreducible group $\lambda(MAN)$ we have $\lambda(N) = 1$ by Lie's theorem. So $\lambda = \mu \otimes \chi : man \mapsto \mu(m)\cdot\chi(a)$ where μ is an irreducible unitary representation of M and χ is a not-necessarily-unitary character on A. Now define

(4.8a) $$\rho = \tfrac{1}{2} \sum_{\Delta_A^+} (\dim G_0^\phi)\cdot\phi \quad \text{linear form on } A.$$

If μ is an irreducible unitary representation of M and η is a complex valued linear form on A, we define

(4.8b) $$\beta_{\mu,\eta}(man) = \mu(m)\cdot\exp((\rho + i\eta)(\log a)),$$

representation of MAN on the representation space V_μ of μ. The above discussion says that

(4.8c) the $\beta_{\mu,\eta}$ are the irreducible complex representations of MAN.

The reason for insertion of ρ in (4.8b) will come out in a moment.

Given $\beta_{\mu,\eta}$ we have an associated complex vector bundle

(4.9a) $$\mathbb{E}_{\mu,\eta} \to G_0/MAN = K/M.$$

Its space of measurable sections is

(4.9b) $$\Gamma(\mathbb{E}_{\mu,\eta}) = \{f\colon G_0 \to V_\mu \text{ measurable: } f(gman) = \beta_{\mu,\eta}(man)^{-1}f(g)\}.$$

Note that a measurable section $f\colon G_0 \to V_\mu$ is specified by $f|_K$; for every $g \in G_0$ has decomposition $g = kan$, and $f(g) = \beta_{\mu,\eta}(an)^{-1}f(k)$. Let dk denote normalized Haar measure on K and (v,v') the unitary inner product on V_μ. Now we have the space of square integrable sections,

(4.10a) $\Gamma_2(\mathbb{E}_{\mu,\eta}) = \{f \in \Gamma(\mathbb{E}_{\mu,\eta}) : \int_K (f(k),f(k))dk < \infty\}.$

It is a complex Hilbert space with inner product

(4.10b) $\langle f,f'\rangle = \int_K (f(k),f'(k))dk.$

G_0 acts on $\Gamma(\mathbb{E}_{\mu,\eta})$ by the algebraic representation

(4.11a) $[\tilde{\pi}_{\mu,\eta}(g)f](g') = f(g^{-1}g').$

We will check that

(4.11b) $\tilde{\pi}_{\mu,\eta}$ restricts to a bounded representation $\pi_{\mu,\eta}$ of G_0 on $\Gamma_2(\mathbb{E}_{\mu,\eta})$

and that

(4.11c) $\pi_{\mu,\eta}$ is unitary if and only if $\eta: A \to \mathbb{R}$ real.

For that we define

(4.12a) $\kappa_g: K \to K$ and $\alpha_g: K \to A$ by

(4.12b) $g^{-1}k \in \kappa_g(k)\cdot\alpha_g(k)\cdot N \subset KAN = G_0.$

If $f \in \Gamma(\mathbb{E}_{\mu,\eta})$, $g \in G_0$ and $k \in K$, now

(4.12c) $[\tilde{\pi}_{\mu,\eta}(g)f](k) = \exp((\rho + i\eta)\cdot\log \alpha_g(k))^{-1} f(\kappa_g(k)).$

The definition of ρ and the trace of A on N^- say, for $F \in L_1(K)$, that

(4.13) $\int_K F(k)dk = \int_K \exp(-2\rho \log \alpha_g(k))F(\kappa_g(k))dk.$

Let $f \in \Gamma_2(\mathbb{E}_{\mu,\eta})$ and $g \in G_0$, and set

$$F(k) = (f(k),f(k)) \quad \text{and} \quad F^g(k) = (f(g^{-1}k),f(g^{-1}k)).$$

Then (4.11a), (4.12c) and (4.13) give us

$$\int_K F^g(k) = \int_K |\exp(i\eta\cdot\log\alpha_g(k))|^{-2}\exp(-2\rho\cdot\log\alpha_g(k))F(\kappa_g(k))dk$$

$$\leqq C_\eta(g)\int_K \exp(-2\rho\cdot\log\alpha_g(k))F(\kappa_g(k))dk$$

$$= C_\eta(g)\int_K F(k)dk,$$

where

$$C_\eta(g) = \sup_{k\in K}|\exp(i\eta\cdot\log\alpha_g(k))|^{-2}.$$

Thus $\tilde{\pi}_{\mu,\eta}(g)f\in\Gamma_2(\mathbb{E}_{\mu,\eta})$, so $\pi_{\mu,\eta}(g)$ is defined, and $\|\pi_{\mu,\eta}(g)\| \leqq C_\eta(g)$. A similar argument shows that $\pi_{\mu,\eta}$ is continuous. That proves (4.11b). For (4.11c) note that η is real if and only if $|\exp(i\eta\cdot\log\alpha_g(k))|^{-2} \equiv 1$, and that the latter holds precisely when $\pi_{\mu,\eta}$ is unitary.

The *principal series* of unitary representations of G_0 is, by definition, the set of equivalence classes of unitary representations $\pi_{\mu,\eta}$ just defined. We have parameterized the principal series by $\hat{M}\times A'$ where $\hat{M}$ is the set of equivalence classes of irreducible unitary representations of M and A' is the real dual space of A.

It is possible to see that two principal series representations $\pi_{\mu,\eta}$ and $\pi_{\mu',\eta'}$ are equivalent if, and only if, $(\mu',\eta') = (\mu\circ\mathrm{ad}(m), \eta\circ\mathrm{ad}(m))$ where $m\in K$ normalizes A (and thus also normalizes M). More precisely, let W_A be the "restricted Weyl group", i.e.,

(4.14a) $\quad W_A = \{k\in K : \mathrm{ad}(k)A = A\}/M.$

Then W_A acts

(4.14b) $\quad$ on A' by $w(\eta) = \eta\circ\mathrm{ad}(k), \quad k\in w;$

(4.14c) $\quad$ on $\hat{M}$ by $w(\mu) = \mu\circ\mathrm{ad}(k), \quad k\in w.$

Now[3] [3]

(4.15) $\pi_{\mu,\eta} \sim \pi_{\mu',\eta'} \Longleftrightarrow (\mu',\eta') = (w\mu, w\eta)$, some $w \in W_A$.

Principal series representations are not automatically irreducible. Bruhat [3] has shown that

(4.16) if $(\mu,\eta) \neq (w\mu, w\eta)$ for $1 \neq w \in W_A$ then $\pi_{\mu,\eta}$ is irreducible.

A number of authors have worked out, for specific groups G_0, which principal series representations are irreducible; cf. Kunze and Stein [15], Knapp and Stein [12], Helgason [11], Wallach [19].

The *spherical principal series* consists of the classes of principal series representations $\pi_{\mu,\eta}$ such that $\mu = 1_M$ trivial representation of M. Irreducibility of all spherical principal series representations was proved by Parthasarathy, Ranga-Rao and Varadarajan [17] for complex G_0, by Kostant [14] in general. Their algebraic methods have been extended to the principal series by Wallach [19]. Also see Zelobenko [24].

Recall (4.10) that the representation space $\Gamma_2(\mathbb{E}_{\mu,\eta})$ is a certain space of L_2-functions on K with values in the representation space V_μ of μ. In other words $\Gamma_2(\mathbb{E}_{\mu,\eta}) \subset L_2(K) \otimes V_\mu$. We will need to identify it as a subspace. For that

(4.17a) if $\kappa \in \hat{K}$ then U_κ denotes its representation space.

Thus the Peter-Weyl theorem for K says

(4.17b) $$L_2(K) = \sum_{\kappa \in \hat{K}} U_\kappa \otimes U_\kappa^* .$$

In particular,

(4.17c) $$L_2(K) \otimes V_\mu = \sum_{\kappa \in \hat{K}} U_\kappa \otimes \{U_\kappa^* \otimes V_\mu\}$$

---

[3] There seems to be some trouble with the argument in [3], but it is fixed up in a forthcoming book by Garth Warner [20].

where the left action of K is κ on the U_κ-factor of each $U_\kappa \otimes \{U_\kappa^* \otimes V_\mu\}$. The right action of M on $U_\kappa \otimes \{U_\kappa^* \otimes V_\mu\}$ is $1_M \otimes \kappa^*|_M \otimes \mu$, so the M-fixed elements are given by $U_\kappa \otimes \{U_\kappa^* \otimes V_\mu\}^M$ there. Now

(4.18) $$\Gamma_2(\mathbb{E}_{\mu,\eta}) = \sum_{\kappa\in\hat{K}} U_\kappa \otimes \{U_\kappa^* \otimes V_\mu\}^M \quad \text{as a } K\text{-module.}$$

Note that this K-module structure does not involve η. Here we recall from (4.12) that the action of G_0 is

(4.19a) $$[\pi_{\mu,\eta}(g)f](k) = \exp((\rho + i\eta)(\log \alpha_g(k)))^{-1} f(\kappa_g(k)).$$

In other words,

(4.19b) $$\pi_{\mu,\eta}(g)f = \exp((\rho + i\eta)(\log \alpha_g(\cdot)))^{-1} \cdot (f \circ \kappa_g)(\cdot).$$

It is the composition with κ_g that mixes the K-summands $U_\kappa \otimes \{U_\kappa^* \otimes V_\mu\}^M$ of $\Gamma_2(\mathbb{E}_{\mu,\eta})$.

We describe the realization of a principal series representation $\pi_{\mu,\eta}$ on the closed orbit in a complex flag manifold.

$X = G/P$ is a complex flag manifold with $G = G_0^{\mathbb{C}}$. Now denote

(4.20a) $$Y = G_0(y_0) \subset X \quad \text{closed orbit}$$

and suppose that the holomorphic arc components $S_{[y]}$, $y \in Y$, have real normalizers given by

(4.20b) $$MAN = \{g \in G_0 : gS_{[y_0]} = S_{[y_0]}\}.$$

Then $S_{[y_0]}$ is a complex flag manifold of $M^{\mathbb{C}}$ and has descriptions

(4.20c) $$S_{[y_0]} = M^{\mathbb{C}}/Q = M/L = MAN/LAN$$

where

(4.20d) $$Q = M^{\mathbb{C}} \cap P_{y_0} \quad \text{has} \quad Q^r = L^{\mathbb{C}} \quad \text{and} \quad L = M \cap P_{y_0}.$$

Here $M^{\mathbb{C}}$ and $L^{\mathbb{C}}$ denote the complexifications of the topological identity components of the compact groups M and L; the other components of M and L are represented by elements of $\exp(iA)$.

We need notation for representation spaces of irreducible unitary representations of L, M and K:

(4.21a) if $\nu \in \hat{L}$ then W_ν is its representation space,

(4.21b) if $\mu \in \hat{M}$ then V_μ is its representation space,

(4.21c) if $\kappa \in \hat{K}$ then U_κ is its representation space.

Let $\nu \in \hat{L}$ and consider the associated M-homogeneous holomorphic vector bundle

(4.22a) $\mathbb{W}_\nu \to M/L = S_{[y_0]}$.

Suppose that ν is chosen so that we can arrange, e.g. by means of the Borel-Weil Theorem, that it is related to a given fixed $\mu \in \hat{M}$ by the condition that the space of holomorphic sections

(4.22b) $H^0(S_{[y_0]}, \mathcal{O}(\mathbb{W}_\nu)) = V_\mu$ as an M-module.

We construct the representation $\pi_{\mu,\eta}$ as follows. Define

(4.23a) $\sigma_{\nu,\eta}$: irreducible representation of LAN on W_ν

(4.23b) by: $\sigma_{\nu,\eta}(lan) = \nu(l)\cdot\exp((\rho + i\eta)(\log a))$.

We have the associated G_0-homogeneous complex vector bundle

(4.23c) $\mathbb{W}_{\nu,\eta} \to G_0/LAN = G_0(y_0) = Y$,

which is holomorphic over every holomorphic arc component of Y because

(4.23d) $\mathbb{W}_{\nu,\eta}|_{S_{[y_0]}} = \mathbb{W}_\nu$.

Now consider

(4.24a) $\Gamma_2^h(\mathbb{W}_{\nu,\eta})$: square integrable partially holomorphic sections.

It consists of all

(4.24b) $f\colon G_0 \to W_\nu$ measurable such that

(4.24c) $f(glan) = \sigma_{\nu,\eta}(lan)^{-1} f(g)$,

(4.24d) $f|_{gMAN}$ specifies a holomorphic section of $\mathbb{W}_{\nu,\eta}|_{S_{[gy_0]}}$,

(4.24e) $\int_K (f(k),f(k))_{W_\nu}\, dk < \infty$.

Our claim is that

(4.25) $\pi_{\mu,\eta}$ is the representation of G_0 on $\Gamma_2^h(\mathbb{W}_{\nu,\eta})$.

To prove (4.25) we realize $\Gamma_2^h(\mathbb{W}_{\nu,\eta})$ as a space of functions on K. The square integrable sections of $\mathbb{W}_{\nu,\eta} \to Y = K/L$ are just the L-fixed square integrable functions from K to W_ν as follows:

(4.26a) $$\Gamma_2(\mathbb{W}_{\nu,\eta}) = \{L_2(K) \otimes W_\nu\}^L = \sum_{\kappa\in\hat{K}} U_\kappa \otimes \{U_\kappa^* \otimes W_\nu\}^L .$$

The holomorphic square integrable sections are those also fixed under the unipotent radical Q^u of the group Q of (4.20d); thus

(4.26b) $$\Gamma_2^h(\mathbb{W}_{\nu,\eta}) = \sum_{\kappa\in\hat{K}} U_\kappa \otimes \{U_\kappa^* \otimes W_\nu\}^{L\cdot Q} .$$

Similarly, the space of (automatically square integrable) holomorphic sections of $\mathbb{W}_\nu \to S_{[y_0]} = M/L$ is

(4.27a) $$\Gamma^h(\mathbb{W}_\nu) = \sum_{m\in\hat{M}} V_m \otimes \{V_m^* \otimes W_\nu\}^{L\cdot Q} .$$

By hypothesis (4.22b) that is V_μ, so

(4.27b) $\dim_{\mathbb{C}}\{V_m^* \otimes W_\nu\}^{LQ}$ is 1 if $m = \mu$, 0 if $m \neq \mu$.

If we denote, for all $m \in \hat{M}$ and $\kappa \in \hat{K}$

$$n(m,\kappa)\colon \text{multiplicity of } m \text{ in } \kappa|_M ,$$

then (4.27b) allows us to re-write (4.26b) as

$$\text{(4.28a)} \qquad \Gamma_2^h(\mathbb{W}_{\nu,\eta}) = \sum_{\kappa\in\hat{K}} U_\kappa \otimes \mathbb{C}^{n(\mu,\kappa)}.$$

As $n(\mu,\kappa) = \dim\{U_\kappa^* \otimes V_\mu\}^M$ by Schur's Lemma, (4.28a) says

$$\text{(4.28b)} \qquad \Gamma_2^h(\mathbb{W}_{\nu,\eta}) = \sum_{\kappa\in\hat{K}} U_\kappa \otimes \{U_\kappa^* \otimes V_\mu\}^M \quad \text{as } K\text{-module.}$$

Comparing (4.12ab) and (4.24c) we see that the action $g\colon f \to f^g$ of G_0 on $\Gamma_2^h(\mathbb{W}_{\nu,\eta})$ is given by

$$\text{(4.28c)} \qquad f^g(k) = \exp((\rho + i\eta)\log \alpha_g(k))^{-1} f(\kappa_g(k)), \quad k \in K.$$

We use (4.28bc) to compute the distribution character of the representation of G_0 on $\Gamma_2^h(\mathbb{W}_{\nu,\eta})$. If ϕ is a C^∞ function of compact support on G_0 then, for $f \in \Gamma_2^h(\mathbb{W}_{\nu,\eta})$ and $k \in K$,

$$\text{(4.29a)} \qquad f^\phi(k) = \int_{G_0} \phi(g) f^g(k)\,dg$$

$$= \int_{G_0} \phi(g)\cdot\exp((\rho + i\eta)\log \alpha_g(k))^{-1}\cdot f(\kappa_g(k))\,dg.$$

Thus

$$\text{(4.29b)} \qquad \langle f^\phi, f\rangle = \int_K \int_{G_0} \phi(g)\cdot\exp((\rho+i\eta)\log \alpha_g(k))^{-1}\cdot(f(\kappa_g k), f(k))\,dg\,dk.$$

Now choose orthonormal bases

$$\text{(4.30a)} \qquad \{u_\kappa^r\} \text{ of } U_\kappa \text{ and } \{v_\kappa^s\} \text{ of } (U_\kappa^* \otimes V_\mu)^M.$$

That defines an orthonormal basis

(4.30b) $\{f_\kappa^{rs}\}$ of $\Gamma_2^h(\mathbb{W}_{\nu,\eta})$ by $f_\kappa^{rs} = u_\kappa^r \otimes v_\kappa^s$.

Now, for each $\kappa \in \hat{K}$, $k \in K$,

$$\sum_{r,s} (f_\kappa^{rs}(\kappa_g(k)), f_\kappa^{rs}(k)) = \sum_{r,s} ((\kappa_g(k))^{-1} \cdot u_\kappa^r, k^{-1} \cdot u_\kappa^r)$$

$$= n(\mu,\kappa) \sum_r (k \cdot \kappa_g(k)^{-1} u_\kappa^r, u_\kappa^r) = n(\mu,\kappa)\chi_\kappa(k \cdot \kappa_g(k)^{-1})$$

where χ_κ is the character of $\kappa \in \hat{K}$ viewed, as usual, as a function on K. Thus (4.29b) says that the representation of G_0 on $\Gamma_2^h(\mathbb{W}_{\nu,\eta})$ has distribution character $\chi_{\nu,\eta}: C_c^\infty(G_0) \to \mathbb{C}$ given by

(4.30c) $$\chi_{\nu,\eta}(\phi) = \int_K \int_{G_0} \sum_{\hat{K}} \phi(g) \cdot \exp((\rho + i\eta) \log \alpha_g(k))^{-1} \cdot$$
$$\cdot\, n(\mu,\kappa) \cdot \chi_\kappa(k \cdot \kappa_g(k)^{-1}) dg\, dk .$$

Recall that (4.18) shows the representation space $\Gamma_2(\mathbb{E}_{\mu,\eta})$ of $\pi_{\mu,\eta}$ to have the same K-module structure (4.28b) as that of $\Gamma_2^h(\mathbb{W}_{\nu,\eta})$, and (4.19) shows the action of G_0 there to be given by the same formula (4.28c) as that of the action of G_0 on $\Gamma_2^h(\mathbb{W}_{\nu,\eta})$. Thus considerations (4.29) and (4.30) are valid for $\pi_{\mu,\eta}$, which consequently has distribution character given by

(4.31a) $$\chi_{\mu,\eta}(\phi) = \int_K \int_{G_0} \sum_{\hat{K}} \phi(g) \cdot \exp((\rho + i\eta) \log \alpha_g(k))^{-1} \cdot$$
$$\cdot\, n(\mu,\kappa) \cdot \chi_\kappa(k \cdot \kappa_g(k)^{-1}) dg\, dk .$$

Thus, from (4.30c) and (4.31a),

(4.31b) $$\chi_{\nu,\eta} = \chi_{\mu,\eta} .$$

Now $\pi_{\mu,\eta}$ and the representation of G_0 on $\Gamma_2^h(\mathbb{W}_{\nu,\eta})$ are K-finite unitary representations with the same global character. Fell's modification [4] of Harish-Chandra's ([6], Lemma 3) (cf. Warner [20], Theorem 4.5.8.1) says that

(4.31c) $\pi_{\mu,\eta}$ is infinitesimally equivalent to: G_0 on $\Gamma_2^h(\mathbb{W}_{\nu,\eta})$.

Extending Harish-Chandra's [5, Theorem 8], it also shows that infinitesimal equivalence is the same as unitary equivalence for K-finite unitary representations. Thus

(4.31d) $\pi_{\mu,\eta}$ is unitarily equivalent to: G_0 on $\Gamma_2^h(\mathbb{W}_{\nu,\eta})$.

Now (4.25) is proved.

The careful reader will note that our proof of (4.25) gives a geometric realization of the complementary series representations of G_0.

Two assumptions were made in the geometric realization just given for principal series representations. They are the hypothesis (4.20b) that MAN is the real normalizer of the holomorphic arc component $S_{[y_0]}$, and the hypothesis (4.22b) that the representation $\mu \in \hat{M}$ is obtained by holomorphic induction of Borel-Weil type from some $\nu \in \hat{L}$. We discuss them separately, observing that both can always be arranged if P is a Borel subgroup of G, and also if $P = (MAN)^{\mathbb{C}}$ complexification of the minimal parabolic subgroup of G_0.

The closed orbit $G_0(y_0) = Y \subset X$ is automatically measurable, hence partially complex and of flag type [21, Theorem 9.12]. Let τ denote complex conjugation of G over G_0 and $P_{y_0}^r$ the reductive part of P_{y_0} relative to the τ-stable Cartan subalgebra $H \subset G \cap P_{y_0}$ given by

(4.32a) $H = (T + A)^{\mathbb{C}}$, T Cartan subalgebra of M.

Let $N_{[y_0],0}$ be the Lie algebra of

(4.32b) $\quad N_{[y_0],0} = \{g \in G_0 : gS_{[y_0]} = S_{[y_0]}\}_{\text{identity component}}$

and $N_{[y_0]}$ its complexification. Then [21, Theorem 9.2] and [21, Corollary 9.11] say

(4.32c) $\quad \tau P^r_{y_0} = P^r_{y_0} \Leftrightarrow P_{y_0} + \tau P_{y_0}$ algebra $\Leftrightarrow N_{[y_0]} = P_{y_0} + \tau P_{y_0}$.

From [21, (9.13b)] and the last paragraph of the proof of [21, Theorem 9.12] we have

(4.32d) $\quad N_{[y_0]} = (M + A + N)^{\mathbb{C}} \Leftrightarrow P_{y_0} \subset (M + A + N)^{\mathbb{C}} \Leftrightarrow N_{[y_0]} = P_{y_0} + \tau P_{y_0}$.

In summary, the hypothesis (4.20b) is satisfied if, and only if, P is conjugate to a subgroup of $(MAN)^{\mathbb{C}}$; i.e., precisely when there are equivariant holomorphic fibrations $G/B \to G/P \to G/(MAN)^{\mathbb{C}}$ for some Borel subgroup B of G. In that case the holomorphic arc components of Y are the fibres of Y over the closed orbit on $G/(MAN)^{\mathbb{C}}$.

Finally we examine the hypothesis (4.22b) that a given $\mu \in \hat{M}$ is obtained from some $\nu \in \hat{L}$ by $H^0(M/L, \mathcal{O}(\mathbb{V}_\nu)) = V_\mu$. If M (and thus also L) is connected, that is just the Borel-Weil Theorem. If $P_{y_0} = (MAN)^{\mathbb{C}}$ then $L = M$ and we may simply take $\nu = \mu$. In general, operating under the assumption (4.20b) discussed above, we proceed as follows. Denote

(4.33a) $\quad F = L \cap \exp(iA)$, so $F = G_0 \cap \exp(iA)$, and

(4.33b) $\quad L^0 \subset M^0$: topological identity components of $L \subset M$, so

(4.33c) $\quad L = F \cdot L^0$ and $M = F \cdot M^0$, F finite central in M.

In (4.33) we are making essential use of the simplifying assumption

$$G_0 \subset G, \text{ i.e., } G_0 \text{ is a linear group,}$$

of this talk. Now further denote

(4.33d) if $\nu \in \hat{L}$ then $\nu^0 = \nu|_{L^0}$; if $\mu \in \hat{M}$ then $\mu^0 = \mu|_{M^0}$.

Then ν^0 and μ^0 are irreducible by (4.33) and Schur's Lemma. Thus $M/L = M^0/L^0$ says that

(4.34) $H^0(M^0/L^0, O(\mathbb{V}_{\nu^0})) = V_{\mu^0} \Rightarrow H^0(M/L, O(\mathbb{V}_\nu)) = V_\mu$,

so (4.22b) is automatic. The matter is slightly more delicate if $G_0 \not\subset G$, e.g., for G_0 the 2-sheeted universal covering group of $SL(3,R)$, where F is replaced by $\mathrm{ad}_G^{-1}(\mathrm{ad}_G(L) \cap \exp_{\mathrm{ad}\, G}(iA))$ and can fail to be central in L.

5. *Harish-Chandra's decomposition of* $L_2(G_0)$

For convenience we recall Harish-Chandra's construction of the series of unitary representations of G_0 corresponding to a conjugacy class of Cartan subgroups. Warning: "Cartan subgroup" means centralizer of a Cartan subalgebra. Here Cartan subgroups are abelian because G_0 is linear, but they need not be connected.

We denote

(5.1a) H : Cartan subgroup of G_0;

(5.1b) σ : Cartan involution of G_0 with $\sigma(H) = H$;

(5.1c) K : fixed point set of σ, maximal compact subgroup of G_0.

Decompose the Lie algebra H of H as

(5.2a) $H = T_H + A_H$ into (+1)- and (-1)-eigenspaces of σ.

That decomposes H as a direct product

(5.2b) $H = T_H \times A_H$; $T_H = H \cap K$ and $A_H = \exp A_H$.

Now as in (4.5) we have a finite set Δ_{A_H} of nonzero real linear functionals on A_H and a direct sum decomposition

(5.3a) $\quad G_0 = Z_H + \sum_{\Delta} G_0^{\phi}, \quad \Delta = \Delta_{A_H}$, where

(5.3b) $\quad Z_H$ is the centralizer of A_H in G_0 and

(5.3c) $\quad G_0^{\phi} = \{u \in G_0 : [a,u] = \phi(a)u \text{ for } a \in A_H\} \neq 0.$

Δ_{A_H} is the set of "A_H-roots" of G_0.

Every $\phi \in \Delta = \Delta_{A_H}$ defines a hyperplane $\phi^{\perp} = \{a \in A_H : \phi(a) = 0\}$, and $A_H - \bigcup_{\Delta} \phi^{\perp}$ is a finite union of convex cones (the *Weyl chambers* of A_H) that are its topological components. Given a Weyl chamber $\mathcal{D} \subset A_H$ we have the corresponding

(5.4a) $\quad \Delta^+ = \{\phi \in \Delta : \phi > 0 \text{ on } \mathcal{D}\}$, *positive* A_H*-root system.*

As before in (4.7), denote

(5.4b) $\quad N_H = \sum_{\Delta^+} G_0^{\phi}$ and $N_H^- = \sum_{\Delta^+} G_0^{-\phi}$,

(5.4c) $\quad N_H = \exp N_H$ unipotent analytic subgroup of G_0,

(5.4d) $\quad P_H$: normalizer of N_H in G_0.

Then P_H is a parabolic subgroup of G_0 with unipotent radical N_H, whose reductive part, the centralizer of A_H in G_0, has H as Cartan subgroup. Thus

(5.5a) $\quad P_H^r = M_H \times A_H$ and $P_H^u = N_H$ where

(5.5b) $\quad A_H$ is the $\mathbb{R}$-split component of the center of P_H^r,

(5.5c) $\quad M_H$ has a compact Cartan subgroup T_H.

There $P_H^r = Z_H = M_H + A_H$ orthogonal direct sum under the Killing form of G_0.

Construction (5.4) of $P_H = M_H A_H N_H$ is based on the choice of positive A_H-root system Δ^+, i.e., on the choice of positive Weyl chamber $\mathcal{D} \subset A_H$. As $\mathcal{D}$ ranges over the set of all positive Weyl chambers of A_H we thus obtain a collection $\{P_{H,\mathcal{D}}\}$ of parabolic subgroups of G_0, whose elements need not all be G_0-conjugate.

We say that two parabolic subgroups of G_0 are *associated* if their reductive parts are G_0-conjugate, i.e., if the split components of the centers of their reductive parts are G_0-conjugate. Given H, now the G_0-association class of P_H is independent of choice of the positive A_H-root system.

Let $Q \subset G_0$ be a parabolic subgroup. We say that Q is *cuspidal* if the derived group $[Q^r, Q^r]$ of the reductive part Q^r has a compact Cartan subgroup. From (5.5) we note that the P_H are cuspidal parabolic subgroups of G_0. In any case, we may conjugate Q in G_0 so that the split component A^Q of the center of Q^r has Lie algebra A^Q in the (-1)-eigenspace of σ, and then

(5.6a) $\quad Q = M^Q A^Q N^Q$ with $Q^r = M^Q \times A^Q$ and $N^Q = Q^u$,

such that the reductive group M^Q has compact center. Now

(5.6b) $\quad Q$ is cuspidal $\Leftrightarrow M^Q$ has a compact Cartan subgroup.

On the other hand,

(5.6c) $\quad M^Q$ has a compact Cartan $\Leftrightarrow \exists H$ with $A^Q = A_H$.

It follows that

(5.6d) $\quad Q$ is cuspidal $\Leftrightarrow Q$ is G_0-conjugate to some P_H.

In summary

(5.7) $\quad H \mapsto P_H$ induces a bijection from the set of all G_0-conjugacy classes of Cartan subgroups to the set of all G_0-association classes of cuspidal parabolic subgroups of G_0.

We note the two extreme cases of (5.7). If H is a compact Cartan subgroup of G_0 then

$$A_H = \{1\}, \quad N_H = \{1\} \quad \text{and} \quad M_H = G_0 \ ; \quad \text{so} \quad P_H = G_0.$$

If H is a maximally $\mathbb{R}$-split Cartan subgroup of G_0 then in the notation of section 4,

$$A_H = A, \quad N_H = N, \quad M_H = M; \quad \text{so} \quad P_H = MAN \quad \text{minimal parabolic.}$$

Another special case of interest is that (cf. [21, pp. 1190-1191]) in which G_0 has just one conjugacy class of Cartan subgroups.

Define a real linear functional ρ_H on A_H by

(5.8a) $\rho_H(\alpha) = \frac{1}{2}\sum_{\Delta^+} (\dim G_0^{\phi})\cdot\phi$, so

(5.8b) A_H acts on N_H^- with trace $-2\rho_H$.

Now suppose that we have

(5.9a) μ : discrete series representation of M_H on V_μ,

(5.9b) η : real linear functional on A_H.

Then we define

(5.10a) $\beta_{\mu,\eta}$: irreducible representation of P_H on $E_{\mu,\eta} = V_\mu \otimes \mathbb{C}$ by

(5.10b) $\beta_{\mu,\eta}(man) = \mu(m)\cdot\exp((\rho_H + i\eta)\log a)$.

The associated complex Hilbert space bundle

(5.11a) $\mathbb{E}_{\mu,\eta} \to G_0/P_H = K/K\cap M_H$

has space of measurable sections

(5.11b) $\Gamma(\mathbb{E}_{\mu,\eta}) = \{f\colon G_0 \to V_\mu$ measurable :

$$f(gman) = \beta_{\mu,\eta}(man)^{-1}f(g)\}.$$

As in section 4, the subspace of square integrable sections is

(5.12a) $$\Gamma_2(\mathbb{E}_{\mu,\eta}) = \{f \in \Gamma(\mathbb{E}_{\mu,\eta}) : \int_K (f(k),f(k))_{V_\mu}\, dk < \infty\}.$$

It is a Hilbert space whose inner product

(5.12b) $$\langle f,f'\rangle = \int_K (f(k),f'(k))_{V_\mu}\, dk$$

is invariant under the action of G_0. As in the case of the principal series, that defines a unitary representation

(5.13) $$\pi_{\mu,\eta} : H\text{-series representation of } G_0 \text{ on } \Gamma_2(\mathbb{E}_{\mu,\eta}).$$

This H-series is parameterized by $(\hat{M}_H)_{\text{discrete}} \times A_H'$; for [9, §11] the distribution character of $\pi_{\mu,\eta}$ does not depend on the choice of positive A_H-root system Δ^+ used in construction of P_H and ρ_H. Moreover, the Weyl group

(5.14a) $$W_H = \{g \in G_0 : \operatorname{ad}(g)A_H = A_H\}/M_H A_H ,$$

finite group acting on

(5.14b) $$\hat{M}_H \text{ by } \mu \mapsto \mu \circ \operatorname{ad}(g) \text{ and } A_H' \text{ by } \eta \mapsto \eta \circ \operatorname{ad}(g),$$

acts trivially on the H-series [9, §11] in the sense

(5.14c) if $w \in W_H$ then $\pi_{\mu,\eta}$ and $\pi_{w(\mu),w(\eta)}$ are equivalent.

An H-series representation is a finite sum of irreducible representations. Harish-Chandra has an irreducibility criterion that is tentative in the sense that he has not written down all the details. The criterion: choose A for a minimal parabolic subgroup MAN of G_0, such that $A_H \subset A$; extend η to A by zero on the orthocomplement of A_H; suppose then that η is not orthogonal to any A-root; in that case $\pi_{\mu,\eta}$ is irreducible.

If H is compact then the H-series is the discrete series. If H is maximally $\mathbb{R}$-split then the H-series is the principal series. Harish-Chandra's decomposition [9, §12] of $L_2(G_0)$ says that the irreducible H-series representations, where H runs over a system of representatives of the conjugacy classes of Cartan subgroups of G_0, form a subset of the unitary dual $\hat{G}_0$ whose complement has Plancherel measure zero.

6. *Integrable orbits and intermediate series*

Fix a Cartan subgroup $H \subset G_0$. We describe the geometric realization of an H-series representation $\pi_{\mu,\eta}$ on a partially holomorphic cohomology space over an appropriate G_0-orbit in a complex flag manifold.

$X = G/P$ is a complex flag manifold with $G_0^{\mathbb{C}} = G$. Suppose that we have an orbit

(6.1a) $\quad X_H = G_0(x_H) \subset X$ measurable and integrable

whose holomorphic arc components have real normalizers specified by

(6.1b) $\quad P_H = M_H A_H N_H = \{g \in G_0 : gS_{[x_H]} = S_{[x_H]}\}$.

Then $S_{[x_H]}$ is an open M_H-orbit on a complex flag manifold $M_H^{\mathbb{C}}(x_H) = M_H^{\mathbb{C}}/Q_H$ of $M_H^{\mathbb{C}}$, so

(6.1c) $\quad S_{[x_H]} = M_H/L_H = P_H/L_H A_H N_H$ where

(6.1d) $\quad L_H$ contains a compact Cartan subgroup of M_H.

We now add the condition that

(6.1e) $\quad L_H$ is compact.

Before proceeding with the realization of $\pi_{\mu,\eta}$ we note the circumstances under which the arrangement (6.1) is possible. From

(6.1a) and (6.1b) we have [21, Corollary 9.11] that P_H is related to the isotropy subgroup P_{x_H} of G at x_H, via the complex conjugation τ of G over G_0, as follows.

$$(6.2a) \qquad P_{x_H} + \tau P_{x_H} = P_H^{\mathbb{C}} \quad \text{and} \quad \tau P_{x_H}^r = P_{x_H}^r .$$

That implies that $P_H^{\mathbb{C}}$ has unipotent and reductive parts

$$(P_H^{\mathbb{C}})^u = N_H^{\mathbb{C}} = P_{x_H}^u \cap \tau P_{x_H}^u \quad \text{and}$$

$$(P_H^{\mathbb{C}})^r = (M_H + A_H)^{\mathbb{C}} = P_{x_H}^r + (P_{x_H}^{-u} \cap \tau P_{x_H}^u) + (P_{x_H}^u \cap \tau P_{x_H}^{-u}).$$

Thus

$$(6.2b) \qquad N_H^{\mathbb{C}} = P_{x_H}^u \cap \tau P_{x_H}^u , \quad L_H^{\mathbb{C}} + A_H^{\mathbb{C}} = P_{x_H}^r , \quad \text{and}$$

$$(6.2c) \qquad M_H^{\mathbb{C}} = L_H^{\mathbb{C}} + (P_{x_H}^{-u} \cap \tau P_{x_H}^u) + (P_{x_H}^u \cap \tau P_{x_H}^{-u}).$$

In particular, the complex flag on which $S_{[x_H]}$ is an open M_H-orbit is

$$(6.2d) \qquad M_H^{\mathbb{C}}(x_H) = M_H^{\mathbb{C}}/Q_H , \quad Q_H^r = L_H^{\mathbb{C}} , \quad Q_H^u = P_{x_H}^u \cap \tau P_{x_H}^{-u} .$$

Now we can prescribe the construction of $x_H \in X = G/P$ so that (6.1) holds:

(i) Let $L_H \subset M_H$ be the Lie algebra of a compact subgroup of M_H that is the centralizer there of a torus. This is always possible, e.g., by choice of L_H as the Lie algebra of a compact Cartan subgroup of M_H.

(ii) Let $Q_H \subset M_H^{\mathbb{C}}$ parabolic subgroup such that $Q_H^r = L_H^{\mathbb{C}}$ and $M_H^{\mathbb{C}} = L_H^{\mathbb{C}} + Q_H^u + \tau Q_H^u$ direct sum. Let $x_H \in M_H^{\mathbb{C}}/Q_H$ denote the identity coset, so L_H is the isotropy subalgebra of M_H at x_H, and $M_H(x_H)$ is a measurable open orbit on $M_H^{\mathbb{C}}/Q_H$ with compact isotropy group L_H.

(iii) Define $P^u = N_H^{\mathbb{C}} + Q_H^u$ unipotent subalgebra of G. Define P to be the normalizer of P^u in G. Observe that

$$P = (L_H + A_H)^{\mathbb{C}} + (N_H^{\mathbb{C}} + Q_H^u),$$

so it contains a Cartan subalgebra $H^{\mathbb{C}}$ of G and it has unipotent radical P^u. Thus [21, Lemma 7.3] P is a parabolic subalgebra of G. Let P be the corresponding parabolic subgroup of G and define $X = G/P$.

(iv) Observe $G_0 \cap P = L_H A_H N_H$, $\tau P^r = P^r$, and $P + \tau P = (M_H + A_H + N_H)^{\mathbb{C}} = P_H^{\mathbb{C}}$. Thus we identify $x_H = 1 \cdot Q_H \in M_H^{\mathbb{C}}/Q_H$ with $1 \cdot P \in G/P = X$, still having $M_H^{\mathbb{C}}(x_H) = M_H^{\mathbb{C}}/Q_H$, and [21, Corollary 9.11] says that (6.1a) and (6.1b) hold. For (6.1e) recall that $M_H \cap Q_H = L_H$ compact.

Conversely, (6.2) shows that this construction of $x_H \in X = G/P$ provides all $x_H \in X = G/P$ for which (6.1) is valid.

We go on to the realization of the H-series representation $\pi_{\mu,\eta}$ on the orbit $X_H = G_0(x_H) \subset X = G/P$ that satisfies (6.1).

Our notation for representation spaces of irreducible unitary representations of L_H, M_H and K is

(6.3a) if $\nu \in \hat{L}_H$ then W_ν is its representation space,

(6.3b) if $\mu \in \hat{M}_H$ then V_μ is its representation space,

(6.3c) if $\kappa \in \hat{K}$ then U_κ is its representation space.

Fix a *discrete* series representation $\mu \in \hat{M}_H$. If $\nu \in \hat{L}_H$ we consider the associated M_H-homogeneous holomorphic vector bundle

(6.4a) $\mathbb{W}_\nu \to M_H/L_H = M_H(x_H) = S_{[x_H]}$.

Suppose that we can find $\nu \in \hat{L}_H$, and an integer $q \geqq 0$, such that the space of square integrable harmonic forms

(6.4b) $H_2^{0,q}(\mathbb{W}_\nu) = V_\mu$ as a unitary M_H-representation space.

We realize $\pi_{\mu,\eta}$ on partially holomorphic cohomology as follows. Define

(6.5a) $\sigma_{\nu,\eta}$: irreducible representation of $L_H A_H N_H$ on W_ν

(6.5b) by : $\sigma_{\nu,\eta}(lan) = \nu(l)\cdot\exp((\rho_H + i\eta)(\log a))$.

The associated G_0-homogeneous complex vector bundle

(6.5c) $\mathbb{W}_{\nu,\eta} \to G_0/L_H A_H N_H = G_0(x_H) = X_H$

is holomorphic over every holomorphic arc component of X_H in X because

(6.5d) $\mathbb{W}_{\nu,\eta}|_{S_{[x_H]}} = \mathbb{W}_\nu$.

The holomorphic tangent space of $S_{[x_H]}$ is given at x_H, in the notation (6.2d), by

(6.6a) $T_{x_H} = Q_H^{-u} = P_{x_H}^{-u} \cap \tau P_{x_H}^{u}$.

There $A_H N_H$ acts trivially, and the action of L_H is unitary because it is the restriction of the adjoint action of G. Thus the antiholomorphic cotangent space to $S_{[x_H]}$ at x_H is also given by

(6.6b) $\bar{T}^*_{x_H} = Q_H^{-u}$ with $L_H A_H N_H$ acting there by

(6.6c) $A_H N_H$ acts trivially, L_H acts through ad_G.

Now consider the space

(6.7a) $\Gamma^{0,q}(\mathbb{W}_{\nu,\eta})$: partially-$(0,q)$-forms in $\mathbb{W}_{\nu,\eta}$.

It consists of all

(6.7b) $f\colon G_0 \to W_\nu \otimes \Lambda^q Q_H^{-u}$ measurable, such that

(6.7c) $\quad f(glan) = [\sigma_{\nu,\eta}(lan) \otimes \Lambda^q \mathrm{ad}_G(l)]^{-1} f(g)$ identically.

Note the consequences of the definition (6.7abc):

(6.7d) $\quad f|_{kP_H}$ is a $(0,q)$-form on $S_{[kx_H]}$ with values in $\mathbb{W}_{\nu,\eta}|_{S_{[kx_H]}}$.

Thus we have a space

(6.8a) $\quad L_2^{0,q}(\mathbb{W}_{\nu,\eta})$: square integrable partially-$(0,q)$-forms in $\mathbb{W}_{\nu,\eta}$

consisting of all $f \in \Gamma^{0,q}(\mathbb{W}_{\nu,\eta})$ such that

(6.8b) $\quad$ the $(0,q)$-form $f|_{kP_H}$ has finite L_2-norm $\|f_k\|$ a.e. in K,

(6.8c) $\quad k \mapsto \|f_k\|$ is measurable and $\int_K \|f_k\|^2 \, dk < \infty$.

$L_2^{0,q}(\mathbb{W}_{\nu,\eta})$ is a Hilbert space with inner product

(6.9a) $\quad \langle f, f' \rangle = \int_K (f|_{kP_H}, f'|_{kP_H}) dk = \int_K dk \int_{M_H} (f(km), f'(km)) dm.$

The operator $\bar\partial$ along the holomorphic arc components is the restriction of the $\bar\partial$-operator of X, so it commutes with the action of K on X_H and $\mathbb{W}_{\nu,\eta}$. Now, applying (3.16) to every holomorphic arc component of our orbit,

(6.9b) $\quad L_2^{0,q}(\mathbb{W}_{\nu,\eta}) = \bar\partial L_2^{0,q-1}(\mathbb{W}_{\nu,\eta}) \oplus \bar\delta L_2^{0,q+1}(\mathbb{W}_{\nu,\eta}) \oplus H_2^{0,q}(\mathbb{W}_{\nu,\eta})$

orthogonal direct sum of closed G_0-invariant subspaces, where

(6.9c) $\quad H_2^{0,q}(\mathbb{W}_{\nu,\eta}) = \{f \in L_2^{0,q}(\mathbb{W}_{\nu,\eta}) : \text{each } f|_{kP_H} \text{ is harmonic}\}$

is our space of square integrable partially-harmonic-$(0,q)$-forms in $\mathbb{W}_{\nu,\eta}$.

We assert that the H-series representation

(6.10) $\pi_{\mu,\eta}$ is unitarily equivalent to: G_0 on $H_2^{0,q}(\mathbb{W}_{\nu,\eta})$.

That will be our geometric realization of $\pi_{\mu,\eta}$.

Looking at $L_2^{0,q}(\mathbb{W}_{\nu,\eta})$ as a K-module, we see that it is given by

(6.11a) $$L_2^{0,q}(\mathbb{W}_{\nu,\eta}) = \{L_2(K) \otimes [L_2(M_H) \otimes W_\nu \otimes \Lambda^q Q_H^{-u}]^{L_H}\}^J$$

where

(6.11b) $J = K \cap M_H$ maximal compact subgroup of M_H.

From (3.16) and (6.9bc), now

(6.11c) $$H_2^{0,q}(\mathbb{W}_{\nu,\eta}) = \{L_2(K) \otimes H_2^{0,q}(\mathbb{W}_\nu)\}^J \text{ as } K\text{-module.}$$

From the hypothesis (6.4) that $H_2^{0,q}(\mathbb{W}_\nu) = V_\mu$, and the Peter-Weyl decomposition of $L_2(K)$, we see that (6.11c) can be re-written as

(6.12) $$H_2^{0,q}(\mathbb{W}_{\nu,\eta}) = \sum_{\kappa \in \hat{K}} U_\kappa \otimes \{U_\kappa^* \otimes V_\mu\}^J \text{ as } K\text{-module.}$$

Note that η does not occur in the K-module structure of $H_2^{0,q}(\mathbb{W}_{\nu,\eta})$.

If $g \in G_0$ we have $\kappa_g : K \to K$ defined by $g^{-1}k \in \kappa_g(k)AN$. Now $AN \subset MAN \subset M_H A_H N_H$. Thus we have

(6.13a) $\kappa_g : K \to K$, $\mu_g^H : K \to AN \cap M_H$ and $\alpha_g^H : K \to A_H \subset A$

such that

(6.13b) $$g^{-1}k \in \kappa_g(k) \cdot \mu_g^H(k) \cdot \alpha_g^H(k) \cdot N_H.$$

Viewing $f \in L_2^{0,q}(\mathbb{W}_{\nu,\eta})$ as a function from K to $L_2(M_H) \otimes W_\nu \otimes \Lambda^q Q_H^{-u}$ as in (6.11a), it follows from (6.7c) that the image of f under $g \in G_0$ is given by

(6.13c) $$f^g(k) = [\lambda(\mu_g^H(k)) \otimes \exp((\rho_H + i\eta)\log \alpha_g^H(k)) \otimes 1]^{-1} f(\kappa_g(k))$$

where λ is the left regular representation of M_H on $L_2(M_H)$. The hypothesis $H_2^{0,q}(\mathbb{W}_\nu) = V_\mu$ says that, on restriction of f to the subspace $H_2^{0,q}(\mathbb{W}_{\nu,\eta})$, and on then viewing f as a function from K to V_μ as in (6.12),

$$(6.14) \qquad f^g(k) = \exp((\rho_H + i\eta)\log \alpha_g^H(k))^{-1}\cdot\mu(\mu_g^H(k))^{-1}\cdot f(\kappa_g(k)).$$

The representation space $\Gamma_2(\mathbb{E}_{\mu,\eta})$ of our H-series representation $\pi_{\mu,\eta}$ has K-module structure $\{L_2(K)\otimes V_\mu\}^J$, which we write via the Peter-Weyl decomposition of $L_2(K)$ as

$$(6.15) \qquad \Gamma_2(\mathbb{E}_{\mu,\eta}) = \sum_{\kappa\in\hat{K}} U_\kappa \otimes \{U_\kappa^* \otimes V_\mu\}^J \quad \text{as } K\text{-module.}$$

Further, from (5.11b) and (6.12ab) we see that the action of G_0 on $\Gamma_2(\mathbb{E}_{\mu,\eta})$ under (6.15) is

$$(6.16) \qquad [\pi_{\mu,\eta}(g)f](k) = \exp((\rho_H + i\eta)\log \alpha_g^H(k))^{-1}\cdot\mu(\mu_g^H(k))^{-1}\cdot f(\kappa_g(k)).$$

Finally, denote the multiplicity

$$(6.17a) \qquad n(\mu,\kappa) = \dim\{U_\kappa^* \otimes V_\mu\}^J.$$

As $\kappa\in\hat{K}$ has finite degree we have

$$(6.17b) \qquad \kappa|_J = \sum_{j\in\hat{J}} n_1(j,\kappa)\cdot j \quad \text{with} \quad \sum_{j\in\hat{J}} n_1(j,\kappa) < \infty.$$

As J is a maximal compact subgroup of M_H and $\mu\in\hat{M}_H$ we have

$$(6.17c) \qquad \mu|_J = \sum_{j\in\hat{J}} n_2(j,\mu)\cdot j \quad \text{with each} \quad n_2(j,\mu) < \infty.$$

Now compute, using Schur's Lemma and the fact that (6.17b) is a finite sum,

$$(6.17d) \qquad n(\mu,\kappa) = \sum_{j\in\hat{J}} n_1(j,\kappa)\cdot n_2(j,\mu) < \infty.$$

In summary, we have checked that

(6.18) $$\dim\{U_\kappa^* \otimes V_\mu\}^J < \infty.$$

Compare (6.12) with (6.15) and (6.14) with (6.16). In view of (6.18) it follows, as in the case of the principal series, that $\pi_{\mu,\eta}$ has the same distribution character as the representation of G_0 on $H_2^{0,q}(\mathbb{W}_{\nu,\eta})$. Both representations being K-finite and unitary, now they are infinitesimally equivalent, and thus unitarily equivalent. That completes the proof of (6.10).

I have indications of, but no pattern for, a "complementary series" of representations of G_0 associated to each nondiscrete H-series. The proof of (6.10) is designed so that it will automatically give geometric realization for any such "complementary H-series" representations of G_0.

Recall the main hypothesis (6.4) in the geometric realization of $\pi_{\mu,\eta}$: that the discrete series representation $\mu \in \hat{M}_H$ could be realized on the space $H_2^{0,q}(\mathbb{W}_\nu)$ of square integrable harmonic $(0,q)$-forms in the M_H-homogeneous holomorphic vector bundle $\mathbb{W}_\nu \to M_H/L_H = S_{[x_H]}$ associated to some $\nu \in \hat{L}_H$.

Making essential use of our simplifying assumption that $G_0 \subset G$, i.e., that G_0 is a linear group, we see that the finite group

(6.19a) $$F_H = K \cap \exp(iA_H)$$

is central in M_H and contained in L_H. Denote

(6.19b) $L_H^0 \subset M_H^0$: topological identity components of $L_H \subset M_H$;

(6.19c) $$L_H^\dagger = F_H L_H^0 \quad \text{and} \quad M_H^\dagger = F_H M_H^0.$$

Harish-Chandra pointed out to me that, looking at the character of a discrete series representation $\mu^\dagger \in M_H^\dagger$, one sees that the induced representation μ of M_H is irreducible; and also that one obtains all discrete series representations μ of M_H in this way. Same situation for $L_H^\dagger$ and L_H.

As $S_{[x_H]}$ is connected and $F_H \subset L_H$, the subgroups $L_H^\dagger \subset L_H$ and $M_H^\dagger \subset M_H$ have the same index r.

Let $\mu \in \hat{M}_H$ discrete. Now

(6.20a) $\quad \mu|_{M_H^\dagger} = \mu_1 \oplus \ldots \oplus \mu_r, \quad \mu_j \in \hat{M}_H^\dagger$ discrete, distinct,

such that

(6.20b) $\quad$ any of the μ_j induces μ.

Further note

(6.20c) $\quad \mu_j|_{M_H^0} = \mu_j^0 \otimes \chi_\mu, \quad \mu_j^0 \in \hat{M}_H^0$ discrete, $\chi_\mu \in \hat{F}_H$.

Similarly, if $\nu \in \hat{L}_H$, then

(6.21a) $\quad \nu|_{L_H^\dagger} = \nu_1 \oplus \ldots \oplus \nu_r, \quad \nu_j \in \hat{L}_H^\dagger$ distinct,

such that

(6.21b) $\quad$ any of the ν_j induces ν, and

(6.21c) $\quad \nu_j|_{L_H^0} = \nu_j^0 \otimes \chi_\nu, \quad \nu_j^0 \in \hat{L}_H^0, \quad \chi_\nu \in \hat{F}_H$.

From $S_{[x_H]} = M_H/L_H = M_H^\dagger/L_H^\dagger = M_H^0/L_H^0$ we now have

(6.22a) $\quad$ if $H_2^{0,q}(\mathbb{W}_{\nu_j^0}) = V_{\mu_j^0}$ then $H_2^{0,q}(\mathbb{W}_{\nu_j}) = V_{\mu_j}$, and

(6.22b) $\quad$ if some $H_2^{0,q}(\mathbb{W}_{\nu_j}) = V_{\mu_j}$ then $H_2^{0,q}(\mathbb{W}_\nu) = V_\mu$.

Thus the question of validity of (6.4) is reduced to the case of connected linear Lie groups with compact center, whence it further reduces to the case of connected linear semisimple Lie groups. That was discussed in section 3.

References

1. A. Andreotti and E. Vesentini, *Carleman estimates for the Laplace-Beltrami operator on complex manifolds*, Inst. Hautes Études Sci. Publ. Math. 25 (1965), 81-130.

2. R. Bott, *Homogeneous vector bundles*, Ann. of Math. (2) 66 (1957), 203-248.

3. F. Bruhat, *Sur les représentations induites des groupes de Lie*, Bull. Soc. Math. France 84 (1956), 97-205.

4. J. M. G. Fell, book on harmonic analysis, in preparation.

5. Harish-Chandra, *Representations of a semisimple Lie group on a Banach space*, I, Trans. Amer. Math. Soc. 75 (1953), 185-243.

6. ————, *The Plancherel formula for complex semisimple Lie groups*, Trans. Amer. Math. Soc. 76 (1954), 485-528.

7. ————, *Representations of semisimple Lie groups*, IV, Amer. J. Math. 77 (1955), 743-777; V, ibid. 78 (1956), 1-41; VI, ibid. 78 (1956), 564-628.

8. ————, *Discrete series for semisimple Lie groups*, II, Acta Math. 116 (1966), 1-111.

9. ————, *Harmonic analysis on semisimple Lie groups*, Bull. Amer. Math. Soc. 76 (1970), 529-551.

10. S. Helgason, *Applications of the Radon transform to representations of semisimple Lie groups*, Proc. Nat. Acad. Sci. U. S. A. 63 (1969), 643-647.

11. ————, *A duality for symmetric spaces with applications to group representations*, to appear.

12. A. W. Knapp and E. M. Stein, *Singular integrals and the principal series*, Proc. Nat. Acad. Sci. U. S. A. 63 (1969), 281-284; II, ibid., to appear.

13. B. Kostant, *Lie algebra cohomology and the generalized Borel-Weil theorem*, Ann. of Math. (2) 74 (1961), 329-387.

14. ————, *On the existence and irreducibility of certain series of representations*, Bull. Amer. Math. Soc. 75 (1969), 627-642.

15. R. A. Kunze and E. M. Stein, *Uniformly bounded representations*, III, Amer. J. Math. 89 (1967), 385-442.

16. M. S. Narasimhan and K. Okamoto, *An analogue of the Borel-Weil-Bott theorem for hermitian symmetric pairs of noncompact type*, Ann. of Math. (2) 91 (1970), 486-511.

17. K. R. Parthasarathy, R. Ranga-Rao and V. S. Varadarajan, *Representations of complex semisimple Lie groups and Lie algebras*, Ann. of Math. (2) 85 (1967), 383-429.

18. W. Schmid, *On a conjecture of Langlands*, Ann. of Math., 1970.

19. N. R. Wallach, *Cyclic vectors and irreducibility for principal series representations*, to appear.

20. Garth Warner, *Harmonic analysis on semisimple Lie groups*, Springer-Verlag, Berlin-Heidelberg-New York, to appear about 1971.

21. J. A. Wolf, *The action of a real semisimple group on a complex flag manifold. I: Orbit structure and holomorphic arc components*, Bull. Amer. Math. Soc. 75 (1969), 1121-1237.

22. ———, II: *Unitary representations on partially holomorphic cohomology spaces*, to appear.

23. ———, III: *Induced representations based on hermitian symmetric spaces*, in preparation.

24. D. P. Zelobenko, *Analysis of irreducibility in the class of elementary representations of a complex semisimple Lie group*, Math. USSR - Izv. 2 (1968), 105-128.

Rutgers University, New Brunswick, New Jersey 08903 *and*

University of California, Berkeley, California 94720

ture Notes in Mathematics

erschienen/Already published

. Werner, Seminar über Funktionen-Algebren. IV, 30 Seiten. M 3,80 / $ 1.10

A. Borel, Cohomologie des espaces localement compacts J. Leray. IV, 93 pages. 1964. DM 9,– / $ 2.60

. F. Adams, Stable Homotopy Theory. Third edition. IV, 78 pages. M 8,– / $ 2.20

M. Arkowitz and C. R. Curjel, Groups of Homotopy Classes. vised edition. IV, 36 pages. 1967. DM 4,80 / $ 1.40

J.-P. Serre, Cohomologie Galoisienne. Troisième édition. pages. 1965. DM 18,– / $ 5.00

H. Hermes, Term Logic with Choise Operator. III, 55 pages. M 6,– / $ 1.70

Ph. Tondeur, Introduction to Lie Groups and Transformation . Second edition. VIII, 176 pages. 1969. DM 14,– / $ 3.80

G. Fichera, Linear Elliptic Differential Systems and Eigenvalue ns. IV, 176 pages. 1965. DM 13,50 / $ 3.80

P. L. Ivănescu, Pseudo-Boolean Programming and Applications. pages. 1965. DM 4,80 / $ 1.40

: H. Lüneburg, Die Suzukigruppen und ihre Geometrien. VI, en. 1965. DM 8,– / $ 2.20

J.-P. Serre, Algèbre Locale. Multiplicités. Rédigé par P. Gabriel. e édition. VIII, 192 pages. 1965. DM 12,– / $ 3.30

: A. Dold, Halbexakte Homotopiefunktoren. II, 157 Seiten. 1966. – / $ 3.30

: E. Thomas, Seminar on Fiber Spaces. IV, 45 pages. 1966. 0 / $ 1.40

: H. Werner, Vorlesung über Approximationstheorie. IV, 184 Sei- 12 Seiten Anhang. 1966. DM 14,– / $ 3.90

: F. Oort, Commutative Group Schemes. VI, 133 pages. 1966. 0 / $ 2.70

: J. Pfanzagl and W. Pierlo, Compact Systems of Sets. IV, es. 1966. DM 5,80 / $ 1.60

: C. Müller, Spherical Harmonics. IV, 46 pages. 1966. – / $ 1.40

: H.-B. Brinkmann und D. Puppe, Kategorien und Funktoren. Seiten, 1966. DM 8,– / $ 2.20

: G. Stolzenberg, Volumes, Limits and Extensions of Analytic es. IV, 45 pages. 1966. DM 5,40 / $ 1.50

: R. Hartshorne, Residues and Duality. VIII, 423 pages. 1966. – / $ 5.50

: Seminar on Complex Multiplication. By A. Borel, S. Chowla, erz, K. Iwasawa, J.-P. Serre. IV, 102 pages. 1966. DM 8,– / $ 2.20

2: H. Bauer, Harmonische Räume und ihre Potentialtheorie. Seiten. 1966. DM 14,– / $ 3.90

: P. L. Ivănescu and S. Rudeanu, Pseudo-Boolean Methods for nt Programming. 120 pages. 1966. DM 10,– / $ 2.80

: J. Lambek, Completions of Categories. IV, 69 pages. 1966. 0 / $ 1.90

: R. Narasimhan, Introduction to the Theory of Analytic Spaces. pages. 1966. DM 10,– / $ 2.80

6: P.-A. Meyer, Processus de Markov. IV, 190 pages. 1967. – / $ 4.20

7: H. P. Künzi und S. T. Tan, Lineare Optimierung großer e. VI, 121 Seiten. 1966. DM 12,– / $ 3.30

8: P. E. Conner and E. E. Floyd, The Relation of Cobordism to ories. VIII, 112 pages. 1966. DM 9,80 / $ 2.70

9: K. Chandrasekharan, Einführung in die Analytische Zahlen- . VI, 199 Seiten. 1966. DM 16,80 / $ 4.70

: A. Frölicher and W. Bucher, Calculus in Vector Spaces without X, 146 pages. 1966. DM 12,– / $ 3.30

1: Symposium on Probability Methods in Analysis. Chairman. Kappos. IV, 329 pages. 1967. DM 20,– / $ 5.50

2: M. André, Méthode Simpliciale en Algèbre Homologique et e Commutative. IV, 122 pages. 1967. DM 12,– / $ 3.30

3: G. I. Targonski, Seminar on Functional Operators and ons. IV, 110 pages. 1967. DM 10,– / $ 2.80

4: G. E. Bredon, Equivariant Cohomology Theories. VI, 64 pages. DM 6,80 / $ 1.90

5: N. P. Bhatia and G. P. Szegö, Dynamical Systems. Stability and Applications. VI, 416 pages. 1967. DM 24,– / $ 6.60

6: A. Borel, Topics in the Homology Theory of Fibre Bundles. pages. 1967. DM 9,– / $ 2.50

Vol. 37: R. B. Jensen, Modelle der Mengenlehre. X, 176 Seiten. 1967. DM 14,– / $ 3.90

Vol. 38: R. Berger, R. Kiehl, E. Kunz und H.-J. Nastold, Differentialrechnung in der analytischen Geometrie IV, 134 Seiten. 1967 DM 12,– / $ 3.30

Vol. 39: Séminaire de Probabilités I. II, 189 pages. 1967. DM 14,– / $ 3.90

Vol. 40: J. Tits, Tabellen zu den einfachen Lie Gruppen und ihren Darstellungen. VI, 53 Seiten. 1967. DM 6,80 / $ 1.90

Vol. 41: A. Grothendieck, Local Cohomology. VI, 106 pages. 1967. DM 10,– / $ 2.80

Vol. 42: J. F. Berglund and K. H. Hofmann, Compact Semitopological Semigroups and Weakly Almost Periodic Functions. VI, 160 pages. 1967. DM 12,– / $ 3.30

Vol. 43: D. G. Quillen, Homotopical Algebra. VI, 157 pages. 1967. DM 14,– / $ 3.90

Vol. 44: K. Urbanik, Lectures on Prediction Theory. IV, 50 pages. 1967. DM 5,80 / $ 1.60

Vol. 45: A. Wilansky, Topics in Functional Analysis. VI, 102 pages. 1967. DM 9,60 / $ 2.70

Vol. 46: P. E. Conner, Seminar on Periodic Maps. IV, 116 pages. 1967. DM 10,60 / $ 3.00

Vol. 47: Reports of the Midwest Category Seminar I. IV, 181 pages. 1967. DM 14,80 / $ 4.10

Vol. 48: G. de Rham, S. Maumary et M. A. Kervaire, Torsion et Type Simple d'Homotopie. IV, 101 pages. 1967. DM 9,60 / $ 2.70

Vol. 49: C. Faith, Lectures on Injective Modules and Quotient Rings. XVI, 140 pages. 1967. DM 12,80 / $ 3.60

Vol. 50: L. Zalcman, Analytic Capacity and Rational Approximation. VI, 155 pages. 1968. DM 13.20 / $ 3.70

Vol. 51: Séminaire de Probabilités II. IV, 199 pages. 1968. DM 14,– / $ 3.90

Vol. 52: D. J. Simms, Lie Groups and Quantum Mechanics. IV, 90 pages. 1968. DM 8,– / $ 2.20

Vol. 53: J. Cerf, Sur les difféomorphismes de la sphère de dimension trois ($\Gamma_4 = O$). XII, 133 pages. 1968. DM 12,– / $ 3.30

Vol. 54: G. Shimura, Automorphic Functions and Number Theory. VI, 69 pages. 1968. DM 8,– / $ 2.20

Vol. 55: D. Gromoll, W. Klingenberg und W. Meyer, Riemannsche Geometrie im Großen. VI, 287 Seiten. 1968. DM 20,– / $ 5.50

Vol. 56: K. Floret und J. Wloka, Einführung in die Theorie der lokalkonvexen Räume. VIII, 194 Seiten. 1968. DM 16,– / $ 4.40

Vol. 57: F. Hirzebruch und K. H. Mayer, O (n)-Mannigfaltigkeiten, exotische Sphären und Singularitäten. IV, 132 Seiten. 1968. DM 10,80 / $ 3.00

Vol. 58: Kuramochi Boundaries of Riemann Surfaces. IV, 102 pages. 1968. DM 9,60 / $ 2.70

Vol. 59: K. Jänich, Differenzierbare G-Mannigfaltigkeiten. VI, 89 Seiten. 1968. DM 8,– / $ 2.20

Vol. 60: Seminar on Differential Equations and Dynamical Systems. Edited by G. S. Jones. VI, 106 pages. 1968. DM 9,60 / $ 2.70

Vol. 61: Reports of the Midwest Category Seminar II. IV, 91 pages. 1968. DM 9,60 / $ 2.70

Vol. 62: Harish-Chandra, Automorphic Forms on Semisimple Lie Groups X, 138 pages. 1968. DM 14,– / $ 3.90

Vol. 63: F. Albrecht, Topics in Control Theory. IV, 65 pages. 1968. DM 6,80 / $ 1.90

Vol. 64: H. Berens, Interpolationsmethoden zur Behandlung von Approximationsprozessen auf Banachräumen. VI, 90 Seiten. 1968. DM 8,– / $ 2.20

Vol. 65: D. Kölzow, Differentiation von Maßen. XII, 102 Seiten. 1968. DM 8,– / $ 2.20

Vol. 66: D. Ferus, Totale Absolutkrümmung in Differentialgeometrie und -topologie. VI, 85 Seiten. 1968. DM 8,– / $ 2.20

Vol. 67: F. Kamber and P. Tondeur, Flat Manifolds. IV, 53 pages. 1968. DM 5,80 / $ 1.60

Vol. 68: N. Boboc et P. Mustată, Espaces harmoniques associés aux opérateurs différentiels linéaires du second ordre de type elliptique. VI, 95 pages. 1968. DM 8,60 / $ 2.40

Vol. 69: Seminar über Potentialtheorie. Herausgegeben von H. Bauer. VI, 180 Seiten. 1968. DM 14,80 / $ 4.10

Vol. 70: Proceedings of the Summer School in Logic. Edited by M. H. Löb. IV, 331 pages. 1968. DM 20,– / $ 5.50

Vol. 71: Séminaire Pierre Lelong (Analyse), Année 1967 – 1968. VI, 190 pages. 1968. DM 14,– / $ 3.90

Bitte wenden / Continued

Vol. 72: The Syntax and Semantics of Infinitary Languages. Edited by J. Barwise. IV, 268 pages. 1968. DM 18,– / $ 5.00

Vol. 73: P. E. Conner, Lectures on the Action of a Finite Group. IV, 123 pages. 1968. DM 10,– / $ 2.80

Vol. 74: A. Fröhlich, Formal Groups. IV, 140 pages. 1968. DM 12,– / $ 3.30

Vol. 75: G. Lumer, Algèbres de fonctions et espaces de Hardy. VI, 80 pages. 1968. DM 8,– / $ 2.20

Vol. 76: R. G. Swan, Algebraic K-Theory. IV, 262 pages. 1968. DM 18,– / $ 5.00

Vol. 77: P.-A. Meyer, Processus de Markov: la frontière de Martin. IV, 123 pages. 1968. DM 10,– / $ 2.80

Vol. 78: H. Herrlich, Topologische Reflexionen und Coreflexionen. XVI, 166 Seiten. 1968. DM 12,– / $ 3.30

Vol. 79: A. Grothendieck, Catégories Cofibrées Additives et Complexe Cotangent Relatif. IV, 167 pages. 1968. DM 12,– / $ 3.30

Vol. 80: Seminar on Triples and Categorical Homology Theory. Edited by B. Eckmann. IV, 398 pages. 1969. DM 20,– / $ 5.50

Vol. 81: J.-P. Eckmann et M. Guenin, Méthodes Algébriques en Mécanique Statistique. VI, 131 pages. 1969. DM 12,– / $ 3.30

Vol. 82: J. Wloka, Grundräume und verallgemeinerte Funktionen. VIII, 131 Seiten. 1969. DM 12,– / $ 3.30

Vol. 83: O. Zariski, An Introduction to the Theory of Algebraic Surfaces. IV, 100 pages. 1969. DM 8,– / $ 2.20

Vol. 84: H. Lüneburg, Transitive Erweiterungen endlicher Permutationsgruppen. IV, 119 Seiten. 1969. DM 10,– / $ 2.80

Vol. 85: P. Cartier et D. Foata, Problèmes combinatoires de commutation et réarrangements. IV, 88 pages. 1969. DM 8,– / $ 2.20

Vol. 86: Category Theory, Homology Theory and their Applications I. Edited by P. Hilton. VI, 216 pages. 1969. DM 16,– / $ 4.40

Vol. 87: M. Tierney, Categorical Constructions in Stable Homotopy Theory. IV, 65 pages. 1969. DM 6,– / $ 1.70

Vol. 88: Séminaire de Probabilités III. IV, 229 pages. 1969. DM 18,– / $ 5.00

Vol. 89: Probability and Information Theory. Edited by M. Behara, K. Krickeberg and J. Wolfowitz. IV, 256 pages. 1969. DM 18,– / $ 5.00

Vol. 90: N. P. Bhatia and O. Hajek, Local Semi-Dynamical Systems. II, 157 pages. 1969. DM 14,– / $ 3.90

Vol. 91: N. N. Janenko, Die Zwischenschrittmethode zur Lösung mehrdimensionaler Probleme der mathematischen Physik. VIII, 194 Seiten. 1969. DM 16,80 / $ 4.70

Vol. 92: Category Theory, Homology Theory and their Applications II. Edited by P. Hilton. V, 308 pages. 1969. DM 20,– / $ 5.50

Vol. 93: K. R. Parthasarathy, Multipliers on Locally Compact Groups. III, 54 pages. 1969. DM 5,60 / $ 1.60

Vol. 94: M. Machover and J. Hirschfeld, Lectures on Non-Standard Analysis. VI, 79 pages. 1969. DM 6,– / $ 1.70

Vol. 95: A. S. Troelstra, Principles of Intuitionism. II, 111 pages. 1969. DM 10,– / $ 2.80

Vol. 96: H.-B. Brinkmann und D. Puppe, Abelsche und exakte Kategorien, Korrespondenzen. V, 141 Seiten. 1969. DM 10,– / $ 2.80

Vol. 97: S. O. Chase and M. E. Sweedler, Hopf Algebras and Galois theory. II, 133 pages. 1969. DM 10,– / $ 2.80

Vol. 98: M. Heins, Hardy Classes on Riemann Surfaces. III, 106 pages. 1969. DM 10,– / $ 2.80

Vol. 99: Category Theory, Homology Theory and their Applications III. Edited by P. Hilton. IV, 489 pages. 1969. DM 24,– / $ 6.60

Vol. 100: M. Artin and B. Mazur, Etale Homotopy. II, 196 Seiten. 1969. DM 12,– / $ 3.30

Vol. 101: G. P. Szegö et G. Treccani, Semigruppi di Trasformazioni Multivoche. VI, 177 pages. 1969. DM 14,– / $ 3.90

Vol. 102: F. Stummel, Rand- und Eigenwertaufgaben in Sobolewschen Räumen. VIII, 386 Seiten. 1969. DM 20,– / $ 5.50

Vol. 103: Lectures in Modern Analysis and Applications I. Edited by C. T. Taam. VII, 162 pages. 1969. DM 12,– / $ 3.30

Vol. 104: G. H. Pimbley, Jr., Eigenfunction Branches of Nonlinear Operators and their Bifurcations. II, 128 pages. 1969. DM 10,– / $ 2.80

Vol. 105: R. Larsen, The Multiplier Problem. VII, 284 pages. 1969. DM 18,– / $ 5.00

Vol. 106: Reports of the Midwest Category Seminar III. Edited by S. Mac Lane. III, 247 pages. 1969. DM 16,– / $ 4.40

Vol. 107: A. Peyerimhoff, Lectures on Summability. III, 111 pages. 1969. DM 8,– / $ 2.20

Vol. 108: Algebraic K-Theory and its Geometric Applications. Edited by R. M. F. Moss and C. B. Thomas. IV, 86 pages. 1969. DM 6,– / $ 1.70

Vol. 109: Conference on the Numerical Solution of Differential tions. Edited by J. Ll. Morris. VI, 275 pages. 1969. DM 18,– / $

Vol. 110: The Many Facets of Graph Theory. Edited by G. Cha and S. F. Kapoor. VIII, 290 pages. 1969. DM 18,– / $ 5.00

Vol. 111: K. H. Mayer, Relationen zwischen charakteristischen Za III, 99 Seiten. 1969. DM 8,– / $ 2.20

Vol. 112: Colloquium on Methods of Optimization. Edite N. N. Moiseev. IV, 293 pages. 1970. DM 18,– / $ 5.00

Vol. 113: R. Wille, Kongruenzklassengeometrien. III, 99 Seiten. DM 8,– / $ 2.20

Vol. 114: H. Jacquet and R. P. Langlands, Automorphic Forms on G VII, 548 pages. 1970. DM 24,– / $ 6.60

Vol. 115: K. H. Roggenkamp and V. Huber-Dyson, Lattices over Or XIX, 290 pages. 1970. DM 18,– / $ 5.00

Vol. 116: Séminaire Pierre Lelong (Analyse) Année 1969. IV, 195 p 1970. DM 14,– / $ 3.90

Vol. 117: Y. Meyer, Nombres de Pisot, Nombres de Salem et Ar Harmonique. 63 pages. 1970. DM 6.– / $ 1.70

Vol. 118: Proceedings of the 15th Scandinavian Congress, Oslo Edited by K. E. Aubert and W. Ljunggren. IV, 162 pages. 1970. DM 12,– / $ 3.30

Vol. 119: M. Raynaud, Faisceaux amples sur les schémas en gro et les espaces homogènes. III, 219 pages. 1970. DM 14,– / $ 3.

Vol. 120: D. Siefkes, Büchi's Monadic Second Order Succ Arithmetic. XII, 130 Seiten. 1970. DM 12,– / $ 3.30

Vol. 121: H. S. Bear, Lectures on Gleason Parts. III, 47 pages. DM 6,– / $ 1.70

Vol. 122: H. Zieschang, E. Vogt und H.-D. Coldewey, Flächer ebene diskontinuierliche Gruppen. VIII, 203 Seiten. 1970. DM 1 $ 4.40

Vol. 123: A. V. Jategaonkar, Left Principal Ideal Rings. VI, 145 p 1970. DM 12,– / $ 3.30

Vol. 124: Séminare de Probabilités IV. Edited by P. A. Meyer. IV pages. 1970. DM 20,– / $ 5.50

Vol. 125: Symposium on Automatic Demonstration. V, 310 pages. DM 20,– / $ 5.50

Vol. 126: P. Schapira, Théorie des Hyperfonctions. XI, 157 pages. DM 14,– / $ 3.90

Vol. 127: I. Stewart, Lie Algebras. IV, 97 pages. 1970. DM 10,– / $

Vol. 128: M. Takesaki, Tomita's Theory of Modular Hilbert Alg and its Applications. II, 123 pages. 1970. DM 10,– / $ 2.80

Vol. 129: K. H. Hofmann, The Duality of Compact Semigroup C*- Bigebras. XII, 142 pages. 1970. DM 14,– / $ 3.90

Vol. 130: F. Lorenz, Quadratische Formen über Körpern. II, 77 S 1970. DM 8,– / $ 2.20

Vol. 131: A Borel et al., Seminar on Algebraic Groups and Re Finite Groups. VII, 321 pages. 1970. DM 22,– / $ 6.10

Vol. 132: Symposium on Optimization. III, 348 pages. 1970. DM 2 $ 6.10

Vol. 133: F. Topsøe, Topology and Measure. XIV, 79 pages. 1 DM 8,– / $ 2.20

Vol. 134: L. Smith, Lectures on the Eilenberg-Moore Spectral Sequ VII, 142 pages. 1970. DM 14,– / $ 3.90

Vol. 135: W. Stoll, Value Distribution of Holomorphic Maps into C pact Complex Manifolds. II, 267 pages. 1970. DM 18,– / $

Vol. 136: M. Karoubi et al., Séminaire Heidelberg-Saarbrücken-S buorg sur la K-Théorie. IV, 264 pages. 1970. DM 18,– / $ 5.00

Vol. 137: Reports of the Midwest Category Seminar IV. Edite S. MacLane. III, 139 pages. 1970. DM 12,– / $ 3.30

Vol. 138: D. Foata et M. Schützenberger, Théorie Géométrique Polynômes Eulériens. V, 94 pages. 1970. DM 10,– / $ 2.80

Vol. 139: A. Badrikian, Séminaire sur les Fonctions Aléatoires aires et les Mesures Cylindriques. VII, 221 pages. 1970. DM 18 $ 5.00

Vol. 140: Lectures in Modern Analysis and Applications II. Edite C. T. Taam. VI, 119 pages. 1970. DM 10,– / $ 2.80

Vol. 141: G. Jameson, Ordered Linear Spaces. XV, 194 pages. DM 16,– / $ 4.40

Vol. 142: K. W. Roggenkamp, Lattices over Orders II. V, 388 pa 1970. DM 22,– / $ 6.10

Vol. 143: K. W. Gruenberg, Cohomological Topics in Group The XIV, 275 pages. 1970. DM 20,– / $ 5.50

Lecture Notes in Physics

Bisher erschienen/Already published

Vol. 1: J. C. Erdmann, Wärmeleitung in Kristallen, theoretische Grundlagen und fortgeschrittene experimentelle Methoden. II, 283 Seiten. 1969. DM 20,– / $ 5.50

Vol. 2: K. Hepp, Théorie de la renormalisation. III, 215 pages. 1969. DM 18,– / $ 5.00

Vol. 3: A. Martin, Scattering Theory: Unitarity, Analytic and Crossing. IV, 125 pages. 1969. DM 14,– / $ 3.90

Vol. 4: G. Ludwig, Deutung des Begriffs physikalische Theorie und axiomatische Grundlegung der Hilbertraumstruktur der Quantenmechanik durch Hauptsätze des Messens. XI, 469 Seiten.1970. DM 28,– / $ 7.70

Vol. 5: M. Schaaf, The Reduction of the Product of Two Irreducible Unitary Representations of the Proper Orthochronous Quantummechanical Poincaré Group. IV, 120 pages. 1970. DM 14,– / $ 3.90

Vol. 6: Group Representations in Mathematics and Physics. Edited by V. Bargmann. V, 340 pages. 1970. DM 24,– / $ 6.60